T0229890

SERIES CONTENTS

SCIENCE AND PHILOSOPHY IN THE TWENTIETH CENTURY

Basic Works of Logical Empiricism

SERIES EDITOR

SAHOTRA SARKAR

Dibner Institute at MIT and McGill University

A GARLAND SERIES IN
READINGS IN PHILOSOPHY

ROBERT NOZICK, *Advisor*
Harvard University

VOLUME

6

THE LEGACY
OF THE
VIENNA CIRCLE

MODERN REAPPRAISALS

Edited with introductions by

SAHOTRA SARKAR

*Dibner Institute at MIT
and McGill University*

Routledge
Taylor & Francis Group

NEW YORK AND LONDON

First published by Garland Publishing, Inc.
This edition published 2013 by Routledge
711 Third Avenue, New York, NY 10017
2 Park Square, Milton Park, Abingdon, Oxfordshire OX14 4RN

Routledge is an imprint of the Taylor & Francis Group, an informa business

Introductions copyright © 1996 Sahotra Sarkar
All rights reserved

Library of Congress Cataloging-in-Publication Data

The legacy of the Vienna circle : modern reappraisals / edited with
introductions by Sahotra Sarkar.
     p.   cm. — (Science and philosophy in the twentieth
century ; v. 6)
    Includes bibliographical references.
    ISBN 0-8153-2267-4 (alk. paper)
    1. Vienna circle. 2. Logical positivism.   I. Sarkar, Sahotra.
II. Series.
B824.6.L43 1996
146'.42—dc20   95–50328
                                                  CIP

SET ISBN	9780815322610
POD ISBN	9780415628365
Vol1	9780815322627
Vol2	9780815322634
Vol3	9780815322641
Vol4	9780815322658
Vol5	9780815322665
Vol6	9780815322672

CONTENTS

METAPHYSICS

EPISTEMOLOGY

SERIES INTRODUCTION

The early years of the twentieth century saw remarkable developments in the sciences, particularly physics and biology. The century began with Planck's introduction of what came to be known as the "quantum hypothesis," followed by the work of Einstein, Bohr, and others, which paved the way for the development of quantum mechanics in the 1920s. It remains the most radical departure from the classical worldview that physics has seen. Not only were some physical quantities "quantized," that is, they could only have discrete values, but there were situations in which some of these values were indeterminate. Perhaps even worse, the basic dynamics of physical systems was indeterministic. The mechanical picture of the world, inherited from the seventeenth century, and already under attack during the nineteenth, finally collapsed beyond hope of recovery. Nevertheless, the new physics was unavoidable. Not only did atomic phenomena abide by its rules, but it provided a successful account of chemical bonding and valency. Meanwhile, in 1905, Einstein's special theory of relativity challenged classical notions of space and time. A decade later, general relativity replaced gravitation as a force by the curvature of space-time. Developments in astrophysics confirmed general relativity's unusual claims.

Also around 1900, biologists recovered the laws for the transmission of hereditary factors, or "genes." These laws, though published by Mendel in 1865, had remained largely unknown for a generation. By 1905, a new science called "genetics" had been created. For the first time, the phenomena of heredity were subsumed under exact (mathematical) laws. In the early 1920s, these laws were used by Fisher, Haldane, and Wright to formulate a quantitative, basically testable theory of evolution by natural selection. Around 1900 it also became clear that the transfer of chromosomes mediated the transmission of hereditary characters from parents to offspring. Between 1910 and 1920, genes were shown to be linearly positioned on chromosomes. The rudiments of a physical account of biological inheritance were in place by the mid-1920s. Eventually this work was integrated with other biological subdisciplines, especially biochemistry (itself largely a turn-of-the-century creation), to generate molecular biology, arguably the greatest triumph of science since 1950.

The philosophical response to the advances of early twentieth century science was schizophrenic. Some philosophers, especially in Germany, ignored scientific developments almost altogether and continued to elaborate extensive metaphysical systems having little contact with the physical world. Collectively, these projects came to be called phenomenology. In sharp contrast, another group of philosophers attempted to reform—or, perhaps, even replace—academic philosophy so as to bring it into consonance with modern science. At times, they claimed to have inherited the mantles of Aristotle and Descartes, Newton and Leibniz, Locke and Hume, Kant and Marx. More often, they claimed to be doing something altogether novel.

Most prominent among the latter group of philosophers were those who called themselves "logical positivists" or "logical empiricists." Many of them were associated, in their early years, with a group that met regularly in Vienna (starting in 1922) and called itself the Vienna Circle. The central figure was Moritz Schlick. (A complete list of members of the Vienna Circle will be found in their 1929 manifesto, which is reprinted in Volume 2.) The members of the Vienna Circle had an almost worshipful attitude towards the new physics though, in general, they seemed to have been completely ignorant of the equally fundamental changes taking place in biology. They were impressed by developments in logic, particularly Whitehead and Russell's attempt to carry out Frege's project of constructing mathematics from logic. Kurt Gödel, a member of the Vienna Circle, though hardly a logical empiricist in his philosophical leanings, probed the foundations of logic and showed that any relatively complex system of mathematics must allow statements to be formulated that can neither be proved nor disproved using formalized rules of proof—this is Gödel's famous incompleteness theorem.

Meanwhile, in Berlin, a smaller group around Hans Reichenbach came to a similar philosophical orientation and concentrated on probing the foundations of physics. In Poland, an eminent group of logicians, with Alfred Tarski as the central figure, began equally important investigations of logical notions. There was considerable intellectual exchange between these different groups. These exchanges led to convergence on many points—the philosophical theses that were most commonly advanced will be described below (and in the introductions to Volumes 1–4).

To return to the historical story, most of the logical empiricists had relatively progressive politics. A few, notably Otto Neurath, were avowed Marxists. Others, including Rudolf Carnap and Hans Hahn, were socialists. With the rise of nazism and fascism in Europe in the 1930s, many of the logical empiricists emigrated to Britain and, especially, to the United States. There they eventually came to establish a temporary hegemony over academic philosophy. Reichenbach moved to the University of California at Los Angeles; Herbert Feigl to the University of Minnesota; and Carnap to the University of Chicago. Meanwhile, during his youthful days, W.V.O. Quine was already preaching the logical empiricist gospel at Harvard. Of the major figures, only Neurath remained in Europe. (Hans Hahn had died in 1934 and Schlick had been murdered in 1936—see the introduction to Volume 2.)

Because of its migration to the U.S., logical empiricism became part of the Anglo-American tradition in philosophy, in spite of its European origins. It is at least arguable that as a movement it matured in the U.S. However, in spite of being relatively organized compared to other philosophical movements, the logical empiricists did not present a unified system of universally held theses—a point that seems to elude their modern critics—though they generally exhibited a coherent attitude to the analysis of philosophical problems. This attitude can be traced back to the 1920s. They *generally* accepted an *a priori* faith in logic, though they were sometimes known to disagree on what logic could be. Other than in logic (and in mathematics, which, for most logical empiricists, could be derived from logic), the logical empiricists endorsed a thoroughgoing empiricism—hence their name. All factual (that is, nonlogical) knowledge was ultimately empirical. A sharp distinction between empirical, *a posteriori*, synthetic claims on one hand and *a priori*, analytic claims on the other was a cherished doctrine for most (but not all) logical empiricists. Its rejection by Quine and others in the 1950s was a significant event in the decline of logical empiricism (see Volume 5).

Any claim that was neither logic nor able to be adjudicated by empirical means was rejected by the logical empiricists as "meaningless" or "cognitively insignificant," whatever its noncognitive (for instance, emotional) appeal. Logic escaped this fate by being true by virtue of meaning (of the logical connectives such as "not" and "and" and operators such as "all," "any," and "some") or of conventions. Mathematics was true because it could be reduced to, or constructed from, logic. Besides logic, the logical empiricists generally did not accept any other normative discipline as consisting of meaningful claims. (Ethical claims, according to some of them, were only devices to evoke appropriate emotive responses from others.)

Given these positions, there did not remain much metaphysics to be done (at least insofar as "metaphysics" was interpreted by the academic philosophers). Some logical empiricists, notably Carnap, claimed to have successfully eliminated metaphysics. In practice, metaphysics was replaced by attempts—rarely profound—at the analysis and interpretation of scientific concepts. Those logical empiricists who were particularly enamored of the technical apparatus of mathematical logic, again, most notably Carnap, interpreted this endeavor as describing the syntax and elaborating a semantics for the language of science. (In the case of logic itself, the logical empiricists achieved some important successes in their interpretive efforts in the 1930s—see Volume 3.) Metaphysics cast aside, the logical empiricists turned to epistemology; in particular, to the possibility of quantifying the extent to which different scientific claims were grounded in experience. The project turned out to be far more complex—and convoluted—than initially envisioned. By the time logical empiricism disappeared as an explicit movement within philosophy, little progress had been made towards this end.

An enumeration of positions advanced—or of successes and failures—only barely captures the spirit of logical empiricism. Within their self-proclaimed framework of accepting only logic and empirical knowledge, they venerated a critical attitude. This included continual self-criticism. Much has been written about the untenability of the doctrines espoused by the logical empiricists—what unfortunately goes unrecognized is that the most severe (and the most relevant) criticisms almost always came from within the movement or, at least, from individuals schooled in the movement (notably Quine). There were significant disagreements among the logical empiricists (for instance, between Carnap and Reichenbach on epistemology). There were also significant disagreements within the Vienna Circle: Kurt Gödel probably rejected most of the tenets in the Vienna Circle manifesto; Karl Menger refused to reject metaphysics on logical grounds (see his paper in Volume 2). These cases, however, may only show that not all members of that circle should be regarded as logical empiricists. Nonetheless, and most importantly, the logical empiricists believed philosophy to be a collective enterprise, like the natural sciences, and one in which progress could be made.

The logical empiricists' domination of Anglo-American philosophy was never complete and whatever hegemony they established was brief. Even within their chosen subdisciplines, such as the philosophy of science or logic or mathematics, their positions came under attack in the 1950s. Cherished doctrines such as the analytic-synthetic distinction were abandoned by a new generation of philosophers. The value of their type of conceptual analysis was sometimes derided by the later Wittgenstein's followers and by the so-called "ordinary language" philosophers. Metaphysics returned with a vengeance and, arguably, the influence of the logical empiricists was largely confined to the philosophy of science after the 1950s. But the 1960s saw logical empiricism under attack even among philosophers of science. It is probably reasonable to say that by around 1970, a new generation of philosophers of science had decided that the analyses offered by the logical empiricists were largely superficial and were to be replaced by more sophisticated work. The most popular position of those days was "scientific realism," a return to exactly the kind of metaphysics that the logical empiricists had found devoid of cognitive content.

Significant interest in logical empiricism resurfaced again in the early 1980s. This did not indicate any general return to the positions the logical empiricists advocated. Rather, the source of the interest was largely historical, part of a desire to understand the history of twentieth-century philosophy. It was aided by a new interest among philosophers in the history of the philosophy of science. Carnap and Reichenbach were probably the only prominent logical empiricists who had continued to be read during the 1960s and 1970s; now the works of Schlick and Neurath, among others, were once again read (and, sometimes, translated into English for the first time). Archives began to be mined to expose the intricate details of the relationships between the logical empiricists, and between them and other social and cultural movements of the 1920s and 1930s. This new work took place not only in the U.S., but also in Austria, Germany, and to a lesser extent, elsewhere in Europe. Slowly, as this historical work has progressed, a more positive philosophical assessment of the movement than was usually found in the 1960s and 1970s has also emerged (Sarkar 1992). These developments are far too recent for any assessment to be made of their lasting value. While the historical interest is neither hard to explain nor appreciate, it is less clear why, but perhaps even more interesting that, this positive reassessment is taking place.

There seem to be at least three reasons for the relatively positive reassessment that deserve mention: (1) since more than a generation had passed between the heyday of the movement and the mid- and late-1980s, the new commentators found it easier to have a more balanced view of both the contributions and the failures of logical empiricism than those—especially in the 1960s— who felt that they had to react to its dominance; (2) historical

exploration—and exegesis—has revealed that the logical empiricists held a variety of views that are both more complex and more
interesting than what their critics attributed to them (see, for
example, Suppe 1974); and (3) arguably, the various alternatives
to logical empiricism as a philosophy of science that were formulated in the 1960s and 1970s have not delivered on their promises.
Going further, and much more controversially, these alternatives
(including scientific realism) have proved less fertile and less
robust than logical empiricism.

In this new intellectual context, it seems appropriate to make
available, to as wide an audience as possible, some of the basic
works of logical empiricism, as well as some of the new commentaries that have followed the renewal of interest in the movement.
Many of the original pieces are not easily available and there is, at
present, neither a detailed history of logical empiricism nor an
annotated guide to its most important writings. An important old
collection is Ayer (1959), which has a fairly comprehensive bibliography of work up to that point. Many valuable collections devoted
to individual figures have been published. Schilpp (1963) collects
many important critical pieces on Carnap, with Carnap's responses. The basic works of Feigl (1981), Hahn (1980), Kraft
(1981), Menger (1979), Neurath (1973, 1987), Reichenbach (1978),
Schlick (1979, 1987), and Waismann (1977) have been published
as part of the Vienna Circle Collection. Collections of articles on
logical positivism from the 1960s and 1970s include Achinstein
and Barker (1969) and Hintikka (1975). Recent works of interest
include Coffa (1991), Haller (1982), Menger (1994), and Uebel
(1991, 1992). However, a detailed history of logical empiricism
remains to be written.

What makes this series different from these works is an
attempt to present a global picture of logical empiricism, including
the influences that led to its initiation and the criticisms that were
responsible for its decline. The emphasis here is on issues rather
than on individual figures even though some of the most influential figures—especially Carnap and Reichenbach—feature prominently. However, for most of the topics treated, all the historically
and conceptually important exchanges on that topic are collected
together. Finally, modern commentaries are also included to bring
the series up to date. In general, complete papers (in English
whenever translations are available) are included over book sections in an effort to present complete arguments as far as possible.
Volume 1 deals with the initial influences on logical empiricism
and with the Vienna Circle period. Volume 2 concerns primarily
the 1930s, when logical empiricism was at its most confident

phase, when its adherents truly believed that they were reforming philosophy for all future times. Volume 3 includes pieces that reflect logical empiricism in its mature phase, after self-criticism and technical developments induced more sophisticated doctrines than those produced in the 1930s. Volume 4 shows how logical empiricism analyzed the special sciences. Volume 5 consists of the most important criticisms of logical empiricism and its responses. It marks the decline of logical empiricism. All of these volumes, except Volume 4, include a concluding section with modern commentaries. Volume 6 consists entirely of these commentaries. Each volume is introduced with an editorial note that puts the contents in perspective. Thanks are due to Richard Creath and Alan Richardson for advice on selecting the pieces for this series, and to Gregg Jaeger for help in assembling them and for commenting on the introductions. Work on these volumes was done while the editor was a Fellow at the Dibner Institute for the History of Science at MIT. Thanks are due to it for its support.

FURTHER READING

Achinstein, P. and Barker, S., eds. 1969. *The Legacy of Logical Positivism*. Baltimore: Johns Hopkins University Press.

Ayer, A.J., ed. 1959. *Logical Positivism*. New York: Free Press.

Coffa, A. 1991. *The Semantic Tradition from Kant to Carnap: To the Vienna Station*. Cambridge, UK: Cambridge University Press.

Feigl, H. 1981. *Inquiries and Provocations: Selected Writings, 1929 – 1974*. Dordrecht: Kluwer.

Hahn, H. 1980. *Empiricism, Logic, and Mathematics: Philosophical Papers*. Dordrecht: Kluwer.

Haller, R., ed. 1982. *Schlick und Neurath—Ein Symposion. Grazer philosophische Studien* 16 –17.

Hintikka, J., ed. 1975. *Rudolf Carnap, Logical Empiricist*. Dordrecht: Reidel.

Kraft, V. 1981. *Foundations for a Scientific Analysis of Value*. Dordrecht: Kluwer.

Menger, K. 1979. *Selected Papers in Logic, Foundations, Didactics, Economics*. Dordrecht: Kluwer.

Menger, K. 1994. *Reminiscences of the Vienna Circle and the Mathematical Colloquium*. Dordrecht: Kluwer.

Neurath, O. 1973. *Empiricism and Sociology*. Dordrecht: Kluwer.

Neurath, O. 1987. *Unified Science*. Dordrecht: Kluwer.

Reichenbach, H. 1978. *Selected Writings, 1909–1953*. Vols. 1, 2. Dordrecht: Kluwer.

Sarkar, S., ed. 1991. *Rudolf Carnap—A Centenary Reappraisal*. Synthese 93.

Schilpp, P. A., ed. 1963. *The Philosophy of Rudolf Carnap*. La Salle, IL: Open Court.

Schlick, M. 1979. *Philosophical Papers*. Vols. 1, 2. Dordrecht: Kluwer.

Schlick, M. 1987. *The Problems of Philosophy in Their Interconnection*. Dordrecht: Kluwer.

Suppe, F., ed. 1974. *The Structure of Scientific Theories*. Urbana: University of Illinois Press.

Uebel, T. E., ed. 1991. *Rediscovering the Forgotten Vienna Circle*. Dordrecht: Kluwer.

Uebel, T. E. 1992. *Overcoming Logical Positivism from Within: The Emergence of Neurath's Naturalism in the Vienna Circle's Protocol Sentence Debate*. Amsterdam: Rodopi.

Waismann, F. 1977. *Philosophical Papers*. Dordrecht: Kluwer.

INTRODUCTION

As was mentioned in the introduction to this series, since the early 1980s, there has been a revival of interest in logical empiricism. The resulting reappraisal of that movement has not led to a revival of logical empiricism or an endorsement of its best-known epistemological (and metaphysical) doctrines, such as those that became the major foci of criticism (the analytic-synthetic distinction and the empiricist criteria for cognitive significance, among others—see Volume 5), at least insofar as those doctrines were presented in their undiluted form in the 1930s. Certainly, the reappraisal has not resulted in any endorsement of the logical empiricists' analyses of the empirical sciences (see Volume 4). Conventionalism in physics continues to be uniformly unpopular. Physicalism, as understood by the logical empiricists, is usually held to be misleading or vacuous in both psychology and sociology. Reductionism in biology (which, in any case, was not an important logical empiricist thesis) is rarely defended.

Nevertheless, it is at least arguable that the postpositivist philosophies of science have not delivered on their promises from the 1960s any more than logical empiricism delivered on its promises from the 1930s. In this context, there is a growing perception that there remains much to be learned from the details of the logical empiricist enterprise, including its mistakes—a theme that will recur in many of the pieces in this volume. Moreover, to some modern commentators, especially those who came to philosophy when logical empiricism's dominance was long gone (including the present editor), it is far from clear that the arguments against logical empiricism that became dogma in the 1960s and 1970s are quite as forceful as they are often made out to be. These arguments may rule out the epistemology of logical empiricism in the 1930s, but are not compelling against the more sophisticated doctrines of the 1950s. Logical empiricism continues to be philosophically interesting, perhaps more so than the post-positivist philosophies of science that replaced it.

At the same time, much of the recent attention to logical empiricism has been historical. This has resulted in the deployment of the usual methods of historical excavation: examination of little-known texts, archival research, and contextual examination of background and influence. The logical empiricists emerge from this work as more fascinating figures than their often ponderous style of writing would suggest. They turn out to be strongly constrained by the history of philosophy they claimed to reject, and certainly not quite as revolutionary as they claimed to be. Nonetheless, a detailed examination of their positions reveals that they were also not quite as naive as their critics in the 1950s and 1960s claimed they were.

The final volume of this series collects some of these recent writings on logical empiricism. The main criterion of selection (besides quality) has been diversity, not only of viewpoint but also of approach. Almost all these pieces have already been put into context in the introductions to the previous volumes of this series, and so the introductory remarks here will be very brief. The first two sections of this volume deal primarily with historical issues; the last two with more philosophical ones. However, as will be indicated below, this division should not be taken too seriously—it was adopted more for convenience of organization than because of any deep commitment to a difference between history and philosophy in this context.

The first section deals with conceptual history. Sauer's paper is on the Kantian background of logical empiricism, as is Richardson's paper in the third section (see also the papers by Haack and Friedman in Volume 1). Ryckman's paper examines the role of implicit definition in the early epistemologies of Carnap and Schlick. Lewis's paper provides a critical examination of Schlick's early work. Haller's paper discusses the meetings between Hahn, Frank, and Neurath before Schlick arrived in Vienna in 1922. Jeffrey's short paper provides an insider's view of Carnap's method of working (Jeffrey was one of Carnap's collaborators on inductive logic.) The emphasis in this paper is on how Carnap pursued philosophy as a collective project.

The second section provides a sample of the new historical agendas that have been used to illuminate the development of logical empiricism. Wartofsky's paper provides a searching examination of logical empiricism as a political movement. Galison's paper explores the Vienna Circle's interactions with the Bauhaus. Stadler's paper analyzes the some of the political and ideological tensions within the Vienna Circle. Finally, Reisch's paper provides the only available detailed history of the unity of science project in the late 1930s. It is a fascinating look at how the project led to the publication of two volumes of the *International Encyclo-*

pedia of Unified Science and then collapsed. (See, also, Holton 1992.)

The third section consists of papers that look mainly at the metaphysical tenets of logical empiricism. Only Coffa's paper is primarily historical: it is a history of the invention and development of formal semantics in the 1930s, with particular attention to Tarski and Carnap. (More personal and historical detail can be found in Coffa 1991.) Putnam attempts a critical exposition of Reichenbach's metaphysics. He provides a discussion of the differences between Reichenbach and Carnap in a partly personal account—Putnam was Reichenbach's student at the University of California at Los Angeles. Shimony's paper is equally personal— he was, briefly, Carnap's student at the University of Chicago. This paper is important because it shows how Carnap's epistemology is supposed to work in actual scientific contexts. It also criticizes Carnap's metaphysical views. Richardson's paper has already been mentioned; it explores all the neo-Kantian influences—not just Cassirer—on the early Carnap. (More detail will be found in Richardson [forthcoming]). Friedman's paper, in spite of its misleadingly general title, only deals with early logical empiricism and provides a summary of the Kantian influences on Schlick, Carnap, and Reichenbach.

The last section consists of papers that analyze the epistemological tenets and methods of the logical empiricists. Only two of these pieces—by Oberdan and Uebel—are basically historical. Oberdan analyzes the nature of observation in Carnap's later work. Uebel's position is radical: not only does he argue that Neurath anticipated Quine's program in naturalistic epistemology (which is relatively uncontroversial), he goes on to say that Neurath anticipated parts of the views of such critics of logical empiricism as Hanson and Kuhn. (Uebel argues his position in more detail in *Overcoming Logical Positivism from Within* [1922]).

Jeffrey's paper in this section provides an assessment of Carnap's later work on inductive logic (a description of that system can be found in Hilpinen's paper in Volume 3). Salmon's paper gives a critical discussion of Reichenbach's approach to the problem of induction. It is, partly, a defense of Reichenbach. Skyrms's paper is about a technical extension of Carnap's inductive logic. It is included here because, besides its intrinsic interest, it shows how even some of the technical work of the logical empiricists continues to be a resource for further inquiry. Finally, Stein's paper provides a searching examination of the cogency of Quine's criticisms of Carnap. He argues persuasively that Quine's arguments against analyticity have less force than customarily thought. He also analyzes Carnap's later program (from the 1950s) in some detail and points out both its shortcomings and its advantages (especially when compared to Quine's epistemological holism). More than anything else, Stein's paper shows that reports of the demise of logical empiricism (see the introduction to Volume 5) are at least somewhat exaggerated.

FURTHER READING

Coffa, A. 1991. *The Semantic Tradition from Kant to Carnap: To the Vienna Station*. Cambridge, UK: Cambridge University Press.

Holton, G. 1992. "Ernst Mach and the Fortunes of Positivism in America." *Isis* 93: 27–60.

Richardson, A. Forthcoming. *Carnap's Construction of the World: The* Aufbau *and the Emergence of Logical Empiricism*. Cambridge, UK: Cambridge University Press.

Uebel, T. E. 1992. *Overcoming Logical Positivism from Within: The Emergence of Neurath's Naturalism in the Vienna Circle's Protocol Sentence Debate*. Amsterdam: Rodopi.

On the Kantian Background of Neopositivism

Werner Sauer

I

According to what may be called the received view of Neopositivism's place in the history of philosophy, the title of this article must sound far-fetched at least. For according to that view Neopositivism, as it flourished in the Vienna Circle and elsewhere, is in a simple and straightforward way antithetical to Kantian philosophy because of its radical denial of the latter's basic thesis of a synthetic *Apriori*, and beyond this also in the wider sense of representing a philosophical tradition entirely heterogeneous from the Kantian one.

This view took shape already in the Vienna Circle's own days, when in its manifesto of 1929, *Wissenschaftliche Weltauffassung. Der Wiener Kreis*, Otto Neurath in an attempt to throw the light of *Ideengeschichte* on the Circle's genesis, traced its teachings back to sources such as the earlier empiricism and positivism from Hume to Mach, the methodology of science from Helmholtz to Einstein, and the development of mathematical logic by the work of, among others, Frege and Russell.[1] The method employed by the new empiricism and positivism of the scientific world-conception, the manifesto then explains its central tenets, is that "of logical analysis", applied with the aim of showing that what can be known *a priori* is only tautologies (analytic statements), and of effecting a "step-by-step reduction" of every empirical concept "to concepts . . . referring to the given itself"; the ultimate result of this reduction process would be the establishment of the *Einheitswissenschaft* in the form of "a constitutional system" of those concepts such that "the constitutional theory", i.e., the theory concerned with the nature of such a conceptual system, as set forth in Rudolf Carnap's *Der logische Aufbau der Welt* (1928), "provides the frame in which logical analysis is applied by the scientific world-conception".[2]

Thus in the manifesto the historical background of Neopositivism is drawn with no reference to the Kantian tradition at all — even though at least when Helmholtz is mentioned, being he after all one of the precursors of Neokantianism, such a reference would only be a matter of course. Kantianism comes into the picture only as an opponent whose basis has been definitely uprooted by the method of the scientific world-conception; and moreover, the negation of that basis figures in the account given as the very cornerstone of the new movement:

> . . . through logical analysis metaphysics not only in the . . . classical sense of the word is overcome, . . . but also the covert metaphysics of Kantian and modern apriorism. The scientific world-conception does not know . . . of any "synthetic judgments *a priori*" as presupposed by Kant's theory of knowledge. . . . It is precisely the rejection of the possibility of synthetic knowledge *a priori* which constitutes the fundamental thesis of modern empiricism.[3]

Weighty protest against this way of viewing the relationship between the two philosophical movements has come from Lewis White Beck, the eminent Kant scholar and historian of philosophy. "The resemblances", he maintains in disagreement with the received view,

> between Kant's program and the program of the school of logical positivism are unmistakable. A history of logical positivism should bring its Kantian provenance to light. The mediating role of the Marburg school of neo-Kantians should be given special attention. . . . While the differences are as obvious as the similarities, they have been so magnified . . . that the resemblances have been ignored and possible historical influences little suspected.[4]

In the following, material will be assembled which supports the thesis Beck has put forward, and in doing so we shall concentrate in systematic respects chiefly on Carnap's before-mentioned book: a restriction surely not wholly arbitrary and unjustified in the face of the central role within the scientific world-conception attributed to the *Aufbau's* constitutional theory in the Vienna Circle's manifesto. Before going into the *Aufbau*, however, some remarks on the two features of the received view as such will be in order.

Topoi 8: 111–119, 1989.
© 1989 *Kluwer Academic Publishers. Printed in the Netherlands.*

1

II

To turn first to the question of the synthetic *Apriori*, we are even in this seemingly so obvious and straightforward case of disagreement well advised to be cautious in order not to fall prey to mere *idola fori*, handy slogans which conceal rather than reveal what the issue really amounts to. Of course, Kant asserts and the Neopositivists deny that there are epistemologically sound statements that are both synthetic and *a priori*. But in fact the matter is not as clear-cut as that; to see this it will suffice to consider briefly Kant's thesis that arithmetic is synthetic.

Kant's claim that "$7 + 5 = 12$" is not analytic but synthetic is the claim that such a formula is not true solely in virtue of the principle of contradiction, or, put differently, that it does not, explicitly or implicitly, possess the form "*AB* is *B*".[5] The polemical side of this doctrine was directed against the Leibnizian account of arithmetical truth. Leibniz not only asserted, as Hume did, what Kant then denied but actually attempted to demonstrate, using "$2 + 2 = 4$" as his example, the reducibility of arithmetical equations to overt identities by means of definitional substitutions only.[6]

Frege, in renewing the quest for a demonstration of the analyticity of arithmetic, agrees with Kant to the point that the literal Leibnizian view is untenable.[7] So he replaced the account of analyticity upon which the issue between Kant and Leibniz was based by a somewhat broader one which expands Kant's truth in virtue of the principle of contradiction into truth in virtue of the general laws of logic (*plus* definitions, of course).[8] However, since Kant characterizes logical truth just as analyticity in terms of the principle of contradiction,[9] we may safely reconstruct his notion of analyticity such that it nominally fits Frege's. Of course, Frege's logic is far more powerful than the logic Kant knew. But again, nor does this fact as such yet create any profound difference in the matter; in the final analysis, the difference that really counts lies rather in a basic epistemological assumption to which, furthermore, the analytic/synthetic distinction is differently related to.

From Kant's account of the distinction it immediately follows that in case a statement is synthetic its truth depends on there being given some object as the extra-conceptual "third thing" in which the statement's concepts are interconnected. Now the basic epistemological assumption alluded to above is, in Kant's case, that objects are given to us never by thought unrelated to

sensibility, or pure reason, but only through sensibility-based intuition, and hence to make an epistemologically sound synthetic statement involves referring, in some way or other, to intuition, be it empirical or pure.[10] Thus the core of Kant's doctrine that arithmetic is synthetic amounts to the two theses that its truths are about *objects* and that, secondly, those objects are supplied by (of course, pure) intuition.

It is the second thesis which Frege attacks. He rejects Kant's epistemological assumption and claims that objects can be given to us by pure reason as well.[11] But from this he does not draw the conclusion that pure reason yields synthetic as well as analytic statements: strictly holding fast to the connection of syntheticity with intuition[12] he takes the opposite line and infers that there are analytic statements involving reference to objects. And he believed to have shown that there are in fact such *logical objects*, out of which then the numbers could be constructed, by way of the principle embodied in axiom (v) of his *Grundgesetze der Arithmetik*, stating, roughly, that iff the same objects fall under both concepts *F* and *G* the value-ranges of *F* and *G* are identical. However, Frege himself was not all too sure as to the logicality of this principle which links a statement about *concepts* to one about *objects*, items he strictly distinguishes from concepts; commenting on the basis of his edifice he writes, "As far as I can see, dispute can arise only about my principle of value-ranges (v) . . . I hold it to be purely logical. In any case, by it the place is marked where the decision must fall".[13]

And it fell. What is here of interest are the conclusions Frege eventually drew from Russell's antinomy which overthrew his axiom (v). The antinomy, he came to be convinced, has shown that by "the logical source of knowledge . . . alone no objects can be given", and hence, that one has "to give up the opinion that arithmetic is a branch of logic".[14] At first sight Frege's total rejection of logicism may appear as an extreme overreaction to the antinomy, but in truth it was just consistent. Frege holds, as Kant did, that universality and abstraction from any particular domain of objects are essential marks of a logical law;[15] but after the antinomy he could have sustained his programme only by imposing certain restrictions upon axiom (v) which, when so modified, he could not have acknowledged as a *logical* principle anymore.

Now Russell, carrying on the logicist programme and, unlike Frege, applying it also to geometry, claimed in 1903 to have demonstrated "that all mathematics follows from . . . logic";[16] however, what he thought to have shown thereby is emphatically not what one would expect. Clarifying his position he says,

> Kant never doubted for a moment that the propositions of logic are analytic, whereas he rightly perceived that those of mathematics are synthetic. It has since appeared that logic is just as synthetic as all other kinds of truth.[17]

And this very peculiar brand of logicism Russell espouses still in 1912: two years after the publication of volume I of his and Whitehead's *Principia Mathematica* he still says that Kant "deserves credit . . . for having perceived that . . . all pure mathematics, though *a priori*, is synthetic",[18] and hence, by implication, that logic is synthetic as well.

The objectionable element of Kant's philosophy of mathematics Russell sees, like Frege in the case of arithmetic, in the connection of mathematics with intuition.[19] But as is already apparent, in the period under consideration Russell does not move on from his rejection to the analyticity thesis, and *insofar* he remains a Kantian. Still, on the deeper level concerning the sources of logico-mathematical knowledge his divergence from Kant is as crucial as is Frege's. The central issue between Frege and Kant was about the question whether by pure reason we can attain to truths about objects, and such truths of reason in the tradition of rationalism is what Russell's logicism is about just as much as Frege's. In Russell's case those objects are universals, i.e., *n*-adic properties, conceived of platonistically; in his own words, "[a]ll *a priori* knowledge" — hence all mathematical knowledge — "deals exclusively with the relations of universals".[20]

So, when the Neopositivists saw in the rise of logicism just a successful assault on the very stronghold of the synthetic *Apriori* in the service of empiricism,[21] a considerable amount of historical repression was involved. The common denominator of the logicism both of Frege and of Russell up to *Principia* was a *rationalistic* conception of logicism opposed not only to *Kantian* Apriorism but even more so to empiricism. To be sure, both of them came to abandon rationalism. But for Frege this was tantamount to abandoning logicism altogether. Russell, under the influence of Wittgenstein, indeed soon came to hold that logic, and hence mathematics, consists only of empty tautologies;[22] yet he remained skeptical as to the possibility of a satisfactory explication of the notion, similar in intention to Kant's analyticity, of tautologousness, still in 1937 candidly

admitting to be "unable to give any clear account" thereof.[23] But without such an account, which, moreover, has not become a more promising project since then either,[24] the theses of the Neopositivists as to what logicism has proved against Kant are necessarily obscure and lacking in sound foundations; moreover, later on we shall have occasion to note *in concreto* convergences with certain Kantian views on the role of formal elements in the fabric of human knowledge hardly reconcilable with the ideas of mere repetitiveness and emptiness associated with the notions of analyticity and tautologousness.

III

Turning now to the other, more general feature of the received view, we find that Neurath's reconstruction, in the Vienna Circle's manifesto, of Neopositivism's historical background did not even represent the *communis opinio* within the movement itself. For when from the Austrian Neurath who in socio-cultural perspective tied the movement's rise to the intellectual climate prevailing in Austria, conspicuous for its lack of any stronger Kantian tradition,[25] we turn to German members of the movement who did not share Neurath's background nor his obsession to purge the movement from any connection to the "German tradition", an entirely different picture emerges.

A striking piece of evidence for this we find in the journal *Erkenntnis*, the joint organ of the Vienna Circle and the Berlin Society of Scientific Philosophy, edited by Carnap and Hans Reichenbach. In volume three of the journal there appeared the correspondence between Friedrich Albert Lange, one of the founders of Neokantianism, and the zoologist Anton Dohrn. In their introduction to the correspondence the journal's editors, i.e., Carnap and Reichenbach, declare as the aim of its publication the furthering of the scientific way of thinking in philosophy, thereby celebrating Lange's *Geschichte des Materialismus* (1866) as an attempt to

> transform scientific thinking into philosophical criticism and, at the same time, to test philosophical thinking at scientific material . . . ; and indeed the historical merit of Lange, and of the Marburg school founded by him, was exactly to have rediscovered the scientist Kant and to have prevailed with him over the metaphysical interpretations of Kantianism.[26]

And concerning the correspondence they write:

> Let us here have speak for us a scientist and a philosopher;
> certainly, the contents of their problems are no longer ours, their
> Kantianism is as far behind us as the physics of the ether and of
> classical mechanics, but their attitude and their striving, their
> earnestness and their affirmation of the scientific way of thinking
> are still a model for us today.

Kantianism as a forerunner, in spirit still exemplary, of the new scientific philosophy: the change in outlook as against that of the Vienna Circle's manifesto is striking indeed. In this view, Kantianism, at least as exemplified by Lange and by the Marburg school of Neokantians, and Neopositivism belong to *one* tradition in philosophy, united by the spirit of scientific philosophizing, but separated by the developments in science from classical to modern physics;[27] and this view of the relationship between the two philosophical movements must have occurred quite naturally to a Carnap and a Reichenbach, reflecting, as it does, their own development, as both of them began their career as philosophers of science within the framework of Neokantianism.

Reichenbach's early Kantianism shall not concern us here,[28] but only Carnap's. In his doctoral thesis on space of 1921, Carnap espoused a Kantian position at least insofar as he regarded the topological properties of perceptual space as synthetic *a priori* in Kant's sense.[29] Taking into account further works of Carnap in that period, we get, in a very brief outline, the following general picture of his early philosophical position. Regarding "the sources of physical knowledge", Carnap contends, "pure empiricism has lost its dominance", since it has increasingly become evident that in "the construction of physics ... non-empirical principles must be employed".[30] In his eyes, the main deficiency of Kant's synthetic *Apriori*, understood as a set of *incorrigible* non-empirical principles of experience, is its being far too rich in content: only the "form factors" of immediate sense-experience, namely "a certain spatial and temporal order, and further, certain qualitative relations of sameness and difference", are synthetic *a priori* in this sense,[31] whereas the non-empirical form factors governing the transition from the "primary world" of immediate sense-experience to the "secondary world" both of common sense and of science are conventional stipulations.[32]

However, conventionalism regards the secondary world is not really a break with Kantian Apriorism as such. Carnap criticizes Neokantianism — clearly with the Marburg school of Hermann Cohen and his followers in mind — mainly on two points: to have overlooked the difference between the primary and the secondary world, and to hold the latter to be uniquely determined by incorrigible form factors which according to him are true only of the primary world. By these criticisms, however, Neokantianism's "true achievement", namely, the demonstration of the object-producing function of thought", Carnap emphasizes, "remains untouched and underlies also our own conception of the secondary world".[33] So one might call his own non-empirical principles of the secondary world as well a relativized synthetic *Apriori*, and he himself refrains from doing so only because he wants to reserve the expression "synthetic *a priori*" for incorrigible principles.[34] And it may be added that with his conception of the secondary world Carnap is even closer to the Neokantians than he himself admits if we take into account the fact, neglected by him, that they themselves were well on the way to liberalize their synthetic *Apriori* in accordance with the evolution of science and to take it in a more relativistic spirit.[35]

Carnap's principal deviation from (Marburg) Neokantianism, then, lies in the notion of the primary world, which notion combines the positivistic idea of the given with the original Kantian idea of synthetic *a priori* forms of intuition.[36] In this combination the Kantian component is dominant, however: the defining mark of the primary world is not experiential immediacy but "necessity of [its] forms",[37] and correspondingly, the givenness of the primary world does not refer to any pure content of experience, this being "a mere abstraction of thought",[38] but to the content already *formed* by those necessary (i.e., synthetic *a priori*) form factors which determine the realm of the primary world.

As Carnap himself says, from 1922 on he was already working on the *Aufbau*, of which a first version was finished in 1925; his Kantian doctrine of synthetic *a priori* forms of experience, however, he abandoned only after he had come to Vienna in 1926.[39] Presupposing the received view, and its characterization, from the Vienna Circle's manifesto up to the present, of the *Aufbau* enterprise as a grand attempt to substantiate "the old empiristic—positivistic programme ... of reducing all scientific concepts to ... what is immediately given in sense-perception",[40] the chronology is bound to cause puzzlement. But perhaps we get a more consistent picture if we look at the *Aufbau* the other way round. To be sure, its published version has

4

officially renounced the synthetic *Apriori*, tersely stating that "in the view of the constitutional theory 'synthetic judgments *a priori*' do not exist at all",[41] and thus accommodating its teaching to what the Vienna Circle's manifesto calls the basic thesis of the new empiristic-positivistic philosophy. In the face of the chronological facts cited, however, it is difficult not to suspect forthwith that this amounts to no more than a mere surface adaption, obscuring but not erasing the *Aufbau's* Kantian heritage; and taking a closer look at certain aspects of the *Aufbau* edifice, we shall indeed find ample evidence for a strong continuance of specifically Kantian patterns in Carnap's thought.[42]

IV

Carnap's aim in the *Aufbau* is to expound his constitutional theory and to sketch the outlines of a constitutional system of all empirical concepts on a solipsistic basis. To constitute a concept F on the basis of the concepts G and H is to state a rule, a "constitutional definition", which allows the transformation of any open sentence containing F into such ones that contain G and H only, and the constitutional basis, the system's "ground elements", is to be taken from the domain of one's own experiences because the *Aufbau* system should satisfy the criterion of epistemic primacy of its basis. Carnap's ground elements, however, are not the traditional sense data, sensations, etc. of empiricism, but what he calls elementary experiences, that is, places in "the stream of experience" in their "totality and undivided unity" which allow only of statements to the effect "that one such place stands to another one in a certain relation".[43] Therefore, besides its ground elements the constitutional basis must also comprise at least one relation; in the *Aufbau* system it is exactly one, the recognition-of-similarity (*Er*) obtaining between elementary experiences.

With this basis, the class *erl* of the elementary experiences and *Er*, Carnap believes to have combined the insights of two different philosophical creeds:

The merit for having uncovered the necessary basis of the constitutional system . . . belongs to two quite different, often mutually hostile philosophical schools. Positivism has emphasized that the only material of knowledge consists in the raw given of experience; there the ground elements of the constitutional system are to be found. Transcendental idealism, especially of Neokantian line, . . . has, however, rightly pointed out that

these elements don't suffice; in addition, orderings must be posited, our "ground relations".[44]

Although from this passage it may appear so, the "positivistic" and the "Kantian" component of Carnap's constitutional basis are not on an equal footing: just as before the content of the primary world was subordinated to its form, so now the material component of the constitutional basis is actually secondary to the formal ordering imposed upon it by the ground relation since subsequently *erl* will be defined (constituted) as the range of *Er*.

In discussing the requirement of introducing at least one ground relation Carnap refers, *inter alia*, to Ernst Cassirer's *Substanzbegriff und Funktionsbegriff* of 1910. In the part of the book he refers to in particular,[45] Cassirer deals with the problem of how science which is essentially abstract can deal adequately with the reality of experience which is essentially concrete. According to Cassirer the entire problem arises, however, only under the misguided presupposition of interpreting the generality of science in terms of the generality of the *genus* concept in traditional logic. The peculiar feature of the concepts of science, Cassirer has already pointed out before, lies in their role "to arrange the 'given' into sequences and to allocate it its fixed position within those sequences".[46] From this point of view there arises no logical gap between the general and the particular, since it is the very task performed by the general concept "to make possible, and to exhibit, the connection and the order of the particular itself". Therefore, the process of ordering the particular along these lines does not destroy its particularity, but leads only to its "dematerialization", as Carnap will say, since it replaces the determination of the particular through a complex of perceptual attributes by its determination through a complex of non-perceptual relations; and in this process of "idealization", the completed determination of the particular through a relation complex functions as a regulative idea which fixes "as infinitely distant point the direction of cognition" and provides us with "the true and full expression of objectivity".[47]

Cassirer's specific version of the common Marburg doctrine that "[t]he entire and indivisible content of thought must itself be a product of thought", as Cohen had put it,[48] finds its close parallel in the *Aufbau's* insistence on the primacy of form over content. In introducing a relation as his system's primitive concept Carnap pursues the aim of closing the gap between the subjectivity of individual experience and the objectivity of scientific reality: if all content of knowledge belongs to the different subjective streams of experience, then the objectivity of knowledge cannot derive from its content but must be somehow grounded in its form. To this end Carnap points out that although "the material of the individual streams of experience . . . is altogether incomparable, . . . certain structural properties are the same for all streams of experience",[49] the basic common structure of all streams of experience being given, of course, by the asymmetry of *Er*. Now, the "structural description . . . constitutes the highest stage of formalization and dematerialization",[50] and therefore, by its complete detachment from any content of experience, also the highest stage of objectification. Hence, any truly scientific statement must be transformable into a structural statement, for "science wants to speak of what is objective", whereas "everything which does not belong to the structure . . . is, in the final analysis, subjective".[51] Thus, at its completion the *Aufbau* system would be a structural description — not to be identified with a full structural determination which of course for Carnap, too, remains an "incompletable task"[52] — of empirical reality in its known entirety, and this by the construction of classes and relations with *Er* as the only non-logical basis of the edifice.

Clearly, there is a close affinity between Carnap's idea of objectification through structuralization and certain essential features of Cassirer's "logical idealism". Of course, Neokantian "logism" is not set on formal but on a transcendental logic as the logic that objectifies experience, whereas Carnap knows of formal logic only. But in fact this difference is not as serious as it looks. In Cassirer's view, the calculus of classes and relations of modern formal logic, Carnap's tool of constitution, *is* a transcendental logic. Pointing out his fundamental agreement with the Russell of 1903 in this respect,[53] he defends its syntheticity; the critical question then is, just as in the case of Kant's categories, on which domain of objects its validity is to be grounded. And the answer to this question is, again in line with Kant but contrary to Russell's platonism, that the forms of synthesis, such as class and relation, provided by modern logic find "their justified application only within empirical science itself", in that only by applying them to the experiential manifold it "becomes possible to speak of a stable and law-like order among appearances and thus of their objective significance".[54]

Although Carnap holds that "logic (including mathe-matics) consists only . . . of tautologies",[55] so that Cassirer's problem of justification would seem not to arise anymore, his approach leads nevertheless in the very same direction. Constitution is an objectifying process carried out by the "*synthetic* means . . . of class and relation".[56] Here the problem has come to the fore alluded to at the end of section II, namely, of how a logic which is tautological can yield tools for an *objectifying synthesis*. The function of formal logic, or more precisely, of its calculus of classes and relations, in the *Aufbau* is exactly that of a transcendental logic in Kant's or Cassirer's sense since it provides an answer to the Kantian question of how objective experience is possible. This question, and its answer, Carnap shares with Cassirer; it is only that in Cassirer the answer to this question is at the same time an answer to the further question of justifying that logic by providing it with objectual content which it, not being analytic in Cassirer's view, is in need of.

Compared with Carnap's earlier position, the most conspicuous change that has taken place is the dissolution of the dualism between the primary and the secondary world. This amounts to an important transformation of the primary world, in which the positivistic idea of the given was anchored, into a Kantian direction, to which we now shall turn.

V

Carnap substantiates the introduction of the elementary experiences as the ground elements by a criticism of the traditional notion of a sense datum, as exemplified by Ernst Mach's notion of a sensation. Such concrete *qualia*, Carnap argues, are "not the given itself but abstractions from it, hence something which is epistemically secondary",[57] thereby citing supporting evidence both from philosophical literature and from *Gestalt* psychology. The impact of the latter on his thought is not very specific, however, for he takes from it only the general thesis of the primacy of the whole over the part in the apprehension of a complex which he radicalizes to his conception of an elementary experience; philosophically, the notion of such undifferentiated totalities yet founding all experiential differentiations goes back, in the last analysis, to Kant's claim that every representation both is an "absolute unity" and contains "a manifold within itself".[58]

Entities of such a kind cannot be objects of immediate cognition; they are "never present in consciousness as a bare, unprocessed material, but in . . . combinations and configurations all along".[59] The primary forming of the elementary experiences is that of *Er*, the proper primitive of the *Aufbau* system. Now it is clear that the items whose structural property of asymmetry *Er* represents are not a given in the usual sense of an object of cognition but cognitive acts, and moreover, rather complex ones if we look at Carnap's explication — not in the proper constitutional language of pure structural descriptions in which there can be no explication of the primitive *Er* but externally, in the "realistic language" — of "*x* stands in *Er* to *y*": "*x* and *y* are elementary experiences which by comparing a memory image of *x* with *y* are recognized as part similar".[60] Thus an instance of recognition-of-similarity contains the following four components: (1) an elementary experience *y*; (2) the retention of an elementary experience *x* by means of a memory image; (3) a comparison of *x*, by way of its memory image, with *y*; and finally, (4) the recognition of the fact that *x* and *y* are partly similar. Now in order to arrive at (4), it is not sufficient that (1) to (3) occur; there must also obtain a certain connectedness among the four components which accounts for their being different components of one and the same act of recognition-of-similarity: required is also the unity of consciousness in them, something akin to what Kant calls the "synthetic unity of apperception".[61] In this way, the basis from which Carnap proceeds in the *Aufbau* is consciousness itself in its "cognitional synthesis", or "apperceptive processing . . . of the given",[62] as he himself puts it in conspicuously Kantian language.

So the relationship of the *Aufbau's* constitutional basis to Carnap's earlier primary world is as follows. Of the latter's forms we still have (part)-similarity and temporal order as the primary forms of the material of recognition-of-similarity: on the level of the realistic mode of expression, they are, like *erl*, already included (temporal order by implication) in the explication of *Er*, and on the level of constitutional language proper, they are, again like *erl*, definable immediately in terms of *Er*.[63] But they have ceased to be primary *simpliciter*, the real starting-point not being the primary world (now: *erl plus* part-similarity and temporal order) as such anymore but its cognitional apprehension; and herein lies the crucial change that has taken place as against Carnap's earlier position.

By now it is clear why Carnap's own characterization, cited in the previous section, of the constitutional basis

of the *Aufbau* system as a combination of a positivistic and a Kantian component is not quite accurate. According to that characterization the positivistic component of the constitutional basis is to be seen in the elementary experiences understood as a "raw given of experience". But in this sense of the word there was no given even in Carnap's former position in which the notion of givenness referred to the complete, composite primary world, the given in the sense of a raw material of experience figuring as an abstraction only. Within the *Aufbau* scheme, however, there is no room anymore even for a given as espoused by Carnap before; to use his earlier language, the primary world has also become "produced in thought", thus in its *modus essendi* merging in the secondary world.

The dissolution of the supposed positivistic component of the constitutional basis Carnap himself states clearly and forcefully when summarizing the affinities between the constitutional theory and Neokantian idealism:

> Constitutional theory and transcendental idealism hold the view in common that all objects of knowledge are constituted (in idealistic language: "produced in thought"); indeed, only as logical forms constructed in a certain way the constituted objects are objects of knowledge. This holds ultimately also of the ground elements of the constitutional system. For they are indeed first presupposed as unanalyzable units, but then in the process of constitution various properties are ascribed to them . . . ; and only by this, that is, also only as constituted objects they become objects of knowledge in the proper sense.[64]

In this passage, the distinction between being given and being constituted which carried the former distinction between primary and secondary world is definitely rejected in favour of its second term; and the notion of the given espoused in it is in full accord with the Neokantian thesis that the given which underlies experience is only a bare *X posited* as the basis of the process of cognition which then becomes increasingly determined in the course of that very process:[65] the Neokantian strands in the *Aufbau*, located principally in the "objectification through structuralization" thesis and in the approach from cognizing consciousness, have come to the fore in all their pervasiveness.[66]

VI

Perhaps it appears to have been not a wholly idle undertaking to look at Neopositivism, in some of its

facets at least, from the unusual perspective of the
Kantian tradition: neglected historical connections may
become visible, and we may find once more that the
actual way of proceeding in the realm of philosophical
thought is much more complex and much less stream
lined than the handy classifications we usually work with
in dealing with philosophy's history would let us suspect.

To concentrate in an attempt like the present on
Carnap's *Aufbau* needs no particular justification, as,
while holding a prominent place within Neopositivism
at a certain stage of its development, it also presents
particular puzzles, systematically as well as in its genesis,
for the received view. A revisionary look at the *Aufbau*,
as here has been attempted,[67] may throw light on certain
features of the doctrine set forth in it, as it is apt to
render intelligible the fact that the book took shape at a
time when Carnap still had quite pronounced Kantian
convictions; and finally, it works also in the other time-
direction. Perhaps in this way we might find it easier to
understand that a philosopher like Nelson Goodman,
who has systematically pursued and developed the
Aufbau programme, has, without any dramatic break in
his outlook, recently arrived at a position which is
openly and admittedly Kantian in some manner:[68] in the
presented view of Carnap's work, it is just that things
have turned full circle.

Notes

Research for this article was assisted by the *Fonds zur Förderung der
wissenschaftlichen Forschung*, Vienna.
[1] "Wissenschaftliche Weltauffassung. Der Wiener Kreis", Otto
Neurath, *Gesammelte philosophische und methodologische Schriften*
(Vienna 1981), vol. 1, 303. The manifesto does not name an author.
Its preface is signed by Rudolf Carnap, Hans Hahn and Neurath; its
actual authors are Neurath, who wrote a first draft of it, Carnap and
Herbert Feigl: for this see Rudolf Haller, *Fragen zu Wittgenstein
und Aufsätze zur Österreichischen Philosophie* (Amsterdam 1986),
193 (n. 8). The account of Neopositivism's historical background in
the manifesto can safely be regarded as being in all essentials due to
Neurath; it reappears, in a somewhat more detailed version, in his
article cited in n. 25 below.
[2] "Wissenschaftliche Weltauffassung" (n. 1 above), 307—8.
[3] *Ibid.*, 307.
[4] Lewis White Beck, *Early German Philosophy. Kant and His
Predecessors* (Cambridge/Mass. 1969), 483.
[5] For Kant's account of the analytic/synthetic distinction see, e.g.,
Kritik der reinen Vernunft, B 190—94, and *Logik*, Academy Edition
of Kant's *Gesammelte Schriften* (Berlin 1902ff.), vol. 9, 111.
[6] Leibniz, *Nouveaux Essais sur l'Entendement Humain*, IV. VII. 10.
For Hume see *An Inquiry Concerning Human Understanding*, Sect.
IV, Part I.

[7] See Gottlob Frege, *Die Grundlagen der Arithmetik* (Breslau
1884), 7.
[8] *Ibid.*, 4.
[9] In *Logik* (n. 5 above), 52—53, Kant states three principles of
logical truth, the other two being the principles of sufficient reason
and of the excluded middle. Elsewhere, however, he indicates that he
regards these two themselves to be application instances of the
principle of contradiction, i.e., to be analytic. See *Preisschrift über die
Fortschritte der Metaphysik*, Academy Edition, vol. 20, 277—78.
[10] For this see *Kritik der reinen Vernunft*, B 194—95, 746—50.
[11] See *Die Grundlagen der Arithmetik* (n. 7 above), 101, 115.
[12] *Ibid.*, 103.
[13] Frege, *Grundgesetze der Arithmetik*, vol. 1 (Jena 1893), vii.
[14] Frege, *Nachgelassene Schriften* (2nd ed. Hamburg 1983), 298—
99.
[15] See, e.g., *ibid.*, 139. For Kant see, e.g., his concise account in
Logik (n. 5 above), 12—13. On the affinity of Frege's conception of
logic in this respect to Kant's, and on the consequences of Russell's
antinomy given this conception, see also Hans D. Sluga, *Gottlob
Frege* (London, Boston, Henley 1980), 108—10.
[16] Bertrand Russell, *The Principles of Mathematics* (7th impression
London 1956), 9.
[17] *Ibid.*, 457.
[18] Russell, *The Problems of Philosophy* (new ed. London, New
York, Toronto 1967), 46—47.
[19] See *The Principles of Mathematics* (n. 16 above), 4, 456.
[20] *The Problems of Philosophy* (n. 18 above), 59. Russell's early
view that mathematics is synthetic *a priori*, and on its relationship to
Kant's philosophy of mathematics see also Ronald Jager, *The
Development of Bertrand Russell's Philosophy* (London, New York
1972), 218—23; Alberto Coffa, "Russell and Kant", *Synthese* 46
(1981); G. G. Taylor, "The Analytic and Synthetic in Russell's
Philosophy of Mathematics", *Philosophical Studies* 39 (1981).
[21] For a good example of this attitude see Rudolf Carnap, "Die alte
und die neue Logik", *Erkenntnis* 1 (1930—31).
[22] See Russell, *Introduction to Mathematical Philosophy* (London
1919), ch. 18.
[23] *The Principles of Mathematics* (n. 16 above), Introduction to the
2nd ed., xii.
[24] For a concise exposition of the difficulties faced by logicism in
this respect, see Stephan Körner, *The Philosophy of Mathematics*
(London 1960), 55—58.
[25] A fact which in the eyes of Neurath made the Austrian intellectual climate a particularly fertile soil for the germination of
Neopositivism. See his "Die Entwicklung des Wiener Kreises und die
Zukunft des Logischen Empirismus" (1936), *Gesammelte . . .
Schriften* (n. 1 above), vol. 2, 676—79. On the early Kantianism in
Austria and on the reasons why it could not consolidate to a stable
tradition, see my *Österreichische Philosophie zwischen Aufklärung
und Restauration* (Würzburg, Amsterdam 1982).
[26] "Dokumente über Naturwissenschaft und Philosophie. Brief-
wechsel zwischen Friedrich Albert Lange und Anton Dohrn",
Erkenntnis 3 (1932—33), 262—63. Next quotation *ibid.*, 265.
[27] A similar view is expressed by Lothar Schäfer, *Karl R. Popper*,
Munich 1988, 47—48.
[28] For Reichenbach's early Kantianism see his *Relativitätstheorie
und Erkenntnis Apriori* (Berlin 1920), 46—47; see also Andreas
Kamlah's notes in Hans Reichenbach, *Gesammelte Werke*

(Braunschweig 1977ff.), vol. 1, 473—74, 476—80, vol. 3, 475—76, and his "The Neo-Kantian Origin of Hans Reichenbach's Principle of Induction", Nicholas Rescher (ed.), *The Heritage of Logical Positivism* (Lanham, London 1985).

[29] Carnap, *Der Raum. Ein Beitrag zur Wissenschaftslehre* (Berlin 1922), 63—67.

[30] Carnap, "Über die Aufgabe der Physik und die Anwendung des Grundsatzes der Einfachstheit", *Kant-Studien* 28 (1923), 90.

[31] Carnap, "Dreidimensionalität des Raumes und Kausalität. Eine Untersuchung über den Zusammenhang zweier logischer Fiktionen", *Annalen der Philosophie und philosophischen Kritik* 4 (1924—25), 106. Although Carnap does not say so expressly, it is, both by his reference on the same page to Kant's "necessary form factors" of experience and by comparison with *Der Raum*, unambiguously clear that he considers the form factors in question as synthetic *a priori*.

[32] *Ibid.*, 106—7. On Carnap's early conventionalism see Hiram Caton, "Carnap's First Philosophy", *Review of Metaphysics* 28 (1974—75), and Edmund Runggaldier, *Carnap's Early Conventionalism* (Amsterdam 1984).

[33] "Dreidimensionalität des Raumes" (n. 31 above), 108.

[34] "Über die Aufgabe der Physik" (n. 30 above), 97.

[35] See, e.g., Hermann Cohen, *Logik der reinen Erkenntnis* (Berlin 1902), 499—500, and Paul Natorp, "Kant und die Marburger Schule", *Kant-Studien* 17 (1912), 209.

[36] Of Carnap's necessary forms of the primary world the third (qualitative relations of sameness and difference) does not, of course, correspond to a Kantian form of intuition, but rather to what Kant, *Kritik der reinen Vernunft*, A 114, calls the "transcendental affinity".

[37] "Dreidimensionalität des Raumes" (n. 31 above), 109.

[38] *Der Raum* (n. 29 above), 39.

[39] Carnap, "Intellectual Autobiography", Paul Arthur Schilpp (ed.), *The Philosophy of Rudolf Carnap* (La Salle/Ill. 1963), 16, 19, 50.

[40] Lothar Krauth, *Die Philosophie Carnaps* (Vienna, New York 1970), 12.

[41] Carnap, *Der logische Aufbau der Welt* (4th ed. Hamburg 1974), § 106.

[42] In the next two sections, I follow closely my "Carnaps 'Aufbau' in kantianischer Sicht", *Grazer Philosophische Studien* 23 (1985), sec. II and III.

[43] *Aufbau*, § 67.

[44] *Ibid.*, § 75.

[45] Ernst Cassirer, *Substanzbegriff und Funktionsbegriff* (2nd ed. Berlin 1923), 292ff. (For Carnap's reference see n. 44 above).

[46] *Ibid.*, 196. Next quotation *ibid.*, 298.

[47] *Ibid.*, 309, 362—63.

[48] *Logik der reinen Erkenntnis* (n. 35 above), 49.

[49] *Aufbau*, § 66.

[50] *Ibid.*, § 12.

[51] *Ibid.*, § 16.

[52] *Ibid.*, § 179.

[53] See Cassirer, "Kant und die moderne Mathematik", *Kant-Studien* 12 (1907), 35—37.

[54] *Ibid.*, 43, 45.

[55] *Aufbau*, § 107.

[56] *Ibid.*, § 69 (emphasis added).

[57] *Ibid.*, § 67.

[58] *Kritik der reinen Vernunft*, A 99. Although Carnap in his references in *Aufbau*, § 67, does not mention this passage directly, he refers to Robert Reininger, *Philosophie des Erkennens* (Leipzig 1911), 370—71, where it is quoted.

[59] *Aufbau*, § 100.

[60] *Ibid.*, § 78.

[61] *Kritik der reinen Vernunft*, B 135.

[62] *Aufbau*, § 100.

[63] *Ibid.*, §§ 109—10, 120. Spatial order has ceased to be such a form because of the nature of the elementary experiences as total presents.

[64] *Ibid.*, § 177.

[65] See, e.g., Natorp, "Kant und die Marburger Schule" (n. 35 above), 207—8.

[66] The idealistic commitments in the *Aufbau* are also pointed out, albeit from quite a different point of view, by Coffa, "Idealism and the Aufbau", Rescher (ed.), *The Heritage of Logical Positivism* (n. 28 above).

[67] Another deviant line in the *Aufbau* is pursued in my "Carnaps Konstitutionstheorie und das Programm der Einheitswissenschaft des Wiener Kreises", *Conceptus* 53/54 (1987).

[68] See Nelson Goodman, *Ways of Worldmaking* (Hassocks/Sussex 1978), especially ch. 1.

Institut für Philosophie
Karl-Franzens-Universität Graz
Heinrichstraße 26
A-8010 Graz
Austria

9

Designation and Convention: A Chapter of Early Logical Empiricism[1]

Thomas A. Ryckman

Northwestern University

We have yet to fully understand the manner or the measure to which logical empiricism emerged as a conventionalist response to both traditional Kantian and empiricist epistemology and to the apparent triumphs of "conventionalist stratagems" (in Popper's aspersive locution) in the foundations of science. By "conventionalism", however, is here understood a broader sense than customary, an extrapolation of views on the foundations of geometry and physics (associated in the first instance with Poincaré) to an encompassing epistemological consideration of the development and validity of scientific concepts generally. In this new construal, the concepts of science are neither derivable from sense experience, nor are they transcendentally valid *a priori* conditions of its possibility. Rather they are "free creations of the human mind" whose provenance is "logically arbitrary", as Poincaré and (subsequently) Einstein put it.

In the initial phases of the period which was to produce the "linguistic turn" in philosophy, Schlick and, some ten years later, Carnap, provide seemingly quite different reconstructive accounts directed towards clarifying and systematizing the role of conventional elements in the formation of scientific concepts. For each, the conventional — reincarnate as the purely formal —paradoxically affords the requisite means to distinguish and privilege scientific knowledge. For Schlick, the precision and, for Carnap, the objectivity, of scientific knowledge are to be accounted for by reconstructive projects aimed at showing how it is possible for the concepts of the sciences to be purely formally determined. In so doing, the intuitive, ineluctably private and incommunicable content or meaning of these concepts is declared irrelevant or epistemologically inert. Of course, fashioning a rigorous separation between the formal and the intuitive, contentual elements of experience proves in each case to be an ideological conceit. But the failure is an instructive one, in particular, in view of the subsequent trajectory of logical empiricism.

In his (1918) Schlick provides the new conventionalism with its characteristic separation of form and content, arguing for a doctrine of concepts as purely formal, implicitly defined **signs** which are coordinated (*zugeordnet*) to objects through the paradigmatically conventional act of designation. In this very influential book, Schlick attempts to provide a general account of scientific knowledge in terms of a

PSA 1990, Volume 2, pp. 149-157
Copyright © 1991 by the Philosophy of Science Association

11

"merely semiotic" doctrine of implicitly defined concepts and their designata, while coupling lengthy critical scrutinies of Machian positivism and Kantian *a priorism*, with a "hypothetical realism" owing much to Helmholtz.

Employing a judo-like maneuver, Carnap in the *Aufbau* holds that logic itself (in the sense of the second edition of *Principia Mathematica*) is conventional; nonetheless, logical structure is to be the reconstructive medium of scientific objectivity, i.e., of the intersubjectivity, of scientific knowledge. Following Poincaré in holding that science treats of, and so describes only relations between sensations, not the private and incommunicable content of the sensations themselves, Carnap's goal is to sketch a method for redefining all the concepts of science using only the resources of pure logic and a sole non-logical relation obtaining among autopsychological experiences. The procedure is targeted at a reconstruction of the sign relation, together with its inherent arbitrariness, the lynchpin of Schlick's semiotic conventionalism. Within the *Aufbau*'s *Konstitutionstheorie*, the designative concept- and relation-signs of science are to be systematically recast *à la* Russell as *Kennzeichen*, indicator-signs ("structural definite descriptions") that unambiguously identify (and thereby "constitute") the objects of the various sciences. In this enterprise at the end of the day, the only remaining non-logical signs are signs for a basic relation holding among autopsychological experiences; ultimately, even these too are to be formally conjured away. After a brief preliminary look at the doctrine of concepts in Schlick's early classic, I will suggest that this 'semiotic' reading of the *Aufbau*, emphasizing its shared problematic with Schlick, effects something of a reconciliation between two heretofore widely varying interpretations of this work current in the literature (cf. Friedman, 1987 and Richardson, 1990).

1. A "Merely Semiotic" Account of Scientific Knowledge

In the *Allgemeine Erkenntnislehre*, Schlick argues that in science (particularly, theoretical natural science) what we find is that, as in ordinary life, knowledge (*Erkenntnis*) is not acquaintance (*Kennen*) but a matter of recognition (*Wiedererkennen*) or rediscovery (*Wiedererfinden*), the finding of one thing in another. As opposed to everyday knowledge, the process of knowing in science has two distinctive features: it is reductive and it seeks to employ a minimum of explanatory principles. But knowledge *qua* recognition assumes a comparison and, in an exact science such as physics, what is compared must be fixed with "absolute constancy" and "determinateness". Only concepts, not ideas, impressions, or intuitive images satisfy this requirement; they alone can be clearly individuated and identified with complete assurance. When concepts in science are clearly defined, this is not due to whatever intuitive meaning or content might attach to them. Rather it is to consider them as mere signs (*Zeichen*) for all the objects whose properties include the defining characteristics (*Merkmale*) of the concept. This is neither to say that concepts are not invariably or often accompanied by intuitive images (*anschauliche Vorstellungen*) nor that concepts are real, in the sense of mental or Platonic entities. A concept is a mere fiction (*blosse Fiktion*), and strictly speaking, only the conceptual function — the designation of an object by a concept, considered solely as a coordination (*Zuordnung*) of concept to object — is real (§5).

Schlick goes on (§7) to urge that Hilbert has accordingly indicated "a path of the greatest significance for epistemology" in showing that concepts can be defined merely by the fact that they satisfy chosen axioms. Thus the method of implicit definition — which, Schlick argues, is profitably extendable outside the realm of pure mathematics to the sciences generally — shows how it is possible to completely determine the content (*Inhalt*) of concepts in a manner which does not fall back on the intuitive (*im Anschaulichen*), and that it is possible to speak of the meaning (*Bedeutung*) of concepts without reduction to intuitive images. With implicit definition, an instrument has been found that enables complete determination of concepts and the attainment of "strict precision" in thinking. Satisfying only the requirement of consistency, the method of implicit definition provides purely formal specification of both the sense (*Sinn*) and the designata (*Bedeutungen*) of concepts in the mathematical sciences; the sense of mathematical concepts is acquired solely in virtue of the particular axiom system, while the *Bedeutung* of such concepts is nothing real, but consists in "a determinate constellation of a number of the remaining concepts". Not mincing implications, this means that mathematics "has only the significance of a game with symbols"; whereas, more generally, in the empirical sciences of nature:

> A system of truths created with the aid of implicit definitions does not at any point rest on the ground of reality. On the contrary, it floats freely, so to speak, and like the solar system bears within itself the guarantee of its own stability (p.35).

In so many words, "the bridges are down" between "concept and intuition, thought and reality".[2]

Due to a "correlation" between judgements and concepts, the result of implicit definition is a network holism of conceptual meaning. On the one hand, "implicit definitions determine concepts by virtue of the fact that certain axioms — which are themselves judgements — hold with regard to these concepts; thus such definitions make concepts depend on judgements". Concepts must occur in several different judgements if they are to have any sense and meaning (*Bedeutung*) at all. On the other hand, "judgements are linked to one another by means of concepts: one and the same concept appears in a number of judgements and thus sets up a relation between them". The mutual dependence of concepts and judgements anticipates the favored metaphor of a later, and more well-known, holist:

> "every concept constitutes, as it were, a point at which a series of judgements meet (namely, all those in which the concept occurs); it is a link that holds them together. The systems of our sciences form a net (*ein Netz*) in which concepts represent the knots (*die Knoten*) and judgements the threads that connect them (p.43).

But it is only in two further sections, on the nature of judgement (§8) and truth (§10), that we begin to see how a semantically self-contained system created through implicit definition is to be tempered with what Carnap will term, in an appropriately vivid phrase, "the blood of empirical reality". Restoration of links to the empirical world is established through acts of designation or coordination (*Zuordnung*). For judgements are also merely signs designating not only relations between objects but that a certain relation obtains between the designated objects. Judgements therefore designate facts or sets of facts (*Tatbestände*) which are presupposed (p.40). However, designation by judgements is doubly holistic: it is only the judgement as a whole that is coordinated to a fact as a whole, whereas which fact a given judgement (proposition) designates is determined solely by the position occupied by the proposition (*Satz*) in our system of judgements (p.62). Moreover, as against the purely formal requirement of consistency to ensure the truth of mathematical theories, an empirical judgement is to be considered true if and only if it "unambiguously designates" (*eindeutig bezeichnet*) or is "coordinated" to (*zugeordnet*) a set of facts. Weaving together the formal mechanism of implicit definition with the conventionalism of designation yields a structural, though holistic, account of truth as *eindeutige Zuordnung*:

...it is the structural connectedness of the system of our judgements which produces the unambiguous coordination and conditions its truth;...(p.62).

Elsewhere, Schlick observes that in these sections of his (1918) scouted above, the
case is made

that such a mapping (*Abbildung*) of the lawfulness of the real (*des Wirklichen*)
with the help of a sign-system comprises in general the essence of all knowledge...(1921 p.156, n.15).

The radically formalist method of determining concepts through implicit definition, together with his thesis of "the merely designative (or semiotic) character of
thinking and cognition," that thinking involves — as does any act of coordination —
"somewhat arbitrary conventions" (1922, p.105), decisively settles, in Schlick's eyes,
the whole question against Kantian philosophy (1918, p.306). Yet Schlick's account
cannot be fully satisfactory, requiring us to accept as unproblematic the considerable
weight placed on the notion of *Zuordnung*,[3] by all appearances a mysterious and
seemingly synthetic act which is difficult, if not impossible, to reconcile with empiricist precepts against synthetic *a priori* principles as well as with strict (non-holistic)
verificationist doctrines of meaning. In his attack on intuition and alleged intuitive
means of acquiring knowledge, Schlick had been content to leave the characterization
of *Zuordnung* as only "a fundamental act of consciousness...a simple ultimate...to
which every epistemologist must, in the end, advance"; in other words, an unex-
plained explainer, a fundamental, not-further-reducible act of thought.[4]

2. The *Aufbau*: Reconstructing Designation Itself

Although he is far more explicit than is Schlick that he is giving a "rational reconstruction" of actual concept formation (cf.§§ 100, 143) Carnap in the *Aufbau* also maintains the purely symbolic (or designative) view of scientific knowledge to confront
"metaphysical" accounts of the acquisition of knowledge through intuition. With
Carnap no less than Schlick, to exclude the intuitive, the subjective and the private, entails upholding the possibility of a purely formal determination of scientific concepts as
requisite to the precision and intersubjectivity of scientific knowledge. For Carnap, the
system of conceptual knowledge that is science (§§ 180,182) is itself symbolic; hence,
an object must be designated by a sign —at least in principle —if it is to be an object of
conceptual knowledge (§ 19). Venturing some rare irony, Carnap even states his agreement with Bergson on the latter's definition of metaphysics as "That science which
wants to get by without symbols", glossing this position as advocating

metaphysics would seek to grasp its object not through the detour (*Umweg*) of
concepts, which are symbols, but immediately through intuition (§ 182).

While in agreement with Schlick on the necessity of a purely formal determination of
scientific concepts, Carnap will offer in the *Aufbau* a completely different method for
its attainment.

Now science is distinguished, preconstructionally as it were, from metaphysics by
the fact that it pursues not problems of essence (*Wesenprobleme*) but only problems of
coordination (*Zuordnungsprobleme*), i.e., of determining which objects or object-
types are associated with each relation. This has been especially emphasized by
Mach, and in several places Carnap repeats Mach's strictures that science seeks only
fixed functional dependences (§§ 21-2, 169). The reconstructive method of
Konstitutionstheorie (hereafter 'KT') is itself part of science, since in order to be able

to make statements about objects at all, these objects must first be constituted, else their names have no meaning (*Sinn*); for this reason indeed, "**the formation of the constitution system is the first task of science**" (§ 179). As part of science, KT is similarly bounded by the injunction to pursue only *Zuordnungsprobleme*, whereas the particular relation of concern to KT is the sign relation (*Zeichenbeziehung*) obtaining between a concept-sign, or a relation symbol, and its designation or *Beudeutung*, i.e. its extension.[5] KT is given the task of rationally reconstructing the sign relation,[6] showing that merely structural statements (*blosse Structurangaben*) can yield a univocal, i.e., unambiguous, sign-indication (*eindeutige Kennzeichnung*) for the domain of objects of science: if no difference in formal sign-indication can be given for two objects, these objects are not distinguishable by scientific methods (§§15-16). Hence identity in science is identity of "relation number" i.e., structural isomorphism (§§11-12). Recalling Reichenbach's puzzlement (expressed in his 1920 monograph on relativity[7]) over the "remarkable fact that, in cognition we carry out a coordination (*eine Zuordnung*) of two sets, the elements of one of which are first defined through this coordination", Carnap maintains that the rational reconstruction of the sign-relation as purely structural indicator signs gives an explanation (*Erklärung*) of this fact. For thereby it is demonstrated how it is possible through a sign-coordination (*Zeichenzuordnung*) to first carry out the individuation of single objects, thus fixing an intersubjective domain of discourse in science which can then be subjected to conceptual treatment (§15).

Picking out individual objects through univocal coordination of signs is precisely what Schlick's otherwise laudable, because purely formal, method of implicit definitions does not allow. Implicit definition Carnap notes here, and argues at some length in an important, but little-known, 1927 paper, succeeds in defining only classes of objects; hence empirical concepts that are implicitly defined are "improper concepts". KT, on the other hand, uses the method of explicit definition which, taken in a "widened sense", comprises both real (eliminative) definitions and Russellian definitions-in-use. For it is only through real definition that "the blood of empirical reality enters and...flows into the most remote veins of the thus-far empty schema" (1927, p.373) of theory created by implicit definition; thereafter, definitions-in-use single out and identify the objects (to be called "quasi-objects" in the *Aufbau*) of the ostensibly different domains of the special sciences (e.g., 'cow' in zoology and in economics, (1927, p.357)). However, these are ultimately eliminable through further logical decomposition in favor of real definitions and symbols of logic alone.[8] As against the unavoidable meaning holism stemming from conceptual determination through implicit definition and the implacable arbitrariness of restoring the linkage to the empirical through mere *Zuordnung* of concept and object, KT's approach begins with the structural relations of the empirical at the ground level and then builds upward. In consequence, it becomes possible to institute a non-network-holist verifiability theory of meaning for the individual statements of a science. Strict verificationism — as a semantic doctrine demarcating the legitimate concepts of science — is to be vindicated.

It is important to see that the derogation of objectivity to purely formal structure and thus to logic — which "**consists only in conventional posits concerning the use of signs**" (§107, original emphasis) — goes hand in hand with the rational reconstruction of the sign-relation. The signs of logic and mathematics do not designate at all; there are no real logical objects. However, empirical concept-signs of the special sciences do purport to designate objects. Within KT, an explicit translation can be formulated which will show that in fact, each of these concept-signs designates only a "quasi object", a logical construction from the sole non-logical basic relation(s) located at one or another level of stratification (corresponding to the various domains of the special sciences). The thesis of KT maintains that any legitimate scientific state-

ment can thus be step-by-step recast as a statement 'about' "quasi-objects" on succes-
sively lower levels, and ultimately can be transformed into a statement containing
only logical symbols (which do not designate) and signs for the basic relation(s)
which "indicate" (§106) or "express a definite (formal and extensional) state of affairs
relative to the basic relation".[9] Since it is possible to recognize whether the basic relation obtains or not between elementary experiences, and because in principle there
are only finitely many elements of the basic relation extension (i.e., combinations of
the elementary experiences), the truth or falsity of every legitimate scientific statement can in principle be ascertained in a finite number of steps.

The *Aufbau* might thus seem to be sounding a Russellian theme about how impov-
erished is the class of "logically proper names", holding that the only genuinely de-
noting (designating) terms are signs for the basic relation, i.e, relations of immediate
acquaintance. Strictly speaking, however, these sole remaining empirical concepts are
held to not really designate objects or to have *Bedeutungen* at all; indeed, to suppose
they do is to mistakenly inquire into the "metaphysical essence" of the object (§161).
Rather the only linkage remaining between concept-sign and object lies not in designation but in a **sententially** holistic "indication of the truth criteria" — the epistemological bedrock of experiential content — "for those sentences in which the sign of
this object can occur" (§§160-1). In this evaporation of the relation of designation
through rational reconstruction, the arbitrariness inherent in designation is systematically replaced by constitution based upon explicit conventions governing logical signs
themselves. The *Aufbau*-project of the constitution of objects — providing unambiguous sign-indications of them — is the first aim of science, and it is attained through
convention (*eine Festsetzung*); the second aim of science is empirical, and lies in determining the various properties and relations of these objects within the special sciences. In the conception of KT, the only two components of knowledge are the conventional (*konventionelle*) and the empirical, separate but equally important (§179).

It is because KT is thus crafted as a **semiotic** doctrine reconstructing concept and
relation **signs** that Carnap can maintain that the program of the *Aufbau* abides
metaphilosophical neutrality: KT is a means of cleaning up the problem of designation,
independently of the realism-idealism dispute. Moreover, in "explaining" the relation
obtaining between sign and designatum through reductive and eliminative explicit definition via constructional chains, ultimately from logical symbols and the basic relation
alone, KT demonstrates as a lemma, as it were, the thesis of unity of science, i.e., that
all and only objects (concepts) of science are constructible — and hence there can be
intersubjective agreement upon **what** is under discussion — within the domain of a
single *Konstitutionstheorie*. If this reading is correct, then we can perhaps locate the
Aufbau's most illustrious philosophical precedent in Leibniz's universal characteristic
(as did Carnap himself (§3) as well as Schlick and Heinrich Scholz in their contemporary reviews of the *Aufbau*), similarly imbued with the spirit of metaphilosophical neutrality in proposing a purely symbolic means of resolving otherwise intractable philosophical disagreements ("*Calculemus!*"). Looking forward from the cusp of the "linguistic turn" a bit into the 1930s, we can identify a continuity with Carnap's method of
syntactical designation via "the formal mode of speech" as a corrective for "pseudo-
object" sentences, as well as with the later metaphilosophy of logical tolerance, and of
linguistic methods for solving problems internal to a chosen framework.

On this interpretation, the *Aufbau* is attempting to simultaneously balance two ostensibly incompatible positions: first, a Russellian and Machian verificationist doctrine of meaning which requires that rational reconstruction of empirical concept formation reflect the epistemic primacy of autopsychological experience and, secondly, a
view that the objectivity of scientific knowledge is accounted for by a shared lan-

guage whose intersubjectivity rests on the possibility of formulating purely structural statements. The equilibrium solution advanced is targeted at the sign relation, whose semantic arbitrariness is systematically reconstituted — conventionally, but not arbitrarily — by pure logic alone. A strict verificationist account, providing for the meaningfulness of all and only scientific statements and couched in the autopsychological terms of (near-) positivist orthodoxy, can be advanced while still securing intersubjectivity through a method that constitutes away the relation of designation, systematically installing purely formal *Kennzeichen*. In this enterprise, logic is a *Hilfsmittel*, but it is more than that: it is the sole locus of structure which alone is communicable. The step-by-step dismantling of the designation relation — down to the sole empirical signs for the basic relation and perhaps even beyond —and its systematic replacement by a purely formal criterion of identity, structural isomorphism, is intended to reconcile the competing demands of verificationism and intersubjectivity. Thus verificationism and formalization go hand in hand; to the extent we maintain the former, we need the latter, so to speak, as compensation. We should not, however, slight Carnap's interest in providing only a **rational reconstruction**: just as "science wants to speak about what is objective" (§16), so too the issue of complete formalization (i.e, including the basic relation(s)) arises only in the abstract, as the limiting point of this reconstruction of scientific intersubjectivity.[10] Of course, the *Aufbau*-project of explicitly defining scientific concepts in the primitive idiom of sense experience flounders long before this point is reached (see note 8).

In this paper, I have argued that despite opposing assessments of the designation relation — an end point for Schlick, a starting point for Carnap — a central link is forged between their major initial works in that each offers a general epistemological program for eliminating all intuitive, imagistic, private and hence incommunicable 'material' or content from scientific concepts. Within a few years, Tarskian semantics would provide an entirely novel framework for discussion of semantic terms such as 'designation' and 'truth'. But we should not lose sight of the fact that much of the ground upon which the "linguistic turn" transpired was already in place, prepared by recognition and accommodation of conventionalism within empiricist doctrines of scientific concept formation.

Notes

[1] I wish to thank Michael Friedman and Thomas Uebel for comments and criticisms of previous drafts, and Alan Richardson for preliminary discussions on these matters.

[2] Of course, the bridges are not really "down"; see the remarks linking "concrete" (ostensive) to "implicit" definitions in §7.

[3] Cf. Coffa (forthcoming, chapter 11).

[4] Schlick 1918, p.326; on the significance of the notion of *Zuordnung* in the epistemology of science in this period, see Ryckman (1991).

[5] As Proust (1989, p.176) notes, the Fregean distinction between concept and object is rendered "a mere matter of words" by Carnap's adherence to a thesis of extensionality.

6 Rolf George's translation of *Zeichenbeziehung* as "designation relation" thus obscures the cardinal point of the interpretation ventured here, that the *Aufbau's* reconstructive project is precisely to 'syntacticize' this semantic relation.

7 Reichenbach (1920, p.38); cited in §15.

8 As has repeatedly been observed, Carnap steps beyond the strict limits of explicit definition in the constitution of physical objects in assigning qualities to space-time points, where this is done via postulates, i.e., implicitly. See Kraft (1950, p.100), Quine (1951, p.40) and Carnap (1967, p.viii).

9 §180; cf. the formulations in §§ 16, 112, 119-21, 144 and 160.

10 Does Carnap's understandable inability to establish "foundedness" as a purely logical concept (§§ 154-5) thus undermine the entire *Aufbau* program (cf. Friedman 1987, pp. 532-3)? If we view his concern with objectivity merely as counterpart to a corresponding concern with strict verificationism, then the failure here plausibly indicates only the *explicative* (to anachronistically retrofit this term) limits of the method of KT.

References

Carnap, R. (1927), "*Eigentliche und Uneigentliche Begriffe*", *Symposion* I, pp. 355-74.

_ _ _ _ _ _. (1928), *Der Logische Aufbau der Welt*. Berlin-Schlactensee,Weltkreis-Verlag.

_ _ _ _ _ _. (1967), *The Logical Structure of the World*, translated by Rolf A. George, with a Preface by Carnap. Berkeley and Los Angeles, University of California Press.

Coffa, J.A. (forthcoming), *To the Vienna Station; Semantics, Epistemology and the a priori from Kant to Carnap*, L. Wessels and G. Steinhoff (eds.). New York and London, Cambridge University Press.

Friedman, M. (1987), "Carnap's *Aufbau* Reconsidered", *Noûs* 21, pp. 521-45.

Kraft, V. (1950), *Der Wiener Kreis; der Ursprung des Neopositivismus. Wien,* Springer-Verlag.

Proust, J. (1989), *Questions of Form; Logic and the Analytic Proposition from Kant to Carnap*, translated by A. Brenner. Minneapolis, University of Minnesota Press.

Quine, W.V.O. (1951), "Two Dogmas of Empiricism", reprinted in *From a Logical Point of View*, 2nd. rev. edition. New York, Harper and Row, 1961, pp. 20-46.

Reichenbach, H. (1920), *Relativitätstheorie und Erkenntnis A Priori*. Berlin, J. Springer.

Richardson, A. (1990), "How not to Russell Carnap's *Aufbau*", *PSA 1990*, vol. 1, pp. 3-14.

Ryckman, T.A. (1991), "*Conditio sine qua non? Zuordnung* in the Early Epistemology of Cassirer and Schlick", *Synthese*, 88.

Schlick, M. (1918), *Allgemeine Erkenntnislehre*. Berlin, J. Springer.

_____. (1921), *Hermann v. Helmholtz Schriften zur Erkenntnistheorie*, P. Hertz and M. Schlick (hg.). Berlin, J. Springer.

_____. (1922), *Raum und Zeit in der gegenwärtigen Physik*. Vierte Aufl. Berlin, J. Springer.

_____. (1929), *Rezension von Carnap (1928)*, *Die Naturwissenschaften*, 27, pp. 550-1.

Scholz, H. (1930), *Rezension von* Carnap (1928), *Deutsche Literatur Zeitung*, Heft 13, Sp. 586-92.

Hidden Agendas: Knowledge and Verification

Joia Lewis

University of San Diego

1. Introduction

Schlick has been accused of a number of philosophical sins over the years, most notably his rather casual, and frequent, traversing of the borders between language, experience, and reality. While we allow our scientists the freedom to roam creatively throughout the peripheral regions of Epistemology and Metaphysics, we are not so tolerant of our philosophers. We know that Schlick gave up the physics laboratory for the philosopher's armchair, and we expect him to stick to a particular position.

Schlick's colleagues in the Vienna Circle were not totally blameless in this regard, either, and it may be that part of the charm of their work is precisely their inability to stick to the positions that solidified over 2500 years of western philosophy. Empiricism and rationalism appear to coalesce in their writings, idealism and realism cohabitate the same theories, the Senses follow the Intellect right out of Plato's Cave and into the light of certainty. We become seasick in Neurath's boat, wondering whether his rebuilders can keep it afloat, or if it will eventually hit the empirical ground after all.

In Schlick's case, we wonder whether he really did 'convert' from realism to something other than realism, and what exactly Wittgenstein had to do with it. We wonder what his realism entailed: in the inimitable words of Alberto Coffa,

In Schlick's hands, realism had been turned from the boring, trivial common sense view that it was before Kant, into an exciting, bold and utterly unbelievable conjecture. ...The world of common sense has been torn to pieces: its time is subjective and transcendentally ideal and so is its space (Coffa, 188).

We also wonder how Schlick could lodge meaning safely within logical relationships in one paragraph and then go on in the next to talk about meaning and empirical content. Michael Friedman has noted (see Friedman, 1983) that Schlick vacillated throughout his philosophical career between a holistic and formalistic account of knowledge and meaning, and an atomistic and foundationalist account, apparently without ever committing himself completely to either one. Tom Ryckman has written

PSA 1990, Volume 2, pp. 159-168
Copyright © 1991 by the Philosophy of Science Association

that Schlick's original linking of thought with reality by means of the notion of *Zuordnung*, or coordination, constituted a very minimalized epistemology:

> It is therefore all the more surprising to find him actively engaged, within a decade or so, in a truly traditional dispute over the "foundations of knowledge" (Ryckman, 56).

This latter dispute over the foundations of knowledge is perhaps what we wonder about most of all: why was the exchange between Schlick and his colleagues in the famous protocol debates of the 1930's so bizarre? Carnap, Neurath and Hempel allegedly found a "detestable metaphysics" in Schlick's comparison of a statement in his Baedecker with the number of spires on a certain cathedral; Schlick accused them of being relativistic and rationalistic, as well as irrationalistic, if they could not allow him to determine the truth of the matter by simply checking his guidebook while standing in front of the cathedral.

What was at issue was Schlick's theory of *Konstatierungen*. These were, according to Schlick, our momentary affirmations of reality, which functioned as the foundation of empirical knowledge. However, their foundational authority was short-lived: as soon as they illicited our assent or dissent of a particular claim, they metamorphosed into hypotheses and joined the ranks of all other claims about empirical reality. Schlick's colleagues had dispensed with the idea that we could ever have an absolutely certain foundation of empirical knowledge, and therefore rejected Schlick's theory on the basis of its metaphysical claim about the possibility of comparing statements to facts. Tom Oberdan has pointed out quite correctly that there was a great deal of misunderstanding going on in the debates: "[Schlick] seems to have assumed—incorrectly—that his commitment to conventionalism was apparent to anyone who knew his work." (Oberdan, 18) If Schlick was indeed as committed to a conventionalist account of language and meaning as his colleagues, what then were they arguing about?

There is a growing literature on this topic, which includes many dissections and interpretations of the discussion between Schlick, Carnap, Neurath, and Hempel, on the nature of basic empirical statements and their role in verification (see Haller, Gadol, *Synthese 64* (1985)). What I would like to focus on specifically is the dual nature of Schlick's major philosophical theses. In the case of Schlick's verificationist theory of meaning, why did he require an account that would be both intersubjectively accessible as well as privately incorrigible? In the case of his 'affirmations of reality,' why did Schlick need to invent, or more modestly, to discover, something that would function as a foundation of knowledge but that would at the same time not clog our systems of conceptual claims with any debris from our perceptual apparatus? I would find this a very frustrating assignment, if asked to create a model fulfilling these terms.

I believe the answer to these questions can be found by exposing the assumptions behind what we can consider to be Schlick's two separate agendas, one formalistic and conventional, the other foundational and empiricist. The agendas are present and fully formed in his earliest philosophical writings. Given that the hypothesis of split personality is not warranted in Schlick's case, the question to ask is, why would he deliberately develop inconsistent lines of thought? What makes the situation interesting is not only that the two agendas are not compatible, which is evident in his early work, but also that Schlick consciously attempted to *make* them work together in his later accounts of meaning and verification. Schlick's conviction that the assumptions in each agenda were equally correct can explain a good deal about the development of his mature theses and is, I believe, responsible for much of the criticism that his work has received, both before and after his death in 1936. Making the two agendas explicit can also provide us with a better framework within which to judge Schlick's work. Finally, the story provides an instructive example of how one's firmest beliefs constrain and dictate one's choice of solutions and perhaps define the problems themselves.

2. Agenda 1

The first agenda had to do with the uncompromising distinction Schlick made between conceptual knowledge and intuitive experience. Athough he subscribed to the general empiricist claim that all knowledge stems ultimately from sense experience, Schlick was always careful to point out the vast difference between precisely formulated and enduring concepts as opposed to imprecise and ephemeral intuitive experiences. Schlick sought to clarify this distinction by means of Hilbert's notion of *implicit definition*, whereby the basic or primitive concepts of a system of truths are defined by virtue of the fact that they satisfy the axioms. He contrasted implicit definitons to *concrete definitions*, in which "the defining terminates when the ultimate indefinable concepts are in some way exhibited in intuition...[they involve] pointing to something real, something which has individual existence." (*General Theory of Knowledge*, abbrev. *GTK*, 37) He described the relationship between concrete and implicit definitions as follows:

it is through concrete definitions that we set up the connection between concepts and reality. Concrete definitions exhibit in intuitive or experienced reality that which henceforth is to be designated by a concept. On the other hand, implicit definitions have no association or connection with reality at all.... A system of truths created with the aid of implicit definitions does not at any point rest on the ground or reality. On the contrary, it floats freely, so to speak, and like the solar system bears within itself the guarantee of its own stability. None of the concepts that occur in the theory designate anything real; rather, they designate one another in such fashion that the meaning of one concept consists in a particular constellation of a number of the remaining concepts. (*GTK*, 37)

We begin, then, by assigning concepts to designate specific experiences. Once we have the concept, we can find an implicit definition for it among concepts already incorporated into our system of knowledge. Thus the relationship between the concepts utilized in concrete definitions and sense-experiences is that of names to the objects they name. Our implicitly defined concepts, however, bear no systematic relationship at all to sense-experience.

Besides the influence of Hilbert, Schlick was also following Planck in his separation of conceptual knowledge from intuitive experience. Schlick accepted Planck's mandate that science should move away from descriptions of reality that originate with our sensations and toward a purely quantitative picture of the world. It is for this reason that Schlick also rejected Mach's restriction of science to descriptions of phenomena within the realm of our sensations. Further, following Planck, Schlick agreed that only abstract quantitative formulations could assure a *unified* picture of the world; such a goal could not be reached by focusing exclusively on sense-dependent descriptions. Schlick identified Planck's distinction between the subjective and qualitative as opposed to the objective and quantitative with his own distinction between concepts and intuition.

Consonant with this first agenda was Schlick's abhorrence of any philosophy which elevated the role of non-conceptualized experience or intuition above that of systematically determined concepts, either with respect to questions of truth or to existence. Schlick considered these positions to be direct attacks on scientific thought, and on any philosophy designed to reflect the scientific spirit. Passages like the following appear repeatedly throughout Schlick's early works:

> the arrogant structures of idealist thought, which created the bitterness and brought the philosophic spirit into discredit, have long since crumbled...these doubts about philosophy rest merely upon hasty judgment, upon deliberate neglect of the ultimate problems of science, in short, upon lack of clarity. (*Philosophical Papers, Vol. I*, abbrev. *Vol. I*, 104)

Clarity, for Schlick, as we have seen, required that knowledge be wholly pruned of its roots in intuitive experience.

3. Agenda 2

Schlick's second agenda had to do with the epistemological foundation of empirical knowledge. Under this agenda are his doctrines of *unique coordination* and *verification*.

Schlick considered the notion of *unique coordination* to be the defining characteristic of *truth*; the method to check uniqueness was then *verification*. In his words,

> truth is defined by a single, extremely simple characteristic: the uniqueness of the correlation of judgments with facts. (*GTK*, 162)

In "The Nature of Truth in Modern Logic" (1910), Schlick likened the notion of judgment to sensation and ideation, in that all three notions involved designation of some sort. He was expanding on Helmholtz's theory that our sensations are signs of things-in-themselves, not in the sense of images of reality, but by giving us the formal features of the otherwise unknowable noumenal realm. According to Schlick, sensations and ideas are "merely signs for the *content* of experience, the world given to us, without regard to the form in which this content makes its appearance" (*Vol I*, 91) But we are aware of order among our experiences, of the fact that "elements of experience are connected in relations to one another...[that] they possess *order* and *form*." (*Vol. I*,91) Another type of sign is therefore required, not one that is given to us but one which we actively employ to designate the relations among our experiences. Judgments fulfill this function. They are "designations of facts, of forms of experience, of the ordering and association of elements of the given." For a judgment to be true, it must isolate a single fact unambiguously, or, in Schlick's words, "*A judgment is true if it univocally designates a specific state-of-affairs.*" (*Vol. I*, 94) The relationship is not one of similarity or picturing; a judgment cannot be "more than a sign in relationship to a set of facts.... A judgment pictures the nature of what is judged as little as a musical note pictures a tone, or the name of a man pictures his personality." (*GTK*, 60) It is the rather the *interconnection* between judgments that allows an adequate designation of reality:

> By virtue of the interconnection of judgments a new truth receives a specific place in the circle of truths; the fact corresponding to this new truth is thereby assigned to the place that, by virtue of the interconnection of facts, it occupies in the domain of reality. ...Hence it is the structural connectedness of our system of judgments that produces the unique coordination and conditions its truth. (*GTK*, 67)

Recall that concepts, for Schlick, were originally connected to experience through concrete definitions. Introducing a concept by means of a concrete definition was a conventional act, "a quite arbitrary stipulation, and consists in introducing a particular name for an object that has been singled out in one fashion or another." (*GTK*, 69) These were contrasted to implicit definitions of concepts within the system of interlocking judgements, which "will then be connected to one another by a system of judgments coinciding fully with the network of judgements that on the basis of experience had been uniquely coordinated to the system of facts." (*GTK*, 70)

Schlick's conviction that this was the case could not have been stronger:

> Obviously, to suppose that the world is intelligible is to assume the existence of a system of implicit definitions that corresponds exactly to the system of empirical judgments. (*GTK*, 70)

So far this fits in with Agenda 1, in that we have only a conventional link between knowledge and experience, through our concrete definitions of concepts. But Schlick's second agenda also had a new nonconventional element.

The system of empirical judgments was uniquely coordinated to the system of facts on the basis of experience. Schlick called the system of empirical judgments *historical* or *descriptive* judgments, and distinguished them from *definitions* and *conventions*. Historical judgments that hold for facts not being immediately observed were *hypotheses*. Schlick noted that this distinction cannot strictly be maintained, however, since "the class of historical judgments dwindles to zero if we consider that strictly speaking it can embrace *only such facts as are immediately experienced in the present moment*." (*GTK*, 73; my emphasis) Schlick thus introduced the idea of *fundamental*, or *perceptual* judgments, to distinguish the judgments embracing "immediately experienced facts" from other descriptive or historical judgments: these were the 'building blocks' of the system, the "propositions...by virtue of which the system rests directly on real facts.... If the whole edifice is correctly built, then a set of real facts corresponds not only to each of the starting-points—the fundamental judgments—but also to each member of the system generated deductively." (*GTK*, 78)

4. Schlick's Early Account of Verification

In Schlick's early account, verification is the method to check for *uniqueness of coordination* between judgments and facts, which, as we have seen, is the defining characteristic of truth. What is involved in verification, according to Schlick, is the identity of two judgments, one of which must be a fundamental or perceptual judgment, as just described.

Schlick first illustrated his account of verification with an example from the history of science, also in his (1910) article, "The Nature of Truth in Modern Logic." He explained that the verification of the existence of the new planet Neptune consisted in the identity that obtained between the prediction that a certain planet would be found and the perceptual judgment made at the time that it was first observed. The prediction was deduced by Leverrier from Newtonion law and observed facts about the perturbed orbit of Uranus. When Galle looked through the telescope and perceived the presence of a planet exactly where it was predicted to be, the perceptual judgment that he formed upon making this observation was identical to the prediction-judgment originally formulated by Leverrier.

Later in the *General Theory of Knowledge*, Schlick presented a more detailed analysis of what happens when a judgment is verified. The procedure to verify an arbitrary assertion about reality, J, is as follows, where J', J", ...are auxiliary assumptions:

(A) Derive J_1 from J and J', where the truth of J' is considered established, for one of the following reasons:

    (i)   J' is an assertion about reality
    (ii) J' is a definition
    (iii) J' is a purely conceptual proposition

(B) Derive J_2 from J_1 and J"... until finally J_n is derived, which is a prediction in the following form:

"At such and such a time and at such and such a place under such and such circumstances such and such will be observed or experienced." (*GTK*, 163)

(C) At the appointed time and place, we make a perceptual judgment, P. If P and J_n are identical then J_n is verified, and so is the original judgment J.

Since both P and Jn designate the same fact, even though we arrived at them through "two entirely different paths," we have, according to Schlick, established a unique correlation. The judgment is therefore true. (*GTK*, 163) Thus, when J_n and P designate the same fact, we may consider the original J to be true. But in order to do this, we must first accept P to be true. Schlick considered the truth of P noncontroversially established as the expression of the fact. In other words, for J_n to be correlated with the fact, it simply had to be identical with the perceptual judgment P that designated this fact. But, if verification is the method by which we establish the unique coordination between a judgment and a fact, thereby establishing the truth of the judgment, how can we then simply accept that the 'perceptual judgments', or judgments of experience are themselves true? The perceptual judgment P, which embraces "immediately experienced facts" is not in need of verification itself, according to Schlick.

Schlick also stated that not only is the original claim J verified, but also the whole chain J_1, J_2, ..., leading up to the prediction-judgment J_n:

since the last member of the chain of judgments led to a unique correlation, we take this as a sign that the other members, hence the starting-point and the endpoint J, also fulfill the truth condition, and we count the entire process as a verification of judgment J. (*GTK*, 163)

An identity of two judgments was also the basis of the verification of purely conceptual or analytical judgments, for Schlick (*GTK*, 166). What is involved in our recognition of an identity in either case is some kind of intuitive process, a mental picturing of the sense of the proposition. But even though empirical and purely conceptual statements share an identity experience at the end of verification, "a vast difference separates these two classes of judgments, an abyss that no logic or epistemology can bridge." (*GTK*, 168) Verification of empirical judgements can only be probable rather than certain. Schlick gave two accounts of the source of the uncertainty of empirical judgments. One considered the inductive problem of knowing the future: this is the problem of relying on the law-like regularities of nature in our assertions about reality. Since a true proposition should be confirmed "always and without exception...what we can infer, strictly speaking, from a limited number of verifications is not absolute truth but only probability (*GTK*, 168)."

The second source of uncertainty cited by Schlick, in a later passage in the *General Theory of Knowledge*, is that associated with the imprecise nature of intuitive experience:

> Due to the fleeting character of experiences, this act of comparing and finding the same is always subject to an uncertainty that, although harmless and of no significance for the practical conduct of science and everyday affairs, is always present theoretically and stands in the way of absolute infallibility. (*GTK*, 342)

Schlick needed a nonconventional link between his perceptual judgments and the facts of experience to make his account of verification work. But verification was supposed to *produce* this assurance that our judgments are true of reality. Schlick knew that he could rely on the certainty produced through manipulations of his *implicitly* defined system of conceptual judgments, and that his originally concretely defined concepts did not contaminate the system with extra-linguistic uncertainty. But his account of *truth* required the confrontation of a perceptual judgement with an extra-linguistic state-of-affairs. The experience linking one's judgments to reality had to provide the basis of one's designation by Schlick's foundational Agenda 2. But, experience with respect to Schlick's formalistic Agenda 1 was not to be trusted; specifically, it was not to be considered knowledge at all. Coffa wrote:

> It would be hard to exaggerate the significance of this difficulty....the link between knowledge and reality depends entirely upon the link between basic statements and reality, and, as Schlick's example illustrates, it was widely assumed that the key to this link was experience. (Coffa, 356)

And we note a profound ambivalence in Schlick's attitude toward experience.

5. Meaning and Affirmation

We can see Schlick's two agendas still working in his later account of verification, which he presented in articles written from the late 1920's to the middle 1930's. What is new, following Schlick's acquaintance with Wittgenstein's work, is a focus on *meaning* as well as on *truth*. Schlick distinguished now between verifiability and verification: *verifiability* refered to a criterion of the *meaningfulness* of a statement; it was a purely philosophical notion having to do with the logical possibility of a corresponding state-of-affairs. *Verification* referred to a procedure for establishing the *truth or falsity* of a statement; it was a scientific activity having to do with the empirical possibility of confirming a particularly occuring state-of-affairs. (see *Vol. II*, (1936), and *The Problems of Philosophy in Their Interconnection*)

Schlick's discussion of meaning in terms of verifiability for the most part follows the holistic assumptions built into his Agenda #1; his discussion of truth in terms of verification follows those of his foundationalist Agenda #2. But in each there is a struggle that is not evident in his earlier work. Schlick raises questions in each account that show that he is trying to accomodate his commitment to a conventionalist account of knowledge along with his conviction that perceptual judgments can be considered true records of "immediately experienced facts."

Schlick wanted to show that what is meaningful is not based on private psychological experiences but on publicly accessible information. He wrote that "it would be nonsense to say 'We can mean nothing but the immediately given'." (*Vol. II*, 462) What is required when we give the meaning of a sentence is a knowledge of the rules that tell us how the sentence is to be used: these include both *ordinary definitions*, in terms other than the term itself, and *ostensive definitions*, which Schlick formerly called 'concrete definitions'. Schlick stated that "there is no way of understanding any meaning without ultimate reference to ostensive definitions, and this means, in an obvious sense, reference to 'experience' or 'possibility of verification'." (*Vol. II*, 458) He admitted that he had insisted both on an "empirical-meaning requirement" and on the fact that "meaning and verifiability do not depend on any empirical conditions whatever, but are determined by purely logical possibilities." (*Vol. II*, 467) He dealt with this apparent contradiction himself by explaining that the problem stems from the ambiguous use of the term 'experience.' It is used in one sense to refer to 'immediate data', and in another sense, the sense in which Hume and Kant used the word, to refer to the process by which we gather information inductively (*Vol. II*, 468). Schlick's 'ostensive definitions' are the link between his two senses of experience: "through them verifiability is linked to experience in the first sense of the word. No rule of expression presupposes any law or regularity in the world...but it does presuppose data and situations, to which names can be attached." (*Vol. II*, 468) So verifiability is not 'independent of experience' with respect to immediate data, but only with respect to experience as inductive learning. In other words, there is no Kantian synthetic *a priori* knowledge about the laws of nature, but an account of meaning must presuppose the immediate data of sense-perception.

In Schlick's ultimate discussions of verification as a criterion of truth, we find assumptions associated with his second foundational agenda. What is new here is his reference to our moments of perceptual certainty as *Konstatierungen*, or 'affirmations' of reality. We know that for Schlick "all verifications terminate in perception" (see above); it is *Konstantierungen* which function as these endpoints, and which fulfill the same role that his perceptual or fundamental judgements did in his early account. They have, according to Schlick, the "positive value of absolute certainty and the negative value of being useless as an enduring foundation." (*Vol. II*, 386) It would be absurd to question the validity of observations made by oneself, in the present, according to Schlick: we may therefore consider our current and subjective experiences as absolute indicators of the truth or falsity of empirical claims. Schlick's colleagues read his 'affirmations' as basic *statements*, subject to all of the criticisms levelled against Carnap's original construal of incorrigible protocol sentences. Neurath had pressured Carnap into a modified version of protocols as statements which were as fallible as any other empirical statements; they were simply *chosen* as basic statements. A number of writers have noted that had Schlick presented his 'affirmations' unambiguously as perceptions rather than as a type of perceptual *statement*, the debates would have taken a different turn, or perhaps not have occurred at all. (see Oberdan, Chisolm and Hilpinen in Haller) Schlick's position is at least consistent with his Agenda #1 if affirmations are merely the perceptual moments that illicit linguistic responses, rather than statements themselves. They remain outside of science for the reason that they remain outside of language in general:

> Science does not rest on them, but leads to them, and they show that it has led aright. They are really the absolutely fixed points; we are glad to reach them, even if we cannot rest there. (*Vol. II*, 383)

The phrase 'absolutely certain foundation' was also guaranteed to ruffle his colleagues' philosophical feathers, and they viewed Schlick as stubbornly hanging on to an outdated correspondence theory of truth. But was this criticism warranted? In what sense do Schlick's affirmations provide an absolutely certain foundation of knowledge?

The question behind the problem of the absolutely certain foundation of knowledge is, so to speak, that of the legitimacy of the satisfaction which verification fills us with. Are our predictions actually realized? In every single case of verification or falsification an 'affirmation' answers unambiguously with yes or no, with joy of fulfillment or disillusion. The affirmations are final. (*Vol. II*, 383)

Schlick's finality has to do with a personal sense of fulfillment or disillusion, not with an incorrigible base of knowledge. While it appeared to others that Schlick retained a Cartesian desire for certainty about the empirical world, Schlick had settled for momentary flashes of perceptual certainty rather than the constant but distant illumination of a self-evident principle.

6. Concluding Remarks

If Schlick's colleagues misunderstood his commitment to a formalistic account of knowledge and meaning, they also misread his account of verification by affirmation as producing certain rather than probable truth. As we have seen, Schlick had given a detailed analylsis of the verification of judgments of reality in his *General Theory of Knowledge* in terms of probability rather than certainty and very likely assumed that his remarks about an empirical foundation of knowledge were read in this light. Schlick considered his role in the debates "nothing but a gentle warning of a true empiricist against certain tendencies towards...a rather dogmatic irrationalistic formulation of positivistic principles." (*Vol. II*], 400) We may note that Schlick preferred the label 'consistent empiricist' to 'logical positivist' (see *Vol. II*, 283). A consistent empiricist would tend to be interested in what we actually do when we say that we know something, and how we got that information. *Konstatierungen* are, after all, not a bad *description* of what happens when we decide to affirm or deny a particular claim. Schlick's ideas do not even sound that far out of line with current views on perception and cognition. With respect to his conviction that the experiential foundation of knowledge could not be considered part of knowledge itself, since it lacked the precision of linguistic concepts, compare the following passage by Patricia Churchland in her book, *Neurophilosophy*::

> Although some cognitive activity probably is understandable as the manipulation of sentential representations according to logical rules, many cognitive processes likely are not. Indeed, the processes underlying sentential representation are surely themselves nonsentential in nature. (Churchland, 452)

With respect to Schlick's apparent ambivalence about the linguistic status of his perceptual judgments, or affirmations, compare this statement by Gerald Edelman in *Neural Darwinism*:

> Perception...is close to the interface between physiology and psychology...[it] involves categorization, a process by which an individual may treat nonidentical objects or events as equivalent. (Edelman, 26)

We have become accustomed to distinguishing between those who deal in empirically equivalent theories and those who have faith in our ability to 'get it right', in our potential someday to fashion a true description of reality. Schlick understood well that knowledge could encompass competing theories (see *Vol. I*, (1915)), but was also convinced that the role of perception in verification was to guide us in theory selection, ultimately to better theories. We tend to read the second sentiment as an override on the first, as if wanting to 'get it right' precludes acknowledging the hypothetical nature of our knowledge. Schlick watched his colleagues suffer from the unique hubris of the logical empiricist, which made it impossible to avoid murdering one's empiricism and marrying one's logic. Perhaps Schlick's hubris was of a different sort. Perhaps what fueled his conviction that both his logical and empirical agendas were correct was his very strong desire for clarity and precision at all costs. Verification, for Schlick, was the only road to clarity and precision on the empirical front; keeping conceptual knowledge systems floating above the empirical ground guaranteed clarity and precision on the logical front. The price for each of these assumptions, however, was to give up absolute certainty about the empirical world. This, as we have seen, Schlick was willing to do.

References

Churchland, P. S. (1986), *Neurophilosophy, Toward a Unified science of the Mind/Brain.* Cambridge: MIT Press.

Coffa, J. A. (forthcoming) *The Semantic Tradition from Kant to Carnap: To the Vienna Station.*

Edelman, G. (1987), *Neural Darwinism, The Theory of Neuronal Group Selection.* New York: Basic Books, Inc.

Friedman, M. (1983), "Moritz Schlick, Philosophical Papers", *Philosophy of Science* 50: 498-514.

Gadol, E. (ed.) (1982) *Rationality and Science, A Memorial Volume for Moritz Schlick in Celebration of the Centennial of His birth.* Wien: Springer-Verlag.

Haller, R. (ed.) (1982), *Schlick und Neurath—ein Symposion, Grazer philosophische Studien* 16/17. Rodopi, Amsterdam.

Oberdan, T., (1989), "Conventionalism in the Protocol Sentence Controversy: The Case for Schlick". (manuscript)

Ryckman, T., (1990), "Conditio Sine Qua Non? The Concept of *Zuordnung* in the Early Epistemologies of Cassirer and Schlick". (manuscript)

Schlick, M. (1985), *General Theory of Knowledge.* La Salle, Illinois: Open Court Publishing Co. Translation by A.E. Blumberg of *Allgemeine Erkenntnislehre,* Second German edition (1925).

_ _ _ _ _ _. (1978), *Philosophical Papers,* Volume I (1901-1922), ed. by H.L. Mulder and B.van de Velde-Schlick, Dordrecht: Reidel.

_ _ _ _ _ _. (1978), *Philosophical Papers,* Volume II (1925-1936), ed. by H.L. Mulder and B.van de Velde-Schlick, Dordrecht: Reidel.

_ _ _ _ _ _. (1987), *The Problems of Philosophy in their Interconnection,* Winter Semester Lectures, (1933-34), ed. by H. L. Mulder, A. J. Kox, and R. Hegselmann, transl. by Peter Heath (Dordrecht: Reidel).

RUDOLF HALLER

THE FIRST VIENNA CIRCLE*

I

The thesis I present for examination is this. Even before the founding of the so-called Vienna Circle around Moritz Schlick, there existed a first Vienna Circle with Hans Hahn, Philipp Frank and Otto Neurath. This circle is of such constitutive importance for the formation of the circle around Schlick that the judgement can be justified that it was really Hans Hahn who founded the Vienna Circle. To draw attention to this I call the one the *first*, the other the *second* Vienna Circle.

Like every theorist, the historian finds no limits for the formation of hypotheses but their intelligibility. As long as his statements are meaningful assumptions, he has fulfilled one of the necessary conditions of research and possesses the right to have his judgement examined. Initially I do not claim more for my considerations.

Before I defend and justify my thesis I wish to say that it is not just *historical* curiosity which suggests or demands an interest in the question, although the current revitalisation of the discussion of the Vienna Circle does also justify such an interest in the history of its origins. Rather, what demands an explanation which goes beyond the mere description of differences and brings genetic aspects into play, is the question of the *philosophical* and *systematic* divergences in the known Vienna Circle. After all, the revitalisation of the discussion of logical positivism does not spring from the investigation of questions of the nature of theoretical and practical rationality, which emerged only later (Mohn 1977, Hegselmann 1979a), but begins at the end of the 1960's (Barker & Achinstein 1969) and finds its clearest testimony in the monumental *Vienna Circle Collection*.

This revitalisation constitutes first of all a parallel to the rising historicism in the new theory of science. Insofar as this historicism was directed against the standard view of scientific theories, as it was supposedly held in the Vienna Circle, one tended, as I must at once stress, to identify the so-called 'received view' with the theories of the Vienna Circle. It would have only been natural to search out these ideas anew in order to become acquainted with their true nature. For in the meantime, due to the quick

95

T. E. Uebel (ed.), Rediscovering the Forgotten Vienna Circle, 95–108.
© *1991 Kluwer Academic Publishers. Printed in the Netherlands.*

development of the analytical theory of science, the picture of logical positivism had become a cliché which rested mainly on a few simplifications. In this picture, the members of the Vienna Circle supported a reductive theory of knowledge according to which all statements were to be reduced to those about *immediate givens,* or at least could be constituted out of elementary experiences; second, the logical empiricism of the Vienna Circle was *ahistorical* and interested only in the logical-analytical and systematic problems of the sciences; third, the Vienna Circle promoted the model of the *cumulative progress* of scientific theories, and thus an unsupportable conception of theory change. On various occasions I have already pointed out that this picture is wrong and requires radical correction (Haller & Rutte 1981, p. viii; Haller 1977a, 1982a [ch. 10], 1982b). The supposed dogmas of logical empiricism were criticised by the logical empiricists themselves, and some members of the Vienna Circle never held theses which would be contaminated by those just mentioned. Bringing this out clearly appears to me to furnish already sufficient reason for renewed concern with the *real* history of the Vienna Circle. A further reason is simply that logical empiricism was for a long time the only philosophical theory which sought to understand the present century's scientific picture of the world given by logic, mathematics, physics, biology and sociology in a unified perspective, and which made the most developed cognitive tool of evolution, namely, scientific methods themselves, the central object of philosophical analyses. In comparison to this epistemological perspective, one of the original motives of the neo-empiricist movement—the anti-metaphysical enlightenment— could not but recede into the background, although it still thoroughly determined the *scientific* world-conception. The role model of Ernst Mach influenced not only the known Vienna Circle, but particularly so the 'first' to which I now turn.

II

In the introduction to a collection of his essays, Philipp Frank described very clearly and in detail how the first Vienna Circle came to be, which themes dominated the discussions, and who were the main participants (Frank 1949a). Frank's report starts from the fact that around the years 1907 to 1912 there existed in Vienna a group of young scientists who met regularly— incidentally, on Thursdays, like the later Vienna Circle—and whose interests were described as essentially methodological and theoretical, but also as prompted by "political, historical and religious problems". The report makes clear that Ernst Mach and the French theorists of science, Pierre Duhem, Henri Poincaré and Abel Rey, were the main authors who formed and influenced the attitude of this group of philosophically interested scientists. These names also emerged, still earlier than in Frank's, in Neurath's descriptions of the history and prehistory of the Vienna Circle. And indeed, it is these names which we meet again as guides for our orientation in our search for the main lines of thought of the first Vienna Circle.

This type of the theory of science was admittedly never wholly excluded in descriptions of the history of logical empiricism, but mostly it was not investigated in the depth it deserved. And what was particularly overlooked was the degree to which the so-called French conventionalists, together with Mach and Popper-Lynkeus, influenced the ideas of the first Vienna Circle. We are in the first decade of our century during which the main works of Duhem and Poincaré were published—and during which the most important of these were translated into German. Thus *La Theorie Physique: Son Object, Sa Structure* was published in 1908 in the translation by Friedrich Adler with a foreword by Ernst Mach, only two years after its first publication in book form in French.[1] It was preceded by the translations of Henri Poincaré's *La Science et l'Hypothése* and *La Valeur de la Science* . Also in 1908 Abel Rey's *La Théorie de Physique chez les Physiciens Contemporains* was published in German translation by Rudolf Eisler.

I believe one can represent the central thesis of these French conventionalists with the picture which Poincaré gave of the various geometries (Poincaré 1902; cf. Haller 1983). Poincaré rehearses hypothetical models in order to show how the different perspectives which we occupy in respect of objects result in non-self-contradictory structures, structures which we construct ourselves inasmuch as it depends upon our decision which assumptions we make and accept. We construct the frame and fit into it what we agree about. This agreement—the truly conventional moment—is, of course, not a factual one and also not the consensus of an ideal discussion round, but is a determination or definition which is viewed as rational yet cannot be justified by logic and experience alone.

What makes possible this path between (or, if you like, beyond) apriori rationalism and empiricism? By not attributing to science the aims usually attributed to it, namely, to provide a description of phenomena that is as comprehensive and as precise as possible and to discover their basic laws.

33

The answer with which Mach opposes the traditional conception of scientific descriptions of the world already moves the value criterion of the conventionalists into the foreground. Science is not *mere* description, but an *economical* description, that is to say, a description of phenomena and their relations which is reduced to simple elements (Haller 1981; cf. Rey 1905 (1908, pp. 68-109)). The imagination, which presents us with the infinite space of possibility, finds itself increasingly limited by the frugality of rational models with respect to a given domain of objects. The 'adaptation of thought to facts', demonstrated by the natural history of evolution, is continued and perfected in theory construction. Of course, never without the price of simplification, schematisation and idealisation of objects.

Often we are offered a wrong picture of the teachings of Mach—not only, by the way, by the opponents of empiricism—as if he had been, first, a pure inductivist and, second, a pure phenomenalist. Both interpretations are totally unsupportable and only serve to confuse. Without any ambiguity Mach stressed again and again the intervention of the principle of economy in the process of ordering experience; translated, this means the intervention of rational operations in sensual experience. Therefore Mach did not recognise the priority of observation to theory, nor that of the subjective to the objective standpoint. So Mach shares with the French conventionalists the point of view that our experiences are not only ordered according to theoretical criteria, but that the invoked ideal of unique determination is afforded only by decisions which we make in the delimitation of the products of our imagination and in simplifications in general. Mach also agreed with Duhem that "a very large number of experimental values may be assigned to a theoretical value", (Mach 1908, p. iv; cf. Mach 1905, p. 349) and that only a provisional validity of our laws of nature already follows from this.

Indeed, this was one of the most essential insights of Duhem: that to one experiential datum one could give various theoretical interpretations, and that the correspondence with experience—the only criterion of truth for a philosophical theory—is determined by, and bears the stamp of, our decisions. Of course, there remains a great difference between the point of view of Mach and that of a natural classification as the reflex of a real existent order, as propounded by Duhem, for nowhere is Mach inclined to accept Duhem's metaphysical superstructure. But in the characterisation of theory construction, in the estimation of the constructive moments of abstraction, of simplification, no difference between the two seems to obtain.

We cannot deny that some of these arguments need be accepted if we do not want to falsify the nature of scientific research. Thus we must admit that without any description at all no fact may be declared to obtain or not to obtain. We refer to facts and events by individuating them by linguistic means, by describing them as 'a so-and-so' or as 'the so-and-so'. And as we know, we can arrive at an infinite series of descriptions for every fact. And, analogously, one and the same fact may be explained by different theoretical sentences. In the pursuit of practical purposes this multiplicity does not interest us very much: we select the simplest and most convenient, depending upon the requirements of the case in question. But we *choose* and can choose. And matters are not different in the theoretical disciplines: only there, we are also interested in the multiplicity of possibilities.

The decisive considerations which Duhem adds to Mach's about the principle of economy are, first, that there is no *experimentum crucis* which could decide the correctness or truth of one of two theories between which we may choose for the explanation of one and the same phenomenon; second, that, in both the testing of an experimental hypothesis and of a theoretical sentence, we confront the *whole* system of observations with that of theoretical sentences.[2] This holism is difficult to formulate theoretically, but it was perhaps the greatest influence upon the first Vienna Circle. And it is this that has been virtually discounted, if not overlooked, in all descriptions of logical empiricism.

It would be, as noted, an exaggeration to think that Mach's theory of elements totally agreed with the theories of the French conventionalists. Rather, it was just from the tension between the positions taken in the theory of science by Poincaré and Duhem on the one hand and Mach on the other that the bundle of questions emerged which determined the problems which occupied the first Vienna Circle.

I believe that the two French theorists of science primarily arrived at a different interpretation of the laws of nature from Mach because they were interested in *other* problems and investigated other constituent elements of scientific theory construction. The starting point of Poincaré's arguments for a conventionalistic interpretation of theories was the comparison of Euclidean and non-Euclidean geometries. Since—presupposing the axiomatic nature of the theories at issue—different axiom systems can be used for the description of the same structures, the essential question had to be concern for the criterion of the selection of theoretical sentences. And the answer was: if the description of a phenomenon can be separated into two components, a theoretical and an empirical one, and if the former depends upon *conventions*, then the selection of a preferred hypothesis cannot be grounded by criteria of coordination, even less so of correspondence. But if logical or ontological rules of preference cannot guide us, then there remain only pragmatic ones, as e.g. simplicity and convenience. Whether a geometrical system is to be preferred thus does not depend only on whether it sufficiently 'approximates' the facts, but whether it possesses properties which make the coordination with observable phenomena more easy than another, alternative system, and whether it can be integrated into the theoretical system so far used.

What we see here is basically another consequence of Mach's idea of economy. For Mach, the simplification or idealisation of a theoretical sentence is guided by the pragmatic criterion of 'adaptation'; similarly, Duhem repeatedly says: "An infinity of different theoretical facts may be taken for the translation of the same practical fact." (Duhem 1906, p. 134.) It is for this reason that a contradiction by experiment does not determine the truth or falsity of a physical hypothesis. The only test of a theory which Duhem declares to be satisfactory is the comparison of "the entire system of the physical theory with the whole group of experimental laws." (Ibid., p. 200.) It is impossible to know which of the premisses of a many-levelled theory has to be revised if an experimental finding contradicts a theory (Poincaré 1902, pp. 151/2). The explanation which Poincaré provides is illuminating. The core of his argument consists in the thesis that it could be possible that the actual relations within a particular domain were conceived of correctly by the theory, but that the contradiction "only exists in the images which we have formed to ourselves of reality." (Ibid., p. 163.)

It seems to me that here an essential aspect of theory construction in general is touched upon: the framework, within which theoretical hypotheses are developed and made precise, forms the frame of an assumption of intelligibility which concerns the total context of the phenomena of the investigated domain. In most cases we are here led by the natural world-conception, the view of common sense, and the pre-formed ways of predication in natural language. But in many areas scientific research 'violates' such a pre-understanding and does not achieve the re-constitution of the natural picture of the world from which it derived its models. That is why the separation of meaning and designation or, in Frege's terminology, of sense and reference, is more fruitful for the reconstruction of these relations than the use of a rough set-theoretic terminology. And that is why, according to Duhem's proposal, one should use and consider in the selection of a hypothesis not only the logical rules and mathematical and physical principles and postulates, but especially also common sense (*'bon sens'*). Since it is in principle impossible to build up a theory from experimental or observational data alone, that is, purely inductively—for the derivation of theoretical sentences always presupposes or implies further theoretical sentences—every observational fact requires also that it be judged with respect to a theoretical framework.

It is in this that the true justification of the holistic conception lies. It is certainly an exaggeration then to hold that for the conventionalists natural science does not represent "a picture of nature but merely a logical construction." (Popper 1935, p. 79.) As I just mentioned, the integration of the conceptual construction is guided by pictures. Rather, what is demanded—and this demand we find in Poincaré as well as in Rey or Duhem—is that a law must always also have an operative sense, if is to be understood as the definition of an object (according to principles of lawlikeness). In other words, the application of a law, e.g. in operations of measurement, decides about its usefulness and appropriateness. And the conventionalists realised, correctly and early on, that not only verification but also falsification cannot constitute a criterion of selection, and therefore of demarcation.

If then there is no logical (deductive) or empirical (inductive) necessity of preference in this most important of steps in theory construction, the selection of hypotheses, if—as in the pragmatic conception of truth—subjective attitudes and circumstances form the criteria, then we must admit the place of randomness and caprice in the formation of theories. Many see in this the danger of a conventionalist interpretation of scientific research. But just this thought was accepted as liberating and used in the first Vienna Circle, and particularly in the philosophy of Neurath. Why?

Before I return to this question in the next section, I wish to consider briefly the third of the earlier mentioned theorists of science, Abel Rey. I do not thereby recollect a pathbreaking thinker, but a synopticist who was important for the members of the first Vienna Circle in a special way, primarily for the reason that he connected a wealth of material from the texts of contemporary physicists with a clear position and a decisive judgement. Many of the themes and—what is more astonishing—many of the theses which were held in the well-known (second) Vienna Circle can be found explicitly in the work of Rey. First, he shares with Duhem the methodological premiss that a theory of science without regard to the history of science is empty, because the change of scientific theories happens only on the basis of their historical forms and can only be reconstructed on this basis. Second, Rey advocates a theory of the *unity* of science which comprehends philosophy within it and determines its tasks.

In Rey's treatise of 1905 we thus meet already the contrast between the cumulative and the non-cumulative interpretation of the progress of knowledge in the exact sciences in the guise of the contrast between mechanists and conventionalists. Rey, who ever anew discusses sceptical objections to the possibility of actual progress in physical knowledge, seeks to show that it is one and the same domain, one and the same reference, to which both these metatheories recur.

The 'empirical rationalism' of which Rey speaks, unites Mach's empiricist adaptation theory of the evolution of cognition with the idea of a new rationalism whose basic thought rests in the fact that objective experience and thought mutually determine each other, or literally, are "mutual functions of each other". (Rey 1905 (1908, p. 363).) Such a convergence of antagonistic directions seems called for already because, in the interpretation of Rey, empiricism leads "necessarily to relativism". (Ibid., p. 308) One reason for this is seen to be the nature of hypotheses themselves: they are revisable by definition. All generalisations, that is to say, all inductive premisses, are supported by experience and can be falsified by experience.

It follows from this that they do not hold definitively and unchangingly; they are capable of development like the hypotheses, which, however fundamental, are always made more precise, complete, better. - Thus they permanently remain open to revision and to the further delimitation of their domain. (Ibid., p. 319.)

Rey always very clearly seeks to specify the historical location of the perspective at which modern natural science has arrived, namely, one beyond dogmatism and scepticism, the two extreme possibilities to which rationalism and empiricism lead, or sink. Rey therefore sees the essential task of the metatheory of the exact sciences in the development of structuralism, of a theory of relations. He remains a Machian insofar as he reduces the necessity, which seems represented in the fact that certain relations do obtain, to habits which evolution formed in us. The rational and the empirical are thus understood as moments of a historical development.

Also from this quarter then, the members of the first Vienna Circle came to be well-acquainted with the problems of a historical theory of science—a fact which apparently has remained unknown to the anti-positivists of our own day.

III

In order to recreate the spirit of the group around Hans Hahn, it is advisable to make reference primarily to writings which derive from the dependence I here postulate, and which originate from the time before Schlick's arrival in Vienna and before the publication of Wittgenstein's logical-philosophical essay.

Philipp Frank—who in 1912 had been called to Einstein's old chair in Prague—stated in his fundamental essay on 'The Importance for our Times of Ernst Mach's Physicalistic Theory of Knowledge' (1917) that, first of all, the main value of Mach's teachings consisted in the defense of the edifice of physics against attack from the outside, not in its productivity for physical research itself. This is a correct remark about the character of foundational investigations: they do not represent inquiries in parallel with object-theoretical investigations, but *meta*scientific inquiries into scientific research, inquiries also intended to analyse and answer epistemological questions about the type of research in question. This main value of Mach's teachings showed itself most clearly in the attempt to "bring about a connection between physics on the one hand and physiology and psychology on the other that is as free of contradictions as possible", a connection that would dispose of the need to switch the entire vocabulary of concepts when crossing the boundaries of these disciplines. That is, Frank correctly sees in Mach's attempt in the theory of elements—which he interprets phenomenalistically—one step towards the unification of theories, towards unified science. To express this in the lively image of Mach's: "As the blood in nourishing the body separates into countless capillaries, only to be collected again to meet in the heart, so in the science of the future all the rills of knowledge will gather more and more into a common and undivided stream." (Mach 1896, p. 261.) This was one of the reasons for Mach's importance for the intellectual life of the day, which Frank stressed. The other consisted in that Mach literally carried the spirit of enlightenment into the 20th century: it was ultimately his critique of superfluous entities which made Mach doubt the existence of atoms.

Frank, the follower of Einstein, naturally regards Mach's critique of the foundation of Newtonian mechanics, the critique of the conception of absolute space, as the prime example of a critique in the spirit of the principle of economy, and a critique which also was directed against a purely mechanistic physics. We must agree with Frank when he explains Mach's scepticism of atomic theory with reference to his critique of the 'auxiliary concepts' of materialism. Since the original enlightenment was bound up with a theory whose 'auxiliary concepts' like 'matter' and 'atom' had remained impure, it was necessary for the *truly* enlightened mind to criticise these concepts. And that had been the way of Mach's great works in the history of science, of which we need mention only his *Mechanics* (1883) and the early essay 'The History and Root of the Principle of the Conservation of Energy' (1872).

Again and again Mach criticised the *basic concepts* of natural scientific research—space, time, substance, causality, etc.—for he aimed to rob them of their metaphysical-mythological mantle and to determine their place in the framework of the theories of natural science, e.g. psycho-physics. And like logical empiricism later, Mach already used an argument from the critique of language in order to eliminate what he deemed metaphysical concepts, namely, the argument of their meaninglessness. According to this argument, the application of a pseudo-descriptive expression does not result in a false, but a *meaningless* sentence. And for the same reason, even the question at issue can be discounted as a pseudo-problem. It is primarily this latter aspect which guides Frank's characterisation of Mach as an enlightenment philosopher *par excellence*: "The task of our age is not to fight against the enlightenment of the 18th century, but rather to continue its work." (Frank 1917, p. 75.)

These then are the two basic tendencies which Frank ascribes to Mach. But these are also the postulates which are upheld as norms for scientific research; Frank always remained a convinced Machian in this sense. Transcending his own position in the theory of science, Mach became the herald of the natural scientific conception of the world, as it was advocated by Frank, Hahn and Neurath. Enlightenment and, for that reason, liberation from metaphysics, was and remained the first and ever essential aim of the scientific enterprise.

This impression is immediately confirmed if we consult Neurath's writings, in particular the longest of those which he dedicated to the history of the Vienna Circle. There we read:

Mach's and Einstein's criticism of Newtonian physics and the new conceptual edifice which resulted therefrom had a particularly specific influence in Vienna. Already as a young physicist, Philipp Frank stood in direct contact with Mach and Einstein. (Neurath 1936a, p. 695.)

Like Frank, Neurath viewed Mach's anti-metaphysics as the sharpest tool in the continuation of enlightenment. And in his justification of the basic empiricist attitude, Neurath, throughout all phases of his development, always referred back to Mach.[3] But however early we find invocations of the authority of Mach, there we also find references to the conventionalism of Poincaré and Duhem, in Frank as in Neurath.[4]

Already in his early work of 1907 on the causal law, Frank formulated a thesis which he himself literally designated as 'conventionalist.'

The law of causality, the foundation of every theoretical science can be neither confirmed nor disproved by experience; not, however, because it is a truth known *a priori*, but because it is a purely conventional definition. (Frank 1907, p. 54.)

Frank proposes the following interpretation of the real content of the law: "If in the course of time the state A of the universe was followed by state B, then whenever A occurs B will follow it." (Ibid.; cf. Haller 1982f.) In other words, Frank accepts Hume's interpretation of causality as regularity and postulates its conventional definition to be a rule of research for natural science.

The ground is here prepared for the model of two languages, which in the course of the development of logical empiricism was varied in different ways. The two languages are identified with the two branches of natural science.

Whereas experimental natural science describes the properties of bodies as given by our senses, and the changes in these properties, the task of theoretical science is to provide bodies with fictitious properties which first of all insure the validity of the law of causality. Theoretical science is not research but a sort of remodeling of nature; it is the work of the imagination. (Frank 1907, p. 57.)

Here the form of the combination of Mach's teaching with those of the French conventionalists becomes clear. The free postulation of assumptions becomes comparable to the conception of Kant's critical idealism insofar as both claim a presupposed framework of experience; the Kantian movement ascribes this to the nature of reason whereas Frank declares it to be a product of human free choice.

Mach's basic idea that assumptions, schematisations and idealisations are ultimately products of the imagination, and that science therefore derives from the same source as the other cultural achievements of humans, is identified (with an emphasis equal to that of the conventionalists) with the conventionalists' idea that these assumptions etc. are the product of logically unconstrained, free choice. This means that the nature which is the object of theoretical science "is not empirical nature but a fictitious nature". This nature is not therefore only the object of human curiosity, but also the product empirical action, and only because of this it becomes transparent, comprehensible. The constructive traits of a theory such as Frank envisaged, whose definiteness is created and creatable, are of course not mirrored in experiment, in experience. For this reason, experience has more play than theory has. These created pictures of the world are therefore never true or false but, depending on their '*Gestalt*', their structure, they are either simpler or more complicated. As is common in conventionalism, the principle of simplicity steps into the place of truth.

In a similar fashion, Wittgenstein says in his *Tractatus* that we have always the possibility to chose an arbitrary form of description, and that one may be simpler than another. He calls these forms 'nets' or 'systems' of descriptions of the world. Accordingly, the causal law is not a law which refers to objects, but something which *shows* itself in the fact that there *are* natural laws, that is, lawlike connections (Wittgenstein 1921, #6.32 to #6.361). Likewise, Wittgenstein at this juncture seems compelled to stress the accidental aprioricity of such a net as the distinguishing characteristic vis-a-vis the otherwise similar Kantian conception.

In his philosophical main work which appeared 25 years after his essay on the causal law, Frank corrects his reference to similarities with the Kantian position. He now says:

> If I had already recognized then that the doctrines of so-called 'idealistic philosophy', even of 'transcendental idealism', are not empirical statements—because they are neither tautological nor do they say anything about the real world—but assertions about a world beyond all human cognition, then I would have placed much less emphasis on the analogy between the conventionalist and the Kantian conception of causality. The analogy exists, to be sure, namely, in that according to both views the causal law does not say anything about the empirical world, but only about how humans conceive of the world.

And Frank continues in the vein typical of the Vienna Circle: "But if one wants to take a position that is conducive to scientific progress it is important to stress the *contrast* to the *Kantian* conception of the causal law." (Frank 1932, p. 242; cf. Stegmüller 1960.)

It is impossible here to follow up these reasons in greater detail, though a separate investigation of the main difference between a conventionalist and a Kantian interpretation is required. What is required for present purposes was said already, namely, that what gets corrected in the course of further development is the supposed or actual similarity with Kant, not the acceptance of the kernel of the conventionalist thesis that general laws have the character of tautologies which are simple and afford perspicious representation. For this reason they were defined as adaptable to the given state of research and revisable.

Now Neurath formulated the principle of this revision very early on. And although he follows mostly in the steps of Duhem and Poincaré he went further, in my estimation, than the other conventionalists. To be sure, both Frenchmen stressed again and again that a law could not collide with experience—the reason why according to Duhem there can be no *experimentum crucis* (Poincaré 1902, pp. 105, 179, 135 ff., 151/2; Duhem 1906, pp. 188ff., 208ff.). And both nevertheless asserted the revisability of every theoretical system which failed to harmonise with experience. But neither of the two thinkers formulated the dichotomy of possibilities so clearly as Neurath—the reason why I called this principle of adaptation in theory change the 'Neurath-principle' (Haller 1979c [ch. 3], 1982a [ch. 10]). One can comprehend this principle only if one comprehends the holistic interpretation of scientific systems which Neurath apparently takes over from Duhem with all of his arguments. Since there can be no univocal description of the facts of experience with reference to theory, the dignification of an isolated sentence as an instance of falsification must be rejected. Similarly, it is not possible to apply a criterion for the selection of a hypothesis in an *experimentum crucis*. Since in the Duhemian holistic conception of theories it is furthermore only a scientific system *in toto* that is accepted or rejected, the question was bound to be raised *how* a revision is possible in the case of an imaginary or actual case of conflict of experiential sentence and system The so astonishingly obvious thesis of Neurath's says that either the system can be changed (the opinion of the French conventionalists) or the sentence itself, because both are of the same status. If one interprets this principle through theoretical naturalism and the perspective of Quine, then it follows that revisions at the periphery of the system are often the ones first thought of, but that in the end it is the experiential sentences which, as 'cornerstones of semantics', help carry the system and which may claim a certain privilege.

As has become plain, these problems did not only re-emerge in the famous protocol sentence debate of the early 1930's, but they accompany the interpretation of theory structures up to the present day.

Without doubt it was due to the development of conventionalist empiricism in the years of the first Vienna Circle that the problem of the interpretation of theories was joined again at the very focus of the position of the members of the first Vienna Circle. When Hans Hahn returned from

Czernovicz and Bonn in 1921 to Vienna, the old group, whose members were also very close in their political convictions, met anew. And it has since been often stressed that the Circle around Schlick was constituted in the main by members of Hahn's mathematical seminar.[5]

I am not of the opinion, which it seems fashionable to suggest these days, that the philosophical divisions in the Vienna Circle, as they became apparent in the protocol sentence debate, derived from the political fronts in the Vienna Circle (which are likely to have existed from the beginning). Rather, these divisions derived from the return to the problems of conventionalism, which had receded into the background due to the discussions of the *Tractatus* and its related epistemology. But it has not been the purpose of this essay to speak of the polarisation of the conception of science in the well-known later Vienna Circle (Haller 1982c [ch. 16]). It will be sufficient if it has become clear what constitutes the basic edifice of the conceptions of the theory of science of the first Vienna Circle. Recognised once, it can be recognised elsewhere again.

Karl-Franzens-Universität, Graz

NOTES

First published as 'Der erste Wiener Kreis', in *Epistemology, Methodology and Philosophy of Science. Essays in Honor of C. G. Hempel on the Occasion of His 80th Birthday, January 8, 1985* (ed. by W. Essler), *Erkenntnis* 22 (1985) 341-358, © 1985, Reidel, Dordrecht. Translated with kind permission of Kluwer Academic Publishers and the author by T. E. Uebel.
[1]   Remember that the first translation of Duhem into English (by P. G. Wiener) did not follow until 1954! And the earliest contributions by J. Agassi, A. Grünbaum, N. R. Hanson which discussed Duhem did not appear until the end of the 1950's. Amongst the descriptions of neo-positivism the contribution of conventionalism was merited a separate section at least in Kolakowski 1971, however.
[2]   Compare Duhem 1906, ch. 6; Rey 1905 (1908, pp. 122ff.). Compare also Grünbaum 1960; Giannoni 1967; Harding 1976; Diederich 1974; Schnädelbach 1971: 165ff.
[3]   See also Neurath 1915b, and the description, rich in historical materials, by Stadler 1982b, pp. 11-125.
[4]   As one can see from the index of Neurath 1981, Mach, Poincaré and Duhem are the most cited names, aside from himself and his contemporaries.
[5]   Compare the eulogy for Hans Hahn by Philipp Frank (1934) where Hahn is called the "real founder of the 'Vienna Circle'"; cf. Haller 1982b and Menger 1980.

RICHARD C. JEFFREY

AFTER CARNAP

It was through his engagement with the program of logical analysis
floated by Russell in 1915 that Carnap began to affect the shape of 20th
century philosophy. The program aimed at bringing to philosophy a
certain method or attitude, resembling that of the sciences in its focus
on progress in solving problems rather than on defense of doctrines.
Progress brought with it the doctrinal flux that soon saw the phenomenalism of Carnap's *Logische Aufbau der Welt*[1] yield to the physicalism
of his *Logische Syntax der Sprache*[2] and saw his early deductivism give
way to the probabilism of his last 25 years, during which his work in
semantics fostered a flowering of modal logic.[3] He welcomed real progress and its attendant doctrinal flux from whatever source, others no
less than himself. The celebrated "Death of Logical Positivism"[4] refers
to particular doctrines (e.g. phenomenological reductionism) and methods (e.g., syntax) that Carnap and his friends abandoned for reasons
rooted in the program itself. Broadcasting out of Europe those of its
participants and associates whom it did not kill, Nazi power propagated
the movement, which grew and changed rapidly in response to hard
challenges. If what fired the *Wiener Kreis* has either died or grown out
of recognition, that's in large part due to the standard of clarity and
care set by works like the *Aufbau*, *Logische Syntax der Sprache*, *Meaning and Necessity*,[5] "Empiricism, semantics, and ontology",[6] and
Logical Foundations of Probability.[7]

Carnap's earliest philosophical work, in the years just after the first
world war, treated space from a Kantian perspective modified by Einstein's recently enunciated general theory of relativity and by his own
characteristic analysis of the disputes over the nature of space as stemming from conflation of different senses of the term (formal, intuitive,
physical). This was his Jena doctoral dissertation, restricting Kant's
synthetic apriorism to the local topology of intuitive space (with the
rest softened to conventionalism), and viewing physical space as purely
empirical.[8]

In Jena he had attended Frege's lectures on logic, but it was after

Erkenntnis **35**: 255–262, 1991.
© 1991 *Kluwer Academic Publishers. Printed in the Netherlands.*

the war, in 1920, that he read Frege's *Grundgesetze der Arithmetik*[9] and was fired by Whitehead and Russell's *Principia Mathematica*.[10] His dissertation had characterized formal space in terms of the logic of relations. Einstein's reduction of the concepts of space and time to concretely conceived local, momentary operations of measurement played a strong rôle with him then as with Russell and with Whitehead, suggesting a general empirical constructivist program in philosophy. In his intellectual autobiography Carnap recalls his excitement in 1921 upon reading Russell's *Our Knowledge of the External World as a Field for Scientific Method in Philosophy*,[11] where "I found formulated clearly and explicitly a view of the aim and method of philosophy which I had implicitly held for some time".[12] According to Russell[13]

This method, of which the first complete example is to be found in the writings of Frege, has gradually, in the course of actual research, increasingly forced itself upon me as something perfectly definite, capable of embodiment in maxims, and adequate, in all branches of philosophy, to yield whatever objective scientific knowledge it is possible to obtain.

The actual research to which he refers is the construction or reduction or logical reconstruction or analysis in *Principia Mathematica* of the numbers and operations of mathematics in terms of purely logical concepts.

The *Aufbau* was Carnap's serious start on the program of completing that analysis by extending it to the empirical domain. In effect, perhaps, it was a start on the heralded, unforthcoming fourth volume of *Principia*. Russell had described *Scientific Method in Philosophy*[14] as "an attempt to show, by means of examples, the nature, capacity, and limitations of the logico-analytic method in philosophy". He continued:

The central problem by which I have sought to illustrate method is the problem of the relation between the crude data of sense and the space, time, and matter of mathematical physics. I have been made aware of this problem by my friend and collaborator Dr Whitehead ... I owe to him ... the whole conception of the world of physics as a *construction* rather than an *inference*. What is said on these topics here is, in fact, a rough preliminary account of the more precise results which he is giving in the fourth volume of our *Principia Mathematica*.[15]

Volume 3 of *Principia* had laid the foundations for the conceptual apparatus of physics in a theory of measurement (part VI, "Quantity"). As with the characterization of cardinal numbers (e.g., of 2 as a property possessed by the empirical property of being an author of *Principia*

Mathematica, or as a set containing the set {Whitehead, Russell}), so the rationals are characterized in empirically applicable terms, as when 2/5 is defined as a relation between relations.[16] That was appropriate since physical magnitudes were analyzed as relations:

We consider each kind of quantity as what may be called a "vector-family", i.e., a class of one-one relations all having the same converse domain, and all having their domain contained in their converse domain. In such a case as spatial distances, the applicability of this view is obvious; in such a case as masses, the view becomes applicable by considering, e.g., one gramme as + one gramme, i.e., as the relation of a mass *m* to a mass *m'* when *m* exceeds *m'* by one gramme. What is commonly called simply one gramme will then be the mass which has the relation + one gramme to the zero of mass.[17]

The fourth volume was to have treated geometry. Russell makes it sound as though the treatment would spell out something like the fourth chapter of *Scientific Method in Philosophy*, "The World of Physics and the World of Sense". Considering Whitehead's views in 1914 of physics, geometry, and sense, and his exclusive responsibility for volume 4, perhaps what he intended for it was an abstract theory that could be specialized to a construction of geometrical concepts out of perceptual ones somewhat as the theory in volume 3 could be specialized to a *Physikalische Begriffsbildung*.

However that may be, the *Aufbau* was Carnap's start on the project of enlarging the *Principia* construction to include geometry, empirical science, and everyday knowledge. It was to be a collaborative effort, to which Nelson Goodman was the earliest and most productive recruit.[18]

The individual no longer undertakes to erect in one bold stroke an entire system of philosophy. Rather, each works at his special place within the one unified science. . . . If we allot to the individual in philosophical work as in the special sciences only a partial task, then we can look with more confidence into the future: in slow careful construction insight after insight will be won. Each collaborator contributes only what he can endorse and justify before the whole body of his co-workers. Thus stone will be carefully added to stone and a safe building will be erected at which each following generation can continue to work.[19]

He was serious about the importance of rationalization, division of labor, but the suggestion of a centralized allotment of special tasks to individuals is a stylistic artifact; the individual chooses his own special place as Goodman did.

Carnap's 1928 preface was a period piece, with Bridgman's operationalistic call to arms the year before:

We should now make it our business to understand so thoroughly the character of our permanent mental relations to nature that another change in our attitude, such as that due to Einstein, shall be forever impossible. It was perhaps excusable that a revolution in mental attitude should occur once, because after all physics is a young science, and physicists have been very busy, but it would certainly be a reproach if such a revolution should ever prove necessary again.[20]

For Carnap as for Russell, scientific method was a datum, something one had the hang of as a scientifically trained person; its rational reconstruction was not high on the agenda as something to be accomplished before the rational reconstruction of our substantive knowledge could begin. Russell offers general remarks about the education of scientific investigators, and maxims claiming for the scientist such characteristics as honesty and open-mindedness, but neither he nor Carnap then essayed an explicit account of scientific method. Both saw the body of work from Frege to *Principia Mathematica* as a historically given basis and paradigm for scientific work in philosophy analogous to the body of work from Galileo to Einstein in physics. But while these histories included revolutionary change, as in the transition from classical to relativistic physics, and cataclysmic collapse of the sort that Russell's paradox produced in Frege's *Grundgesetze* construction, Carnap's 1928 statement envisaged only slow, irreversible progress. He would soon see his mistake, with Gödel's[21] incompletability proof for the *Principia* project, and Tarski's[22] deflationary rehabilitation of the concept of truth.

In *Logische Syntax der Sprache* Carnap dropped the phenomenalistic reductionism of the *Aufbau* in favor of a version of the physicalism that Neurath had been urging. The project remained that of logical analysis:

That part of the work of philosophers which may be held to be scientific in its nature . . . - consists of logical analysis. The aim of logical syntax is to provide a system of concepts, a language, by the help of which the results of logical analysis will be exactly formulable. *Philosophy is to be replaced by the logic of science* — that is to say, by the logical analysis of the concepts and sentences of the sciences, for *the logic of science is nothing other than the logical syntax of the language of science.*[23]

This was the work Quine read as it issued from Ina Carnap's typewriter in 1932 during his "rebirth in central Europe".[24]

Carnap's last big philosophical project finally addressed the nature of scientific method, in an attempt at rational reconstruction of our way of building scientific and everyday knowledge on experience. This took to a pure extreme the idea, subscribed to by many Bayesians, that it is the difference between your experience and mine that separates your probability function from mine – an idea often characterized by the vaguer saying that there are no unconditional probability judgments, but all are relative to background experience. Characteristically, Carnap took this idea quite seriously, and formulated it simply and clearly: the current probability that any individual i attributes to any hypothesis h in her or his language ought to be $c(h, e_i)$, where c is a "logical" probability function, the same for all rational agents. This constant c is the logical ingredient in logical empiricism, while the variable e_i is the empirical ingredient. Put baldly in Carnap's way the idea is dismissed by Bayesians to whom garbled expressions of it sound like obvious truths. Here as in the *Aufbau*, Carnap was conscious and careful in making moves that others whistle over hastily, eyes averted from the void below.

Like the *Aufbau*, this was the start of a long-range project for which Carnap hoped to recruit collaborators. That was the point of the *Studies in Inductive Logic and Probability*[25] series, to expedite cooperation among scattered collaborators through quick publication of new work on the project. He was optimistic and (Ayer's term) serene, first to last. He died with his logical boots on, at work on the project.

What's wrong with logical empiricism? Quine offered a double answer in "Two dogmas of empiricism": belief in the analytic-synthetic distinction, and reductionism, "the belief that each meaningful statement is equivalent to some logical construction upon terms which refer to immediate experience".[26] For these, Quine substituted the belief that our overall system of (yes/no) judgments is an economizing response to the totality of irritations of our sensitive surfaces. These irritations are not captured by protocol sentences recording sensations, but have their effects here and there in the totality of our assents to and dissents from sentences of various sorts. The account he gives of the economy of thought involves a circle of terms noticed by Charles Chihara:[27] projectibility, similarity, simplicity. Quine sees that as a virtuous circle, unlike the vicious circle so carefully traced in "Two dogmas". But with Chihara, I find the view clearer from a Bayesian, probabilistic perspective – if not quite the one Carnap was mapping in his last 25 years.[28]

But surface irritations aren't enough. We are active animals; we process our inputs in view of what we take ourselves to be doing as we receive them. We are primates who don't swing from trees but do get sensory inputs on the fly, as an integral part of processes like walking, driving cars, conversing, catching balls, writing, etc. There is no hint of that in the common philosophical examples – observing hard brown tables, sensing yellow patches, "here now headache", and the rest. Mach's illustration, the view from his left eye as he lies on a couch, sums it up.[29] Here's where Anscombe and Hampshire on knowledge without observation are right on the money.[30] It's not just surface irritations that provide our sense of what we're doing and trying. That's the big truth in pragmatism as I see it. Rejecting reductionism won't get us out of the way of our own feet until we also reject the view from behind Mach's moustache.

It strikes me that in adopting Russell's program, Carnap adapted it quite characteristically. Just as he advocated artificial languages, consciously constructed by us to serve our purposes, so his philosophical method was synthesis, not analysis – a fact better understood by detractors dismissing him as a mere engineer than by some of his friends. He saw meanings as human artifacts, but had no reverence for traditional modes of conceptual production and their attendant mythology, for the lore of our fathers. He thought it practical, and essential for progress, to select and abide by linguistic rules that fit our purposes. Carnap was an activist, not only in relation to language but in his insistence on human agency as a prime epistemological perspective. This pragmatism or epistemological activism grew during the work on inductive logic to which he devoted much of his last quarter century, in the course of which he came to see rational deliberation as the primary context for his notion of inductive probability.[31]

By continuing in the contentious scientific spirit that Carnap and Russell urged upon us we can get further than they did; that was the idea all along. What we embrace is not a body of philosophical doctrine but a *de facto* method that still wants definition, "explication". Carnap made a good start on that. Where would he have ended? There is no fact of the matter. The next steps are for us, not the "we" of that springtime:

We cannot close our eyes to the fact that philosophico-metaphysical and religious movements opposed to such a stance have again gained strong influence, especially today. What gives us hope that nonetheless our call for clarity, for a science free of metaphysics, will be heard? It is the insight, or to put it more carefully, the faith, that those opposing powers belong to the past. We feel an inner kinship between the attitude underlying our philosophical work and the intellectual attitude that now makes itself felt in entirely

different walks of life; we feel this attitude in artistic movements, especially in architecture, and in movements striving for meaningful forms of personal and communal life, of
education, and of secular organization in general. We feel all around us the same basic
attitude, the same style of thought and action. It is the mentality that seeks clarity
everywhere while recognizing that the fabric of life cannot be fully seen through. It
attends to the design of details and the great lines of the whole, to the bonds between
people and the free development of the individual. The faith that the future belongs to
this mentality sustains our work. (Vienna, May 1928, Rudolf Carnap[32]).

NOTES

[1] Felix Meiner: Vienna, 1928, Hamburg, 1961. Transl. Rolf George, *The Logical Structure of the World*, Berkeley and Los Angeles: University of California Press, 1967.
[2] Julius Springer, Vienna, 1934. Translation by Amethe Smeaton, *The Logical Syntax
of Language*, London: Kegan Paul, Trench, Trubner & Co., 1937.
[3] In *J. Symbolic Logic* 11 (1946) 33–64 Carnap announced soundness proofs for the
simplest (S5) system of propositional modal logic and for quantified S5 relative to a
semantics in which models are represented by the state descriptions of his *Introduction to
Semantics* (Harvard University Press, Cambridge, Mass, 1942). He left the completeness
question open, as did Stig Kanger in *Provability in Logic* (Stockholm 1957), the first
publication proposing relational model theories. In *J. Symbolic Logic* 24 (1959) Saul
Kripke published a completeness proof for quantified S5 (pp. 1–15), and announced
results for other systems (pp. 323–324) with proofs appearing in *Z. mathematische Logik
und Grundlagen der Mathematik* 9 (1963) 67–96, where footnote 2 gives some early
history – as do Hintikka and Kripke in *Acta Philosophica Fennica* 16 (1963) 65–94.
[4] For which many claim reponsibility, starting with Karl Popper. See *The Philosophy of
Karl Popper*, La Salle, Illinois: Open Court, 1974, vol. 1, pp. 69–71.
[5] Chicago and London: University of Chicago Press, 1947, 1956.
[6] *Revue International de Philosophie* 4 (1950) 20–40.
[7] University of Chicago Press, Chicago, 1950, 1962.
[8] *Der Raum, ein Beitrag zur Wissenschaftslehre*, Kant-Studien, No. 56, Berlin, 1922.
[9] 2 vols., Jena, 1893, 1903.
[10] 3 vols., Cambridge, England, 1910–1913; 2nd ed., 1925–1927.
[11] Open Court, Chicago and London, 1915.
[12] *The Philosophy of Rudolf Carnap*, P. A. Schilpp (ed.), Open Court, La Salle, Illinois,
1963, p. 13.
[13] *Op. cit.*, p. v.
[14] The title of *Our Knowledge of the External World as a Field for Scientific Method in
Philosophy* as abbreviated on the spine and cover of the first edition.
[15] *Scientific Method in Philosophy*, pp. v–vi. Russell's emphases.
[16] An example outside physics: the relation 2/5 holds between the grandparent relation
and the great-great-great grandparent relation. See Willard van Orman Quine, "Whitehead and modern logic", *The Philosophy of Alfred North Whitehead*, P. A. Schilpp (ed.),
Northwestern University Press, 1941, p. 161.
[17] *Principia Mathematica*, vol. 3, p. 233.
[18] See *The Structure of Appearance*, Cambridge, Mass., 1951: Harvard University Press.

Goodman reports (p. vii) beginning work on the project within a year or so of the Aufbau's publication.

[19] *The Logical Structure of the Word*, pp. xvi–xvii.

[20] Percy Bridgman, *The Logic of Modern Physics*, New York: Macmillan, 1927.

[21] "Über formal unentscheidbare Sätze der *Principia Mathematica* und verwandter Systeme I", *Monatshefte für Mathematik und Physik*, 1931.

[22] Summaries in Polish and in German were published in 1931 and 1932. The full paper appeared in German translation much later: "Der Wahrheitsbegriff in den formalisierten Sprachen", *Studia Philosophica* 1 (1936) 261–405. Carnap would offer this as prime evidence of the need for an international language. (He had in mind something like Esperanto or Ido, not English.)

[23] *The Logical Syntax of Language*, p. xiii, Carnap's emphases.

[24] See *The Philosophy of W. V. Quine* (ed. L. E. Hahn and P. A. Schilpp), Open Court, La Salle, Illinois, 1986, p. 12.

[25] Berkeley, Los Angeles and London: University of California Press, vol. 1 (Rudolf Carnap and Richard Jeffrey, eds.) 1971, vol. 2 (Richard Jeffrey, ed.) 1980.

[26] First paragraph of "Two dogmas of empiricism", W. V. Quine, *From a Logical Point of View*, Cambridge, Mass: Harvard University Press, 1953.

[27] "Quine and the confirmational paradoxes", *Midwest Studies in Philosophy*, vol. 6, ed. P. French, T. Uehling, Jr. and H. Wettstein, University of Minnesota Press, Minneapolis, 1981, pp. 425–452.

[28] I collaborated with Carnap as a fellow-traveller – toward a destination closer to Bruno de Finetti's: see "Reading *Probabilismo*", *Erkenntnis* 31 (1989) 225–37.

[29] See Figure 1 in Ernst Mach, *Beiträge zur Analyse der Empfindungen*, Jena, 1886; translation by C. M. Williams, *Contributions to the Analysis of Sensations*, La Salle, Illinois: Open Court, 1896.

[30] E.g., see sec. 11.9 of G. E. M. Anscombe, *Intention* (Oxford: Basil Blackwell, 1957; 2nd ed., Ithaca, N.Y.: Cornell, 1963).

[31] See "Inductive logic and rational decisions", *Studies in Inductive Logic and Probability*, vol. 1 (fn. 25 above).

[32] *Aufbau*, preface. I thank Frank Döring, Michael Friedman, Saul Kripke, and Michaelis Michael for setting me straight at various points in this paper.

Department of Philosophy
Princeton University
Princeton, NJ 08544
U.S.A.

POSITIVISM AND POLITICS
THE VIENNA CIRCLE AS A SOCIAL MOVEMENT

Marx W. WARTOFSKY
Boston University

Introduction: Some Methodological and Historical Preliminaries:

At the end of the Pelopponesian wars, Athens found itself in a crisis: economic, social, political, ideological. The old Gods had failed; the civic virtues of Periclean Athens, its culture and arts, its political and military structure, had not been able to stave off the utter defeat by the Lacedaemonians. Sparta had conquered, destroyed the walls to the Piraeus, cut down the orchards, subjected the city to a terrible siege and a great plague. The once great center of science and literature had lost its faith in the old religion, the old virtues. There followed a period of internal violence, revolution and terror, the rule of the Thirty Tyrants and the ensuing democracy, by whose *Gerousia* Socrates was condemned to death. Plato, who had grown up during this period, saw his task clearly: first, he considered a career as an active participant in social and political life, as a legislator perhaps, helping to reconstruct the political greatness of pre-war Athens. But then, dismayed by the examples of political practice, he saw the task of reconstructing a new system of politics in a theoretical way: a model Republic embodying rationality in its very principles. Here, authority would derive not from force of arms, nor from the old Gods, but from Reason itself; and this authority would be invested in those who understood it best — the philosopher-kings. Reason and the social good were not separate, but one. The good would be attained by the exercise and the authority of reason. And reason was science: the knowledge of principles from which there followed deductively those particular consequences which could serve as grounds for rational decisions which would achieve a reconstruction of the community destroyed by the war and by the ensuing rule of the oligarchs and the crowd. Plato's *Republic* is such a

53

rational utopia. It is clearly intended as a theoretical work with practical political consequences: not as a blueprint for an actual form of state, but rather as a construction of the norms and possibilities which could be envisioned in making decisions about matters of politics and statehood.

At the end of the First World War, Austria found itself in a crisis: economic, social, political, ideological. The old Gods had failed; the civic virtues of Habsburg Vienna, its culture and arts, its political and military structure had not been able to stave off utter defeat by the Allies. The once-great center of sciences and literature had lost its faith in the old religion, the old virtues. There followed a period of internal violence, revolution and terror, and the ensuing democratic republic. Otto Neurath who had grown up during this period, saw his task clearly: first, as an active participant in social and political life, helping to reconstruct a new, socialist society to replace the old exploitative one. But then, dismayed by the failure of the program for full socialization, and brought to trial for his role in the revolutionary Bavarian Soviet, he saw the task of reconstructing a new social system in a theoretical way: a model economy embodying rationality in its very principles; and more generally, the development of the scientific conception of the world which was appropriate to the construction of a new society. The scientific world conception and the social good were not separate, but one. The social good would be attained not simply by the authority of a scientific sociology and a scientific political economy on a materialist basis, but rather with the help of these means. Neurath's theoretical-scientific work was intended to have practical consequences: not as a blueprint for an actual form of social-economic life, but rather as a construction of the alternative empirical possibilities which could be envisioned hypothetically as a basis for making decisions about how to best satisfy the social needs of food, clothing, housing, work, and culture.

It should be clear what I intend by this somewhat strained analogy, especially for a conference on the Vienna Circle, in which Otto Neurath is one of the main subjects of discussion. Now, Plato was no positivist, and certainly no empiricist; nor was Plato's *Academy* a forerunner of the *Wiener Kreis*. Athens in the early decades of the Fourth Century B.C. was not Vienna in the early decades of the Twentieth Century A.D. And Neurath was, of course, no Platonist, either in philosophy or in politics. Yet there is a common element: in both cases, what was then taken to be the most advanced philos-ophical-scientific conception of the world was proposed, in a context of radical social change, as a guide to politics or to the organization of social life, and as a means of establishing the practical rule of reason. There is yet another similarity: In its later development as a movement, Platonism came to be transformed into an other-worldly doctrine, a dualism of Spirit and Flesh, of the Heavenly and the Earthly City. It was divorced from its this-worldly origins and motives, as a model of *Sophrosyne*, of practical wisdom and sound-ness of mind, and became instead the domain of philosophers who were no longer engaged in the practical precincts of science and politics. It became theological, dogmatic and formalistic. So too, in its later development, logical empiricism as a movement became divorced from its practical origins — even from the living practices of science, as its theoretical idealizations came to be more and more formalistic.

The disanalogy between the two movements certainly runs deep. Yet, Neurath's project was decidedly like Plato's: to put science or rationality in the service of social transformation, by constructing models of possible forms of social life which would satisfy certain ends. But this is not my topic in this paper, except incidentally. I will not deal with Neurath's proposals for a scientific sociology on a materialist basis, nor with his planning theory, nor with his theory of the democratization and humanization of knowledge — that is, I do not plan to discuss Neurath's *particular* proposals for the role of science in social life. Rather, what I want to focus on in this paper is the more general question of the connection between the positivism of the Vienna Circle — the "scientific conception of the world" — and politics.

How should one proceed in such a task? Can one do justice to the ideational or philosophical content of a movement of thought such as Platonism or logical positivism and at the same time understand this very content itself in the framework of its social and historical setting? Or is it necessary to separate the philosophical analysis and criticism of concepts and theories, as an analytical task, from the analysis of the sociological and historical contexts in which such ideas developed, or found their application? Or is it the case that the norms of a philosophical analysis or critique of ideas are incom-mensurable with those of a social or historical analysis or critique?

That, for example, in the first case, we are concerned with truth, validity, rightness; and in the second, only with descriptive adequacy in matters of fact, and with norms of value which do not bear on the "internal" questions of the validity or truth of theories, and which are therefore "external" to the philosophical analysis?

In this question as to how to proceed, we already encounter the philosophical and methodological content of the Vienna Circle's own thought, i.e., the methodological separation of "internal" and "external" questions, whether in their earlier Viennese forms, or in the forms in which they came to be expressed later, e.g. by Reichenbach and by Carnap, (e.g. in Reicherbach's formulation of the demarcation between contexts of discovery and contexts of justification, or in Carnap's distinction between "internal" and "external" questions). In short, we cannot proceed without getting entangled in the very issues which were internal to the philosophy and the method which the Vienna Circle developed. It therefore seems that we shall not be able to give an analysis of the Vienna Circle as a social movement without at the same time coping with the methodological views of the movement itself in these matters. So, in a sense, the very investigation itself is reflexive, and in a way, self-referential, if not also paradoxical: the consideration of the Vienna Circle as a social movement involves a consideration of what the Vienna Circle philosophers themselves thought about the role of their philosophy, or about the Scientific Conception of the World, with respect to its social genesis or its role in the larger world of scientific and social practices.

It may be useful at the outset to make clear what my own approach will be. It seems to me that the traditional divisions or demarcations between "internal" and "external" contexts are fundamentally mistaken, whether we take these divisions in their historiographical interpretations ("internalist" and "externalist" histories) or in their normative-philosophical connotations (e.g. "discovery" vs. "justification"). I would argue that we cannot begin to understand the genesis or the forms of Platonism, *philosophically*, without recognizing the *Problematik* which Plato posed in the social context of his time, i.e. without understanding why, in his view, rational order was taken as universal, necessary, eternal and unchanging, whether in mathematics or in politcs, whether in art or in social life. For Plato's project, Reason and the Good had to be identified as essential forms beyond the flux of mere sense experience, opinion or everyday politics. Plato's aim was political; but also philosophical, in the sense that politics was to be modelled on rational-scientific principles, as Plato understood them. Thus, in order to understand Platonism philosophically, it needs to be studied not only in the mode of philosophical analysis, or simply in terms of an "internalist" *Ideengeschichte*, however important such approaches may be. It also has to be studied in its own context, by an attempt to reconstruct the *Problematik*, the motives and the self-understanding of Plato's time. Nor can one then claim that all of this is irrelevant to whether or not what Plato and the Platonists said was *true*, or whether the arguments given are *valid*. For I would argue that we cannot begin to understand Plato's claims for his arguments without some understanding of what the claims *were intended to assert*, and why the arguments *were thought to be valid*, without some under-standing of the *self-understanding* which Plato, or the later Platonists brought to their project. We cannot therefore give a context-free analysis or critique of Platonism without giving, at the same time (either tacitly and unwittingly, or explicitly and critically) *some* interpretation of the system of thought within which these claims and arguments have their place. Similarly, in order to understand the philosophical content of the Vienna Circle, it needs to be seen also in its original milieu, in terms of the motives and the self-understanding of its constituents, and against the background of its own time. For logical positivism or logical empiricism, like Platonism, was not only an intellectual movement, but also a social movement, and needs to be understood in this way.

To this end, I will consider the Vienna Circle as a social movement in two senses, which are closely related: *first*, I want to consider the sense in which the Vienna Circle, — or, more correctly, some leading members of it — took their development of a scientific conception of the world as an historically-situated political project, i.e. as a struggle against reactionary forces not only in intellectual life, but also in political and social life in Austria; and more specifically, as a contribution to the movement of social and political change toward a socialist society; *second*, I want to consider the Vienna Circle as *itself* the object of a social and historical analysis as an intellectual movement. That is, I want to examine the Vienna Circle "*als soziale Bewegung*," and also to consider the movement from the point of view of "*Sozialforschung*."

I also want to make clear what I will *not* do in this paper: this is not a study of the historical facts or the social features of the Vienna Circle: such a study would require a much more sustained paper than I can give here and more knowledge than I have. (And such studies are in fact under way, in the works of such younger Austrian scholars as Elisabeth Nemeth, Karola Fleck and Friedrich Stadler[1]). However, I will argue that it is just such historical and social analyses that are needed in order to achieve an adequate *philosophical* assessment and understanding of the *Wiener Kreis*, for the reasons I gave earlier.

2. *The Vienna Circle as a social movement: A first approach*

The view of the Vienna Circle as a social movement is certainly not new. Nor was it a secret. This *self*-characterization of the movement was openly articulated — and debated — within the Circle. Thus, it was Neurath (and together with him, to some extent, also Carnap, Hahn, Olga Hahn-Neurath, Philipp Frank) who understood the movement in this way, though others like Schlick strongly disagreed, as we shall see. It is also clear that some of the early American students of the new scientifc conception of the world understood it in this way as well, e.g. Ernest Nagel and Albert Blumberg. So too, Sir Karl Popper writes, in his memoir of Neurath:

"I had myself long been deeply critical of Marxism with its historical prophecies ... and these critical ideas had led me to become interested in problems of the theory of knowledge and of the philosophy of science, and their bearing on political problems. I was therefore very much interested to find that Neurath also felt that such philosophical problems are important for politics. He later published some articles alluding to the significance of the theory of knowledge for politics; and he mentioned the

1. Elisabeth Nemeth, *Otto Neurath und der Wiener Kreis – Revolutionäre Wissenschaft als Anspruch*, Frankfurt and New York: Campus Verlag, 1981; Karola Fleck, *Otto Neurath: Eine biographische und systematische Unter-suchung*, Inauguraldissertation, Geisteswissenschaftliche Fakultät, Universität Graz, 1979; Friedrich Stadler, "Aspekte des Gesellschaftlichen Hintergrunds und Standorts des Wiener Kreises am Beispiel der Universität Wien," *Wittgenstein, der Wiener Kreis und der Kritische Rationalismus. Akten des Dritten Internationalen Wittgenstein Symposiums.* Ed. by H. Berghel et al. Wien: Hölder-Pichler-Temsky 1979, pp. 41-59.

philosophy of the circle around Schlick and Hahn as promising in this respect.... I have little doubt that it was Otto Neurath who, with the hope in mind of a philosophical reform of politics, attempted to give the circle of men around Schlick and Hahn a more definite shape; and thus, it may have been he, perhaps more than anybody else, who was instrumental in turning it into the 'Vienna Circle' "[2]

Furthermore, at least two of the three figures listed in the the Circle's "Manifesto" (*The Scientific Conception of the World of the Vienna Circle*, 1929) — namely Bertrand Russell and Albert Einstein — may be said to have held similar views about the relevance of scientific rationality to the solution of problems of social life. (About the third — Wittgenstein— the case concerning his views of the relevance of philosophy for social reform is, I believe, more complex than it is ordinarily made out to be). It is worth citing the well-known last paragraph of the 1929 pamphlet (written by Neurath and edited and cosigned by Carnap and Hahn) to catch the flavor of this view:

"Thus, the scientific world-conception is close to the life of the present. Certainly it is threatened with hard struggles and hostility. Nevertheless there are many who do not despair but, in view of the present sociological situation, look forward with hope to the course of events to come. Of course not every single adherent of the scientific world-conception will be a fighter. Some, glad of solitude, will lead a withdrawn existence on the icy slopes of logic: some may even disdain mingling with the masses and regret the 'trivialized' form that these matters take on spreading. However, their achievements too will take a place among the historic developments. We witness the spirit of the scientific world-conception penetrating in growing measure the forms of personal and public life, in education, upbringing, architecture and the shaping of economic and social life according to rational principles. *The scientific world-conception serves life, and life receives it.*"[3] (emphasis in original)

Apart from this understanding of the social nature of the movement by some of its leading members, its American students and critical

2. Karl Popper, "Memories of Otto Neurath," in Otto Neurath, *Empiricism and Sociology*, ed. M. Neurath and R.S. Cohen, Dordrecht and Boston: D. Reidel, 1973, p. 54.

3. *Wissenschaftliche Weltauffassung. Der Wiener Kreis*, Wien: Artur Wolf Verlag, 1929, p. 30. Reprinted in Otto Neurath, *Empiricism and Sociology*, op. cit. pp. 317-18.

philosophical colleagues like Popper, it was also seen as a social movement by its political enemies — the right-wing Austro-German nationalists, the reactionary Catholic-clerical establishment in Vienna and in the University, the proto-Nazis and anti-semites — who saw the Vienna Circle and its ideas as "red" and as "Jewish," and therefore as a social and political threat and not only as an intellectual one.

In short, one cannot begin to give an account of the Vienna Circle without seeing it not only as a movement for a scientific world conception in terms of its logical, epistemological and methodological content, but also as a movement which conceived of its theoretical contributions as being in the service of social reform, and as, in significant measure, allied with the left social movements of its time.

However, it is a matter of great importance to the assessment of the movement that many of its members did not understand it in this way, or were at most only tacitly cognizant of this aspect of it. Clearly, there was no requirement on the members of the Circle to accept such a view. And we know that Schlick did not even like the idea that the scientific and methodological principles of the *Wiener Kreis* had become organized in this way. Herbert Feigl, in his account of this period, writes for example:

> "... in 1929, a slender pamphlet, *Wissenschaftliche Weltauffassung. Der Wiener Kreis*, had been composed by Carnap, Neurath and Hahn, aided by Waismann and myself. This was as it were our declaration of independence from traditional philosophy. We presented this pamphlet to Schlick upon his return from Stanford. Schlick was moved by our amicable intentions; but as I could tell from his facial expression, and from what he told me later, he was actually appalled and dismayed by the thought that we were propagandizing our views as a system or 'movement.' He was deeply committed to an individualistic conception of philosophizing, and while he considered group discussion and mutual criticism to be greatly helpful and intellectually profitable, he believed that everyone should think creatively for himself. A 'movement' like large-scale meetings or conferences, was something he loathed."[4]

It is also interesting to note that, in this very article by Feigl ("The *Wiener Kreis* in America," 1969), the only mention Feigl makes of

---

4. Herbert Feigl, *Inquiries and Provocations: Selected Writings 1929-1974*, ed. Robert S. Cohen, Dordrecht and Boston: D. Reidel, 1981, pp. 70-71.

any sort concerning the social motivations or role of the movement, is the rather oblique sentence: "It was especially Neurath and Frank who envisioned and worked for a new era of enlightenment, propagating the Vienna form of positivism."[5] (This can hardly be called evidence for any social, not to speak of socialist, consciousness with respect to the movement!) Yet, for all this, one cannot ignore the conscious orientation of Neurath and some of the others in this regard; nor can one ignore the historically objective features of the movement's origins: its embattled intellectual position in a context where this position was itself taken to be political both by sympathizers and antagonists; nor (as I hope to show) can one ignore the fact that the movement, by virtue of its intellectual, philosophical-theoretical content itself, came to be important in its social effects, *malgré lui.* What does appear clear is that the movement was social in a problematic way; and it is this very problem that I would like to address.

Thus, I would argue that it is a matter of plain historical fact that the Vienna Circle was (and to an important extent, took itself consciously to be) a social movement. However, this fact was to become *almost totally forgotten*, in a moment of abstraction which itself requires an historical explanation. Logical empiricism became a movement almost exclusively understood in the context of a logic of science and of a linguistic philosophy, both of which were taken as ahistorical and socially context-free.

Now we may present an apparent paradox, and a problem: the social character of the movement in its early history is not only *not* widely known— especially in the Anglo-American scientific and philosophical world where the movement had its greatest effect — but one may also say that the movement *lost* this character as it developed, especially after it was dispersed and took root in England and the United States. But further still — and here is the apparent paradox and the problem — the further logical positivism moved away from those earlier social-political commitments and motivations that it did have in significant measure, the *more* it became the target of a radical critique from the left. It was accused, from a variety of left positions — orthodox Marxist, neo-Marxist (Critical

5. Elizabeth Nemeth, op. cit.

Theorists of the Frankfurt School), and later by the New Left social theorists and critics — of being a major support for, and expression of just those social-forces — e.g. the ruling class, bourgeois ideology, authoritarianism — which Neurath and others had seen as the enemies of that very enlightenment which was represented by the scientific conception of the world. Worse yet, on philosophical as well as ideological grounds, logical positivism and logical empiricism were charged with just that mystification of the social world, and of the sciences, which the Vienna Circle had originally combated in its struggle against reactionary and obscurantist idealist philosophy and "metaphysics." The enlightenment Humanism which had marked the social posture of the Vienna Circle now came under criticism as an anti-Humanist positivism and scientism. The very hallmarks of the scientific philosophy — its logical-analytical rigor, it emphasis on mathematization or formalization, its original emphasis on the quantitative or exact sciences as the models of rationality, its reduction program, its elimination of metaphysics, its transformation of ontological questions into linguistic questions concerning choice among alternative language frameworks — all of these, in one or another way, came to be viewed by the critics as marks of an anti-humanist and anti-social outlook, and indeed, as giving aid and comfort to the enemy.

One might have expected this reaction from just those philosophical quarters which had been sharply criticized by the Vienna Circle originally, namely the phenomenological movement, the *Geisteswissenschaft* theorists of the Dilthey tradition, the *Verstehen* sociologists; and of course, one would have expected such a criticism from the quarters of traditional metaphysics, systematic philosophy and philosophical idealism. But now the charge was being made from the left that positivism in its modern guise as logical empiricism represented not a scientific world conception, but rather the modern form of "bourgeois idealism," i.e., a formalist, anti-materialist view which degenerated into a subjectivist phenomenalism or a relativist conventionalism; or else that it represented "scientism," the ideology of a reductive, technocratic, oppressive and authoritarian view of human nature and society.

Thus, on the one hand, Orthodox Marxism charged logical empiricism with being a form of subjective idealism; on the other hand, humanist Marxism identified the movement with a reductive scientism, physicalism and behaviorism. Furthermore, as so-called "Western Marxism" adopted elements from the phenomenological and the *Verstehen* traditions, and as these movements of thought themselves developed a social-political orientation after the Second World War, the critique of positivism became more general and widespread. (This may be seen, for example, in the post-war development of the Frankfurt School, in the phenomenological Marxism of Sartre and others, and interestingly enough, in the action theoretical approaches which, starting either from Wittgenstein's later writings, or from the renewed interest in the tradition of the *Geisteswissenschaften* and in historicism, came to ally themselves with social criticism). This combination of the predominant political quietism of the movement, in its Anglo-American development, and the growing criticisms from the left resulted in a sharply changed conception of the social character of the movement.

The Vienna Circle had always emphasized formalism — i.e. the value of rigorous logical-linguistic representation of its analyses and arguments, as well as the constructive task of developing formal languages and studying the foundations of mathematics and logic — as a means of criticism and clarification. But on Neurath's view, and that of others, this formalism was in the service of enlightenment, as an instrument against obscurantism. Now, the formalism seemed to proceed in a way more and more detached from those contexts of social and scientific practice which the original movement, insofar as it was socially oriented, had intended to serve.

Further, one of the hallmarks of the scientific conception of the world, and certainly central to Neurath's view, was the importance of prediction in the sciences. The social value of prediction was seen by Neurath as a means of applying rational controls over hypothetical plans of action — whether in research in the sciences, involving the use of experiment, or in social planning, (or in Neurath's own term, "social engineering"). Neurath, in fact, saw the great value of American behaviorist psychology in just this element of control over the boundary conditions of behavior, so that it could be shaped in accordance with decisions as to the preferred end-results. Indeed, this element of prediction for the sake of rationally controlled action is one of the oldest features of positivism, historically. Auguste Comte expressed it in his aphorism: *"Voir pour prévoir; prévoir pour agir."* But now, the critics saw in this project of "social engineering" the grim echoes of Stalinist authoritarianism — the "engineering of human souls" — or on the other hand, the means of social coercion typical of capitalism, which reduced human beings to objects of manipulation by a one-sided "instrumental rationality." That is to say, the *instrumentarium* of rationality could be used for purposes of domination, as well as for enlightenment; or more critically still, it was argued — e.g. by the Critical Theorists — that such modes of scientific rationality are *intrinsically* dominative and oppressive, and that the positivism which gave rise to these conceptions was itself the social and ideological outgrowth of the alienated modes of social life which require manipulation and domination of some people by others, i.e. of the working class, or other oppressed groups by the ruling classes. In this view, the scientific conception of the world of the *Wiener Kreis* and of its followers was seen as inherently "one-dimensional," in Marcuse's phrase, and as reflecting theoretically the forms of social and class oppression characteristic of capitalism.

Now all this may be seen as a social or political critique of a social or political movement; and thus it presupposes the *political* efficacy of logical positivism or logical empiricism. But this criticism from the left came *not* when the movement was more or less explicitly motivated by social concerns in the larger world, but precisely when the movement had lost this character, and when it had turned in on itself within the narrower "internalist" precincts of a technical philosophy of science. The charge, then, must be interpreted in one of two ways; *either* it was an accusation that the movement had *lost* its social commitment, and had shown itself to be irrelevant to larger, or at least more practical concerns — that it had given up the struggle, so to speak; and therefore deserved to be criticized for its asocial and apolitical character, and that therefore it was turning attention away from just those applications of the scientific conception of the world which had originally motivated Neurath and the others; or, it was an accusation that the movement had become *very* socially efficacious in precisely those terms which had originally been envisioned — namely, that as a technical philosophy of science, it was having a profound effect on the norms and values of scientific practice itself, and through this, on social conceptions within the culture at large (e.g. through the proliferation of "positivist" psychology and social science which imitated the reductionist model of physics); but that this social effect was not that of a liberating or humanist enlightenment, as originally intended. Instead, it was in the service of domination and control, or of the "mystification" of social reality as a means of maintaining the status quo against criticism. Thus, the criticism runs in two opposite directions: it is either that The Vienna Circle had failed in whatever original aims it had as a social movement; or that it had succeeded only too well in its social efficacy, in a direction opposite to that originally envisioned, but one which nevertheless developed on the very grounds of its original positivism.

It would indeed be paradoxical if both of these criticisms were right; but no paradox ensues if either one or both of the criticisms are wrong. The easy solution to the problem presented by the left critique of the logical empiricist movement is therefore to discount both criticisms by denying their premises: In the first case, one would argue that the Vienna Circle, and its later development in logical empiricism, was *not* a social movement in the proposed sense to begin with, and therefore could not have failed in a project it didn't undertake. (This would eliminate the critique from the left that claims that it went wrong). In the second case, one would argue that neither the Vienna Circle nor its successors had *any* social effect whatever, either in the domain of scientific practice, or in matters of social ideology or political practice. In effect, this would be to disclaim the general view that the Vienna Circle was, in fact, a social movement. And since I would argue that it clearly was, this resolution of the question seems to me to be untenable. A less frontal approach would be to say that both criticisms fall short of the mark, because both misinterpret or distort the actual content of the philosophy of the *Wiener Kreis*, from tendentious points of view. Therefore, one could argue against the left critics that the Vienna Circle was, in fact, a social movement but *not* of the sort described by the left critics; and that what it failed in was not its *own* project, but rather in theirs. That is to say, even in its most explicit expression, by Neurath, the movement did not intend to achieve either the goals of a Marxist humanism, or of an orthodox Communism. At best, then, this criticism may be taken as a claim that the Vienna Circle, though it was a social movement, should have had a different character than it had, i.e. one that conformed to the goals which the critics themselves would have posed. But this is straightforward external political criticism, and doesn't yet bear on the ideational or philosophical content of the movement, and certainly not on its core in the philosophy of science. The second criticism from the left does bear on this question, however: Was the effect of the *Wiener Kreis*, in philosophical terms, to foster a subjective phenomenalism, or a relativistic conventionalism, or a technocratic ideology of science, or a formalism insensitive to the living contexts of scientific and social practice? Here, the question becomes philosophically interesting. For in point of fact, it was not the external critics from the left who levelled the sharpest criticism of *these* features of the movement, but rather *the logical empiricists themselves*, together with their internal critics *within* the disciplines of logic, linguistic philosophy, and the philosophy and history of science. In short, the severest criticism of the *philosophical* shortcomings of the Scientific Conception of the World did not come from the left critics, but from the inside of the movement, and from its near periphery.

The fact is that the "external" critics from the left, while they carried on an ideological criticism of the movement, were either not competent or not willing to carry out a philosophical critique. Let me give some examples: Insofar as the Marxist humanists or the Critical Theorists attacked "positivism" and "scientism," they in fact adopted the very view of science which they attributed to the logical empiricists, and accepted it as a valid description of science. They then made science, *under this description*, the object of their attack. However, it was those relatively "internal" critics who attacked the specific methodological formulations of the program for the "logical reconstruction of science," who rejected this account of science and of scientific method, and who finally undermined the formalist view of science.

Similarly, the Leninist Marxists who led the attack on the tendencies toward subjectivist phenomenalism and relativistic conventionalism that they attributed to the epistemology of the Vienna Circle did so in the name of a realist epistemology of the sort proposed in Lenin's *Materialism and Empirio-Criticism* (a work in which Philipp Frank had already been criticized). But this epistemological realism remained at the level of a naive reflection theory of knowledge, and of an ahistorical, and hardly dialectical essentialism, with respect to its ontology (at least in that version of it which derived from Lenin's *Materialism and Empirio-Criticism*; his later *Philosophical Notebooks* suggest a more qualified view under the influence of Lenin's reading of Hegel). In its details, this remained a *rhetorical* realism, and it failed to do justice to the problems of contemporary science. But in fact the sharpest critique of phenomenalism and of subjectivism came from within the *Wiener Kreis*, e.g. in Neurath's argument for physicalism, or in Carnap's later adoption of a physicalist language, or in Feigl's "hypercritical realism." And, as is well known, the criticisms by Quine and Popper and the dismantling of the verificationist theory of meaning and of truth were crucial in undermining some of the philosophical foundations of the logical empiricist view. This critique can hardly be said to have come from the left! Further, insofar as the reduction program itself fell prey to devastating criticisms, these came, in large part, from within the movement itself, as self-criticisms and revisions, e.g. in the work of Carnap and Hempel. Insofar as this critique came from without, it was most effectively carried out by philosophers and historians of science like Hanson, Kuhn, Feyerabend and others, who stood close to the logical empiricist tradition in many ways.

Now this may be no more than to say that the most effective critique of a philosophy is an internal critique — i.e., one which examines the premises and the arguments of a philosophical view from within the self-understanding of that view, and thus is able to show that it is, in the end, internally incoherent or crucially weak in its own conclusions. The replacement, or revision then takes place in response to the desiderata which that philosophical view itself had set forth. The external critique, by contrast, demands a more radical rejection of the premises of the criticized view, either on the grounds that they are just false, or that their consequences are false or unacceptable. But if one is to examine the Vienna Circle as a social movement, in a critical way, one cannot avail oneself of *only* one or the other of these approaches. That is to say, one cannot get at a social-critical understanding of the Vienna Circle either by an internal critique of its logical or philosophical coherence, or by a simple approbation or rejection of its premises, or their consequences. What is at issue is something more complex and subtle, namely, the ways in which the "internal" ideational-philosophical content bears upon the "external" social content and role of the movement.

Let me put this in the simplest terms: what we have here is a case study in the problem of the relation of theory to practice, and more particularly, of the relation of a technical epistemological and methodological theory of science to social practice, (including here both scientific practice as well as the wider contexts of political practice). I cannot attempt to carry out such an analysis of the Vienna Circle as a whole, in this paper, of course. But I can attempt to formulate the problem in such a way that it will suggest how one might go about such a task. I have attempted to discuss the preliminary considerations for such an approach in the foregoing section. What I propose to do, in the balance of this paper, is to focus on the case in which this problem is most explicitly articulated in the literature and history of the Circle, namely, in the case of Neurath. For it was he more than any other figure who posed this very question for himself and for the movement. My critical assessment of the nature of the Vienna Circle as a social movement starts therefore from a consideration of his views; and concludes with some theses about the failure of this social movement precisely as it relates to the logical positivist and logical empiricist theory of science, and to its scientific conception of the world.

3. Neurath on the relation of the "Wissenschaftliche Weltauffassung" to social practice: A second approach.

Of all the members of the Vienna Circle, it was Otto Neurath who articulated most clearly the social-critical role of the "Scientific Conception of the World." Though his views underwent some development, it may be said that he presented a fairly consistent position of this question. It has recently been studied in some detail in Elisabeth's Nemeth's excellent work, *Otto Neurath und der Wiener Kreis – Revolutionäre Wissenschaft als Anspruch.* Neurath was quite clear that the scientific world view was to be seen not only in terms of intellectual or scientific enlightenment, but also as one which contributed to the social and political struggle for a socialist society. Thus, he writes, in an embattled spirit: (in 1931)

> "Just as in certain circles the scientific world-view is spreading ever more widely, so the leading academic sociologists in Germany are strongly metaphysical in outlook, or at any rate, they are far removed from any endeavor intended to further a unified science on a materialist basis. This advance of metaphysics becomes understandable if we take a look at those who hold such views. On the whole, the representatives of metaphysically directed sociology are at the same time representatives of the ruling order. Most governments favor metaphysically inclined

scholars, even theologizing ones, especially those who wish to establish this outlook in the field of sociology .... Conversely, the revolutionary masses of workers and the groups attached to them become more vigorous through anti-metaphysical physicalist sociology, and above all the fight against metaphysics and theology meant the destruction of the bourgeois world-order; this struggle plays a great role as a means of propaganda, today as well as in the time of Marx and Engels."[6]

The language of traditional *Klassenkampf* and of *agitprop* may seem blunt and strained here, but it gives some measure of Neurath's militancy. Nor is this an isolated expression. In an earlier work, Neurath wrote:

> ... scientific thinking is safeguarded best by the proletariet which can use it to best advantage, whereas among its adversaries an unscientific attitude often strengthens the front-line. Scientific attitude and solidarity go together. Whoever joins the proletariat can say with justification that he joins love and reason."[7]

Clearly, then, Neurath held that the scientific world-view was not a matter of mere theory, but that it had social-practical consequences, just as did anti-scientific, or "metaphysical" world views. It might be thought that Neurath was speaking here of a scientific sociology alone, since his criticism of "metaphysics" is aimed at Dilthey, Weber and Spengler, among others. But Neurath includes *all* of science and mathematics in his conception of what helps the proletariat in its revolutionary activity. Thus, it is not simply that a scientific *social* theory plays a liberating, or organizing role in the class struggle, but that all forms of scientific rationality do so. Thus, even in the concluding paragraph of the 1929 Manifesto of the *Wiener Kreis*, as we have seen earlier, allusion is made to the contributions to "life" and to "historic developments" even of those who "lead a withdrawn existence on the icy slopes of logic." And in 1928, Neurath had written:

> "Nothing would be further amiss than to think that a Marxist-minded representative of the proletarian class struggle would respect only such scientific work which relates directly to the strategy of the class struggle. It is precisely Marxism that uncovers indirect relations and detours, and thus might 'ascertain that cultivating pure logic and the most general

6. Otto Neurath, *Empiricism and Sociology*, op. cit., p. 356.
7. Ibid., p. 252.

problems of mathematics and physics is especially favorable to revolutionary thinking. The Marxist will tend to regard it not as a mere accident that among the representatives of just these abstract disciplines ordinarily thought to be impractical, there are so many socialists as well as bourgeois in opposition, as for instance the English logician and mathematician Betrand Russell or the German physicist Albert Einstein. A cultivation of this kind of scientific thought seems almost a form of dissolution of metaphysical and half-theological thought, which under many guises and masks is more alive today among the bourgeoisie than two generations ago .... The cultivation of scientific, unmetaphysical thought, its application above all to social occurrences, is quite Marxist."[8]

It is quite clear that Neurath himself saw the requirement for direct application of the scientific world conception to social contexts. In his work, whether *Empiricism and Sociology* (1931) or his earlier "Character and Course of Socialization," (a technical report given to the Munich Workers' Council, 1919) or his "Utopia as a Social Engineer's Construction" (1919) or his activity in establishing the Social and Economic Museum in Vienna, and in developing the modern method of visual display, the *Isotype* project — in all of these ways, Neurath saw the close connection of theory and practice, and moreover, between *scientific* theory and *social* practice. In his 1933 essay, "Museums of the Future", he wrote:

"How to organize human life socially — that is the great question which people are asking today with ever greater insistence. The social museum is the museum for our time. And this is its twofold task: to show social processes, and to bring all the facts of life into some recognizable relation with social processes."[9]

And on the relation of practical activity and engagement to the scientific task of theory-formation Neurath is also explicit. The scientist or theorist in any field depends on the existence of certain practical social or technical conditions for the development of his or her work. In *Empirical Sociology* (1931), Neurath wrote:

"The dependence of social processes on forecasts plays a smaller part for us than does the dependence of forecasts on social processes. We see that the likelihood of obtaining some piece of knowledge depends on certain social changes. If one strives to have a certain line of study

8. Ibid., p. 295.
9. Ibid., p. 220.

followed by hundreds in unanimous cooperation (which might be a prerequisite for achieving certain insights or creating certain modes of thought) one can often count on success only if certain social transformations are already beginning.

"All this leads to an intimate linking of theory and practice. Theory itself is a form of practice, is part of it. Who needs splinters must either wait for the splitting of wood to happen or do it himself. Moreover, sociological insight into contemporary conditions is best gained by someone who is closely linked with the social structure of the present. In physics too, a close link with technical practice produces a stimulus. This is even more so with sociology. The scholar himself is part of the scene."[10]

It is clear, then, that for Neurath, the Scientific World Conception of the Vienna Circle was seen as contributing to an enlightenment whose goals were not simply theoretical, but socially practical as well. It did so, in his view, in all of its contexts: whether in the applied contexts of sociology, social forecasting, social planning, or in the more theoretical and esoteric reformulations and reconstructions of the natural sciences, or in its most formal mathematical-logical or epistemological analyses. Having said this, however, one needs to go further: just how did Neurath see this relation of the theoretical (scientific-philosophical) enlightenment to the contexts of social practice? It is not enough to say, as I have done thus far, that the Vienna Circle *was* a social movement, and understood itself to be one, to a significant extent. What *kind* of a social movement was it, in the eyes of its leading proponent? How did Neurath in fact conceive of the relation between scientific world-view and practical social-political reform or revolution?

What is crucial here, it seems to me, is the context of Austrian Marxism itself, and of Second International Marxism in general, for it was here that the relation of science to socialism was formulated in a distinctive way related to Neurath's own project. Central here, as a tenet of the vigorous positivism and anti-metaphysical stance of this interpretation of science, is the sharp distinction between judgments of fact and judgments of value, between the proposing of hypothetical alternatives and the choice among them, between planning and policy choice. David Hume may not be taken to be the foremost progenitor of Second International Marxism, or of Austro-Marxism,

10. Ibid., p. 406.

but his distinctions found their living embodiment there. Hume held that "When reason has laid out all the alternatives, 'tis sentiment which decides." "Sentiment" here is of course neither mere feeling, nor is it sentimental. For the Marxism of the Second International, of Kautsky, of Hilferding, of Bernstein, the science of society yielded the value-free laws of political economy, as, in effect, descriptive laws of a fully objective sort. Socialism as a social and political movement was constituted by the attitude or stance one took toward these laws. What was "custom" in Hume became the social decisions of a class-conscious proletariet. But the separation of fact and value was maintained. In Austrian Marxism — the "Second-and-a-half International" as it came to be called — a strong strain of this sort was also present. For Neurath too, it was the political process of democratic decision-making by the workers, by the "masses," which was responsible for choosing which among the scientifically worked out alternatives in policy and planning were to be chosen. Science provides the hypothetical imperatives (in Kant's sense: If you want X, then you must do Y). An enlightened and free people, disabused of the ideological mystifications of reactionary metaphysics and theology, would then be the court of last resort for what *ought* to be done, in consciousness of their interests, material, cultural, intellectual. Neurath's *Isotype* project itself was aimed at this sort of an enlightened and well-informed populace, which could then make its decisions on the basis of clearly understood facts, especially statistical facts.

But for Neurath, the facts never dictated a decision; political judgment like scientific judgment, was not algorithmic, but rather involved decision or choice in the face of insufficient knowledge. And this insufficiency in Neurath's view is not simply a matter of ignorance to be made good at the limit, but is rather endemic to the nature of our scientific knowledge itself. Thus, in a very early essay, "The Lost Wanderers of the World and the Auxiliary Motive" (1913) Neurath writes:

> Whoever wants to create a world-view or a scientific system must operate with doubtful premises. Each attempt to create a world-picture by starting from a *tabula rasa* and making a series of statements which are recognized as definitively true is necessarily full of trickeries. The phenomena that we encounter are so much interconnected that they cannot be described by a one-dimensional chain of statements. The correctness of each statement is connected to all of the others. It is

impossible to formulate a single statement about the world without making tacit use at the same time of countless others. Also, we cannot express any statement without applying all of the preceding concept-formation .... We can vary the world of concepts present in us, but we cannot discard it. Each attempt to renew it from the bottom up is by its very nature the child of the concepts at hand."[11]

In this statement of the implicitly contextual-historical nature of thought, Neurath foreshadows the holist conventionalist notions which are later developed by Quine and by Goodman, and which emphasize the "web" or "system" of statements which is presupposed by any truth-claims; and he also echoes, albeit not in any explicit or conscious fashion, the holism of Hegelianism, (a thought which would hardly be congenial, e.g., to either Quine or Goodman). Neurath also says, in an instrumentalist vein: "Our thinking is a tool. It depends on social and historical conditions.One should never forget this."[12] But Neurath's argument here (and one which he repeats much later in this critique of Popper's *Logic of Scientific Discovery* (in "Pseudorationalism and Falsificationism" (1935)) is that any algorithmic view of decision-procedures is in effect pseudo-rationalistic, and akin to superstition, in that it seeks some external authority for the resolution of doubt which will yield a "release from the feelings of displeasure" which doubt raises. Instead, says Neurath, in the early essay, "Rationalism sees its chief triumph in the clear recognition of the limits of actual insight."[13] Therefore, the rational procedure is to give this recognition its full due, to admit the weakness of our insight, and to decide by lot (as a more congenial method than by instinct, or by omens, or by the "pseudorationalist" method of deciding what is correct after weighing all the alternatives. Now this sounds at odds with what Neurath says elsewhere — e.g. "Wherever there is a clear question, there is a clear answer" — but this requires a fuller treatment of Neurath's views than I can give here. Suffice it to say that there is a strong element of conventionalism in Neurath which denies the strong linkage between what we can know and what we ought to do: facts do not dictate values,

11. Otto Neurath, *Philosophical papers, 1913-1946*, ed. R.S. Cohen and M. Neurath, Dordrecht and Boston: D. Reidel, 1983, p. 3.
12. Ibid., p. 45.
13. Ibid., p. 6.

and *oughts* don't fall out as deductive consequences of *is's*.

In practice, this meant for Neurath the sharp separation of economics from politics. Social engineering requires both: the designer of the system and the political chooser and executor of it. Their separation means, in effect, that designers shouldn't be choosers. (In fact, Otto Bauer's defense of Neurath at his trial as head of the Planning Commission of the Munich *Räterepublik* was precisely that Neurath's role was simply that of a planning technician, and not that of a political decision-maker). This separation of technical from political judgment has its dark side, as we know from recent European history. But in Neurath's version of it, it meant that the rationality of science did not dictate decisions, and that the appropriate process of social decision-making could therefore not have the "clean" character of an algorithmic procedure.

From all this, we may begin to reconstruct what Neurath thought of the role of the Scientific World Conception in its relation to social practice. The scientist-philosopher, the theoretician, is a servant of the political and social decision-making process. He is, in effect, its civil servant, who dispassionately provides it with the information necessary for its policy making, or its actions. In the grand and most militant version of this scheme, one may say that Neurath conceived of the social role of the new scientific enlightenment as the civil service of the revolution. The revolution is to be made by the workers; but it is to be enlightened by the Scientific World Conception. In the more sober contexts of its Viennese ambience, and the actual political content of the movement, it might more aptly be characterized as a "Municipal" movement, closely linked to the politics and social character of the city, to its concern with civic matters of housing, social needs, educational reform in the old center of the Habsburg empire moving rapidly into the new world. This is not to belittle the character of the Vienna Circle insofar as it had a socio-political motive, but to assess its character and its limits against the background of its time. The Vienna Circle was a movement of a transitional period, which had one foot in the Habsburg past and one foot in the new world, in a city which was an island of urbanism in an agrarian-peasant sea. The relation between the positivism of the movement and its politics is a complex one, which I have only touched here. Karl Korsch might have shed a great deal more light on it, had he been able to undertake his planned project of a work on the relation between Marxism and Logical Empiricism.[14] Unhappily, he was not able to do this because his health failed. Robert S. Cohen broke the ground more than twenty years ago with his masterful essay "Dialectical Materialism and Carnap's Logical Empiricism."[15] New studies of the movement, in particular those concerning its social and historical contexts, like those of Nemeth, Fleck and Stadler mentioned earlier, promise much for a revised and more adequate understanding of the relations between positivism and politics in the Vienna Circle. The recent publication of Neurath's work, in German and in English[16] now makes widely available the remarkable thought of this great thinker, and reveals aspects of the movement which had been lost to view or ignored. All this makes possible a new understanding not only of the Vienna Circle and of the logical empiricist movement, but of many contemporary issues concerning the relations between science, philosophy and politics.

14. Cf. Karola Fleck, op. cit., p. 34.

15. Robert S. Cohen, "Dialectical Materialism and Carnap's Logical Empiricism," in P. Schilpp, ed., *The Philosophy of Rudolf Carnap*, (The Library of Living Philosophers), Lasalle, Ill.: Open Court, 1963, pp. 99-158.

16. Otto Neurath, *Gesammelte philosophische und methodologische Schriften*, ed. R. Haller and H. Rutte. 2 vls. Wien: Hölder-Pichler-Tempsky 1982.

Aufbau/Bauhaus: Logical Positivism and Architectural Modernism

Peter Galison

1. Introduction

On 15 October 1929, Rudolf Carnap, a leading member of the recently founded Vienna Circle, came to lecture at the Bauhaus in Dessau, southwest of Berlin. Carnap had just finished his magnum opus, *The Logical Construction of the World*, a book that immediately became the bible of the new antiphilosophy announced by the logical positivists. From a small group in Vienna, the movement soon expanded to include an international following, and in the sixty years since has exerted a powerful sway over the conduct of the philosophy of science as well as over wide branches of philosophy, economics, psychology, and physics.

I would like to thank Nancy Cartwright, Robert S. Cohen, Richard Creath, Lorraine Daston, Arnold Davidson, Alain Findeli, Peter Frank, Caroline Jones, J. B. Kennedy, Cheryce Kramer, Timothy Lenoir, and Hans Sluga for many helpful conversations. My thanks as well to the Archives of Scientific Philosophy in the University of Pittsburgh Libraries (especially to Gerald Heverly); to the Peirce Edition Project at the Center for American Studies (especially to Janine Beckley and Christian Kloesel); to the Special Collections at the University of Chicago Library; to the University Library Special Collections Department, The University of Illinois at Chicago (especially to Mary Ann Bamberger); to the Bauhaus-Archiv, Museum für Gestaltung (especially to Frau Stolle and Dr. Magdalena Droste); to the Archives of the History of Art, Getty Center for the History of Art and the Humanities, Los Angeles (especially to Stephen Nonack); and to the Vienna Circle Archive, Vienna Circle Foundation (Amsterdam) (especially to Anne Kox) for their help and permission to cite material held at these locations. This work was completed with the support of the Center for Advanced Study in the Behavioral Sciences and the National Science Foundation.

Critical Inquiry 16 (Summer 1990)
© 1990 by The University of Chicago. 0093-1896/90/1604-0007$01.00. All rights reserved.

The site of Carnap's lecture that day, the Dessau Bauhaus, was a stunning building designed by Walter Gropius and dedicated just three years earlier. Protected by its flat roof and glass walls, the artists, architects, weavers, and furniture designers had made the school a citadel of high modernism. It was here that Carnap addressed an enthusiastic audience on "Science and Life." "I work in science," he began, "and you in visible forms; the two are only different sides of a single life."[1] In this paper I will explore this "single life" of which the new philosophy and the new art were to be different facets; in the process, I hope to cast light on the shared modernist impulses that drove both disciplines in the interwar years.

Any attempt to link philosophy and art in the interwar period must go further than merely identifying parallelisms between movements. In fact, core members of the logical positivist and Bauhaus groups self-consciously sought to articulate a view of the world in which both would play essential roles. Though on opposite political poles of the Vienna Circle, the philosophers Otto Neurath and Ludwig Wittgenstein each spent years pursuing architectural concerns. Throughout their writings Carnap, Neurath, and others singled out modern architecture as the cultural movement with which they most identified; their interests were reciprocated as the logical positivists were more prominent as visitors to the Dessau Bauhaus than members of any other single group outside art and architecture. Further, the two movements faced the same enemies—the religious right, nationalist, anthroposophist, *völkisch*, and Nazi opponents—and this drove them even closer together, toward the conjoint life they had in mind. Both enterprises sought to instantiate a modernism emphasizing what I will call "transparent construction," a manifest building up from simple elements to all higher forms that would, by virtue of the systematic constructional program itself, guarantee the exclusion of the decorative, mystical, or metaphysical. There

1. Rudolf Carnap, lecture notes for his Bauhaus lecture, "Wissenschaft und Leben," prepared 1 Oct. 1929 and delivered 15 Oct. 1929, transcription from shorthand by Gerald Heverly, Carnap Papers in the Archives of Scientific Philosophy, University of Pittsburgh Libraries, University of Pittsburgh (hereafter abbreviated CP, PASP), document RC 110-07-49. Quoted by permission of the University of Pittsburgh. All rights reserved. Translations are my own unless otherwise noted.

Peter Galison is associate professor in the departments of philosophy and physics at Stanford University, where he co-chairs the program in the history of science. His primary interest is in the history and philosophy of experimentation, the subject of his *How Experiments End* (1987) and *Big Science: The Growth of Large-Scale Research*, edited with Bruce Hevly (forthcoming). His current project is entitled *Image and Logic: The Material Culture of Modern Physics*.

was a political dimension to this form of construction: by basing it on simple, accessible units, they hoped to banish incorporation of nationalist or historical features.

From simple observation reports ("protocol statements") and logical connectives (such as "if/then," "or," "and"), the logical positivists sought to ground a "scientific," antiphilosophical philosophy that would set all reliable knowledge on strong foundations and isolate it from the unreliable. Since all valid inferences would be built out of these basic statements, the sciences would be unified by their shared starting points. For their part, the Bauhäusler hoped to use scientific principles to combine primitive color relations and basic geometrical forms to eliminate the decorative and create a new antiaesthetic aesthetic that would prize functionality. So close had the two groups come in their shared vision of modernism that, when the Bauhaus reconvened as the New Bauhaus in Chicago after fleeing the Nazis, the New Bauhaus imported the Vienna Circle's logical positivism as a fundamental component of its basic design program.

The modernism of the Bauhaus spanned many styles, political orientations, leaders, and artists—from their almost expressionist pre–World War I efforts in Weimar to the Marxist and technical orientation of the Dessau years. Similarly, as the logical positivism movement spread, it gained strength by enlisting the cooperation of a myriad of philosophical groupings, from American pragmatists to Polish logicians.[2] By the late 1940s and early 1950s, the impact of both tendencies was vast, but diluted. Here I focus on the late 1920s and early 1930s, a time when the Vienna Circle had just begun its most vigorous and productive phase, and the Bauhaus had recently planted its new roots in Dessau. During this period the connecting links between art and philosophy were real, not metaphorical, as artists and philosophers were bound by shared political, scientific, and programmatic concerns. No doubt by casting a wider net one could find other "affinities" between bits of philosophy and morsels of modern art, music, and literature. But it is in the later interwar period that the modernism of the Bauhaus and the Vienna Circle self-consciously reinforced each other, and in so doing began to articulate a common vision of what both called a modern "form of life."

2. Their polemic included a story of the group's origin in their manifesto, and this story has been repeated in "histories" that have been written (mostly by adherents to the movement) ever since. The standard sources for these recollections/histories are: A. J. Ayer, "The Vienna Circle," in Ayer et al., *The Revolution in Philosophy* (London, 1956), pp. 70–87; Carnap, "Intellectual Autobiography," *The Philosophy of Rudolf Carnap*, ed. Paul Arthur Schilpp (La Salle, Ill., 1963), pp. 3–84; Joergen Joergensen, "The Development of Logical Empiricism," in *Foundations of the Unity of Science: Toward an International Encyclopedia of Unified Science*, ed. Otto Neurath, Carnap, and Charles Morris, 2 vols. (Chicago, 1970–71), 2:847–946; Victor Kraft, *The Vienna Circle: The Origin of Neo-Positivism, A Chap-*

A reconstruction of this modernist form of life would serve two purposes. First, it would afford us a wider cultural understanding of both the philosophical and architectural movements. In particular, it could give a deeper significance to the attempt by philosophers of science to construct a "modern" view of the world. My hope is that by tracing the real links between the Vienna Circle and the Bauhaus, light will be shed on a central strand of canonical high modernism, revealing how each discipline used the other to legitimate its then radical endeavor.[3] Second, it is by now clear that both logical positivism and modernist architecture have come to occupy a central and disputed territory between left and right in the rich discussions of postmodernism. Though it is not my primary task here, it may be that by locating the philosophers of the Vienna Circle within a modernist cultural matrix, we will be better able to see what is and what is not an alternative to their political and philosophical vision.

2. Aufbau and Bauhaus

To an astonishing degree, modern philosophy of science traces its heritage to the Vienna Circle, a small philosophical group comprised

---

ter in the History of Recent Philosophy, trans. Arthur Pap (New York, 1953); Hans Reichenbach, *The Rise of Scientific Philosophy* (Berkeley and Los Angeles, 1951); Herbert Feigl, "The Wiener Kreis in America," in *The Intellectual Migration: Europe and America, 1930–1960*, ed. Donald Fleming and Bernard Bailyn (Cambridge, Mass., 1969), pp. 630–73; John Passmore, "Logical Positivism," in *The Encyclopedia of Philosophy*, ed. Paul Edwards, 8 vols. (New York, 1967), 5:52–57; Philipp Frank, *Modern Science and Its Philosophy* (1949; New York, 1955); and Neurath, *Le Développement du Cercle de Vienne et l'avenir de l'empirisme logique*, trans. du générale Vouillemin (Paris, 1935). These histories essentially provide a list of names, an intellectual constellation that in the receptive environment of liberal Vienna gelled into a commitment to science and to positivism. The list draws from all of history: in classical antiquity the Epicureans are seen as precursors, in the Middle Ages the nominalists, and in more recent times in Vienna the work of Ludwig Boltzmann and Ernst Mach. Their chosen forebears include Francis Bacon, Thomas Hobbes, John Locke, David Hume, Jeremy Bentham, John Stuart Mill, Herbert Spencer, René Descartes, Auguste Comte, Jules-Henri Poincaré, Gottfried Wilhelm Leibniz, and many others. Above all, the group autobiography continues, it was the recent work of the logicians Bertrand Russell, Gottlob Frege, and Ludwig Wittgenstein that, when mixed with the indigenous positivist tradition of Vienna, was to create logical positivism.

   3. In the last decade or so there have been a number of fine studies of the cultural setting of the Vienna Circle. See, for example, Elisabeth Nemeth, *Otto Neurath und der Wiener Kreis: Revolutionäre Wissenschaftlichkeit als Anspruch* (Frankfurt and New York, 1981); Friedrich Stadler, "Aspekte des gesellschaftlichen Hintergrunds und Standorts des Wiener Kreises am Beispiel der Universität Wien," in *Wittgenstein, der Wiener Kreis und der kritische Rationalismus*, ed. Hal Berghel, Adolf Hübner, and Eckehart Köhler (Vienna, 1979), pp. 41–59; and Rudolf Haller, "New Light on the Vienna Circle," *The Monist* 65

mostly of outsiders to philosophy that met regularly during the 1920s. Moritz Schlick, the Austrian aristocrat who stood at the center of the group known initially as the Verein Ernst Mach, had done his doctoral thesis in theoretical optics under the guidance of the physicist Max Planck in Berlin. Other members included Hans Hahn, a mathematician, and Philipp Frank, a theoretical physicist. Neurath was a dynamic, politically committed sociologist who came to the group with an interest in everything from museums to history, philosophy, and physics. Carnap had pursued experimental physics before turning to philosophy; he joined the rest of the Circle in 1926 after being in contact with both Schlick and Neurath. Others came from history, engineering, science, and philosophy. The group rapidly augmented its influence beyond Austrian borders by allying itself with Hans Reichenbach's movement for "Exact Philosophy" in Berlin and similarly oriented efforts in Poland, the United States, Great Britain, and Scandinavia. Throughout its existence, the Vienna Circle conceived of itself as modern and scientific, as a movement that would tear apart the stagnant, pointless inquiry that called itself philosophy. In the place of traditional philosophy the Circle wanted to erect a unified structure of science in which all knowledge—from quantum mechanics to Marxist sociology and Freudian psychology—would be built up from logical strings of basic experiential propositions.

Neurath and Carnap together forged many of the Vienna Circle's most self-consciously modern texts. The first surviving letter from Neurath to Carnap is dated October 1923. Neurath, addressing Carnap as "sehr geehrter Herr Doktor," wrote in the hopes of meeting Carnap to discuss their common interest in the correspondence between concrete reality and mathematical logic. "War and revolution," he said, "have torn chasms that have not yet healed, and it will require some more time before the ease, so necessary for wisdom . . . can be fully regained." Neurath was in a position to know. He had served as a technical expert in the finances of the left-wing revolutionary Munich government; after its defeat he spent a year and a half behind bars. "But," he now wrote Carnap, "I want with this letter to begin to weave a band; my wife and I would be delighted if you could participate in this knotting. Who knows, it might become a real carpet!" Extending a

---

(Jan. 1982): 25–37. Relatively little exists on Neurath's efforts in the visual arts. The following items are extremely helpful: Robin Kinross, "Otto Neurath et la communication visuelle," in *Le Cercle de Vienne: doctrines et controverses,* ed. Jan Sebestik and Antonia Soulez (Paris, 1986), pp. 271–88; Paul Neurath, "Souvenirs des débuts des statistiques illustrées et de l'isotype," in *Le Cercle de Vienne,* pp. 289–97; and *Arbeiterbildung in der Zwischenkriegszeit Otto Neurath–Gerd Arntz,* ed. Stadler (Vienna and Munich, 1982).

hand to Carnap's "neighboring" *Freideutschen*, Neurath added that they too could participate in the reconciliation.[4]

The letter is revealing not only about Neurath but also about the circle of philosophers he would soon join. Above all, it illustrates the manner in which formal, mathematical philosophy could serve as a bridge over a political divide, with Neurath and Carnap on the left and other members of the Circle on the right. During the revolution Neurath clearly allied himself with the workers' cause, but always in his capacity as a neutral, scientific expert. Even when reporting to the Munich Workers' Council in January 1919, Neurath introduced his summary by reminding the audience that the considerations he would discuss regarding social configurations, shelter, food, clothing, and working time were "unpolitical." Elsewhere that same year he described the social engineer as the direct analogue of the mechanical engineer: both transformed the world through scientific work, through the systematic analysis of modern statistics.[5] Evidently his stance of neutral engineering appealed to those in charge, for shortly after Kurt Eisner (minister-president of the Bavarian revolutionary government) was killed in February 1919, Neurath was asked to be president of the central planning office for Bavaria. "I accepted," he recounted a few months later, "stressing that I wished to be an unpolitical administrator" (*ES*, p. 21).

Neurath's scientism—his faith in the neutral, binding threads of statistics, physics, and logic—was key to the consolidation of the Verein Ernst Mach. But even as the Verein was in its infancy, Neurath continued his "unpolitical" technical social work and revealed a deep interest in workers' housing, art, and architecture. For Neurath, mass accommodation had several important political functions: it met the immediate material needs of the workers; it encouraged a collective form of life; and it served to build, sector by sector, Neurath's ultimate goal of full socialization of the economy. By the early 1920s Neurath had become a central figure in the housing movements in and around Vienna, drawing him into the circle of politically engaged modern artists and architects.[6] To Franz Roh, an art critic and close friend, Neurath wrote in 1924: "Just now I am dictating letters about the re-

4. Neurath to Carnap, 19 Oct. 1923, CP, PASP, document 029-16-07, p. 2. Originals of this and all letters by Neurath cited in this article are held by the Vienna Circle Archive, Vienna Circle Foundation (Amsterdam).

5. Neurath, *Empiricism and Sociology*, trans. Paul Foulkes and Marie Neurath, ed. Marie Neurath and Robert S. Cohen (Dordrecht and Boston, 1973), pp. 151–52; hereafter abbreviated *ES*.

6. For an excellent overview of Neurath's involvement in the housing movement, see Robert Hoffmann, "Proletarisches Siedeln—Otto Neuraths Engagement für die Wiener Siedlungsbewegung und den Gildensozialismus von 1920 bis 1925," in *Arbeiterbildung in der Zwischenkriegszeit*, pp. 140–48.

design of workers' housing through propaganda . . . 25,000 apartments were just built. . . . Can you send me information about graphics, color lithography, pictures, etc. concerning such worker housing?"[7]

Beginning at just this time in Germany and Austria, such public mass accommodations were increasingly identified both with Gropius's Bauhaus and with the liberal and left-wing municipal governments that supported the huge building projects.[8] At the time Neurath wrote to Roh, Gropius himself had just spoken in Vienna, leaving Neurath dissatisfied. But while he complained that Gropius had failed to bring sufficiently new ideas to Vienna, Neurath nonetheless told Roh of his outrage at attempts to abolish the Bauhaus on political grounds.[9]

Neurath's concern was not misplaced: nationalists blasted Gropius's Weimar Bauhaus for its left-wing orientation. Not long after Neurath heard Gropius lecture in Vienna, negotiations between the Bauhaus and the Weimar government collapsed, and in 1925 the socialist government of Dessau successfully courted the displaced artists. But turmoil continued to surround the Bauhaus, inside as well as outside its walls. Neurath's suspicion of some of the more conservative—that is, mystical and mythical—components of the Weimar Bauhaus program were shared by some of the artists within the Bauhaus itself. With the move to Dessau and pressure from various sides, including the spartan geometrists of the De Stijl, the Bauhäusler began a profound shift away from the mystical and toward the streamlined and industrial. This change was surely reinforced by the presence in Dessau of the big industrial concerns of Agfa, the Junkers aircraft plant, and factories for the production of gas and chemicals. Reflecting the new priorities, the teaching staff of the Dessau Bauhaus altered their titles from "masters" to "professors," and replaced graphic design with advertising. Their espousal of everything technical and scientific became ever more pronounced; art would act like science and serve as an initiator in the cycle of industrial production.[10]

Nothing pleased Neurath more than this new, scientific turn. When the Dessau Bauhaus opened in December 1926, Neurath was there, and he wrote about the occasion in the journal *Der Aufbau*. Cele-

7. Neurath to Franz Roh, n.d. [probably 1924], Correspondence and Miscellaneous Papers of Franz Roh, Archives of the History of Art, Getty Center for the History of Art and the Humanities, Los Angeles; hereafter designated as Roh Collection.
8. On the role of mass housing as a vehicle for the display of radical architecture, see Barbara Miller Lane, *Architecture and Politics in Germany, 1918–1945*, rev. ed. (1968; Cambridge, Mass., 1985), pp. 58, 84ff; hereafter abbreviated *AP*. According to Lane, it was the Bauhäusler's involvement in the design of these *Siedlungen* that so politicized the movement of left-wing architecture.
9. See Neurath to Roh, n.d. [probably 1924], Roh Collection.
10. Gillian Naylor, *The Bauhaus Reassessed: Sources and Design Theory* (New York, 1985), pp. 124–28; hereafter abbreviated *BR*.

brating the renunciation of ornamentation and decoration of every sort, he gently chided the Bauhaus for relying too much on the *style* of modernism and not sufficiently on its practical implications: "When will the modern engineers run the Bauhaus?" Insofar as the Bauhaus followed a technical, socially driven agenda, Neurath believed, it would serve the great revolution associated with the new form of societal and personal life [*Neugestaltung des gesellschaftlichen und persönlichen Lebens*]. Since he believed that "artists were leading the battle for a spiritual liberation from the past," the Bauhaus's cultural role could not have been greater.[11] This, in the end, was for Neurath the real import of the Bauhaus. Anyone wanting to "enter the promised land" liberated from the past "will seize upon the formation of the new form of life [*Gestaltung des Lebens*] as a technical achievement. This is the thrust of the Bauhaus, unfettering the liveliest discussion, and most vigorous efforts on all sides" ("NB," p. 211).

In his focus on a technically grounded new form of life, Neurath's language was at one with that of the radical architects and their defenders, who never tired of insisting that architectural novelty would underwrite a broader reformation of social and political existence. This was the true import of their new mode of building. The idea also permeated Neurath's writing and was stated most clearly in his 1928 book *Lebensgestaltung und Klassenkampf* [*Form of Life and Class Struggle*], where the philosopher insisted that it was the architect "more than any other creative person" who could anticipate and so shape the future form of life [*Lebensform*] (*ES*, p. 257).[12] Since rationality and scientificity were to characterize the revolutionary proletariat orientation, the architecture of modernity demanded rationality and functionalism. Modern architecture, Neurath believed, could both reflect and shape "the spirit of modern times." Again and again, he argued that "significant movements of the age" striving to shake loose of the past would ignore the example of the Bauhaus only at their peril. On his mind must surely have been his own less well known but equally messianic defense of the Vienna Circle ("NB," p. 211).

The notion that technical innovation could alter the form of life lay deep in the political ideology of left-liberal modernism, especially in architecture. In a remark of 1923 that was typical of his arguments ever since the end of the war, Gropius contended that the new modern architecture would actually produce "'a complete spiritual revolution in the individual'" and a "'new style of life'" (*AP*, p. 67). Defending the

11. Neurath, "Das Neue Bauhaus in Dessau," *Der Aufbau* [Vienna] 1, no. 11/12 (1926): 210–11; hereafter abbreviated "NB."

12. See also the German version of Neurath's *Lebensgestaltung und Klassenkampf, Gesammelte philosophische und methodologische Schriften*, ed. Haller and Heiner Rutte, 2 vols. (Vienna, 1981), 1:235–36.

new architecture, the mayor of Frankfurt am Main, Ludwig Land-mann, insisted that "'our new era must create new forms for both its inner and its outer life'" (*AP*, p. 90). Indeed, the claims for a reforma-tion of life based on modern principles of science became a common slogan of the left-leaning architects in post-World War I Germany—and an irritant to those on the right, who were determined to preserve a *völkisch* life form, imbued with history, nationalism, and racial iden-tity.

Gropius himself began to speak with growing conviction about the science of art as well as architecture: "The Bauhaus workshops are essentially laboratories in which prototypes suitable for mass produc-tion and typical of their time are developed with care and constantly improved."

> Only by constant contact with advanced technology, with the diversity of new materials and with new methods of construction, is the creative individual able to bring objects into a vital relationship with the past, and to develop from that a new attitude to design, namely:
> Determined acceptance of the living environment of machines and vehicles.[13]

Most important, Gropius created a new architecture department under the direction of Hannes Meyer, who, while continuing the scien-tific orientation of the earlier Dessau Bauhaus, to the dismay of some of his colleagues, put his materialism up front.

> "Building is not an aesthetic process. . . . Architecture which 'con-tinues a tradition' is historicist . . . the new house is . . . a product of industry and as such is the work of specialists: economists, statisti-cians, hygienicists, climatologists, experts in . . . norms, heating techniques . . . the architect? He was an artist and is becoming a specialist in organization . . . building is only organization: social, technical, economic, mental organization." [*B*, p. 180]

Here was a man after Neurath's own heart; at last there was a powerful Bauhäusler who put engineering before aesthetics. Instead of the backward-looking "historical" buildings, Meyer wanted the standard-ized, worker-oriented housing project. With Gropius, Meyer actually built such mass housing in the Törten district of Dessau (1926–28).[14]

Typical of Meyer's technocratic ambitions were his and Hans

---

13. Gropius, "Dessau Bauhaus—principles of Bauhaus production," sheet published by the Bauhaus in March 1926; cited in Frank Whitford, *Bauhaus* (London, 1984), p. 206; hereafter abbreviated *B*.

14. See *BR*, pp. 138–42.

Wittwer's 1927 plan for the League of Nations Palace in Geneva. As
their entry to the design contest, the two Swiss architects submitted the
drawing shown in figure 1. They built up the building from geometrical cells, importing all of the latest technological innovations:
paternosters, escalators, express elevators, moving sidewalks, and
automobile access. A new world order could not possibly be "squashed
into a structure of traditional construction; [there could be] no column-
stuffed reception room for tired sovereigns; instead, hygienic work
spaces for active representatives. No labyrinthine corridors for the
labyrinthine intrigues of the diplomats; instead, open glass rooms for
the public affairs of open men. The constructional arrangements of the
League of Nations arise through goal-directed *invention* [Erfindung]
and not through stylistic composition."[15] Architecture, Meyer believed,
would help reconstitute international relations by reorganizing the
material world in which they were conducted, just as new mass housing
would reform working-class forms of life.

Meyer's radical shift toward the rational and scientific antagonized
even some who were broadly sympathetic to that turn in art and architecture, such as László Moholy-Nagy, who had headed the Bauhaus's
metal workshop. In January 1928 Gropius quit, and despite the resignation of Moholy-Nagy and the resistance of others, Meyer took over and
led architecture to center stage. One of his first moves was to invite
guest lecturers in sociology, physics, and philosophy to the Bauhaus to
set the tone of scientific progressivism.[16]

Meyer's fascination with the scientific and the technical led him to
invite Herbert Feigl, a founding member of the Vienna Circle, to the
Bauhaus as the official representative of what the Circle called their
"new scientific world conception." Feigl spent a week (3–10 July 1929)
lecturing and getting to know Wassily Kandinsky, Paul Klee, and
others.[17] Apparently his visit was a smashing success, as a few weeks
later Carnap wrote to Neurath: "just received a very friendly letter
from Hannes Meyer. I'm to come for a week to lecture at the Bauhaus
on the scientific world conception. Feigl's efforts seem not yet to have
sated them, rather only to have agreeably whetted their appetite. In
principle I've said I'll go."[18] Meanwhile the Bauhaus asked Reichenbach
to come lecture at Dessau; Reichenbach was the chief Berlin ally of the

---

15. *Hannes Meyer 1889–1954, architekt urbanist lehrer* (Berlin, 1989), p. 105; hereafter abbreviated *HM*.

16. An excellent collection of Meyer's views can be found in Meyer, *Bauen und
Gesellschaft: Schriften, Briefe, Projekte* (Dresden, 1980).

17. Feigl to Reichenbach, 1 July 1929, Hans Reichenbach Papers, PASP, document
HR 014-06-11.

18. Carnap to Neurath, 25 Aug. 1929, CP, PASP, document 029-15-02.

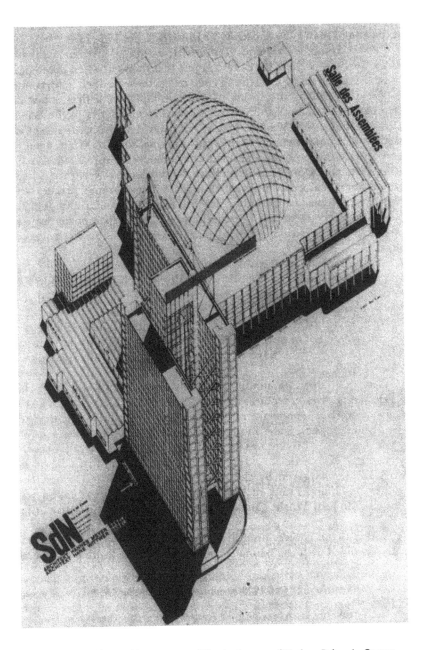

FIG. 1.—Hannes Meyer, proposal for the League of Nations Palace in Geneva, 1927. From *Hannes Meyer 1879–1954, architekt urbanist lehrer* (Berlin, 1989).

Vienna Circle.[19] Neurath himself was invited to lecture at the Bauhaus at the end of May 1929 and again in 1930 (*HM*, pp. 177–78).

For Carnap, Feigl, and Neurath, the timing of their Bauhaus excursions was perfect; they came at just that moment when the logical positivists were doing everything they could to bring their efforts into public view. During the spring of 1929 they printed a flyer soliciting membership in their Verein Ernst Mach: "To all friends of the Scientific World View!" "We live in a critical spiritual [*geistigen*] situation! Metaphysical and theological thought is taking hold in certain groups; astrology, anthroposophy and similar movements are spreading. On the other side: ever more conscious efforts for a scientific world view, logical-mathematical and empirical thought." The Verein's project was grandly ambitious as it sought (in words by then standard among the radical architects) to use the methods of "modern empiricism" to reform not only public but also private forms of life [*Lebensgestaltungen*].[20]

In the logical positivists' attempt to create a new form of life that necessarily extended beyond one's specialty, the positivists were in full accord with the Bauhäusler. Given Neurath's involvement with the Bauhaus controversy and his stated admiration for the architects' leading role in cultural reform, it is perhaps understandable that the Verein's statement of purpose affiliated the logical positivist movement "with wide circles who have trust in the scientific world conception." All were invited to join.[21]

The first project announced for this new widened public of the Verein was a series of lectures to include mathematics, astronomy, sociology of science, modern architecture, and (of course) arguments against metaphysics. Soon another flyer appeared announcing a series of four lectures. This one was headed: "Friends of the Scientific World Conception!" Of particular interest to us is that the very first talk, on 19 April 1929, was given by the Austrian architect Josef Frank, the brother of Philipp Frank of the Vienna Circle. His presentation was entitled "Modern World Conception and Modern Architecture" (fig. 2).[22]

It was an apt choice. Josef Frank was deeply, if sometimes ambivalently, involved with the Bauhaus and was one of the leading architects

19. Carnap diary, 21 Oct. 1929, transcription by Heverly, CP, PASP, document RC 025-73-03.

20. Verein Ernst Mach flyer, "An alle Freunde wissenschaftlicher Weltauffassung!" [before April] 1929, CP, PASP, document 029-30-01.

21. Ibid.

22. Verein Ernst Mach flyer, "Freunde wissenschaftlicher Weltauffassung! . . . Vorträge," n.d., CP, PASP, document 029-30-02.

VEREIN ERNST MACH

SITZ: WIEN. I. BEZIRK, WIPPLINGERSTRASSE 8. III III 331 TELEFON Nr. U 24-3 10

Freunde wissenschaftlicher Weltauffassung!

Der Verein Ernst Mach, der sich zur Aufgabe gestellt hat, wissenschaftliche
Weltauffassung zu fördern, veranstaltet in den nächsten Monaten nachstehende

VORTRÄGE
und zwar:

Freitag, den 19. April 1929

    Prof. JOSEF FRANK: Moderne Weltauffassung und moderne Architektur.

Freitag, den 10. Mai 1929

    Univ. Prof. HANS HAHN: Überflüssige Wesenheiten (Occams Rasiermesser).

Freitag, den 24. Mai 1929

    Bez.-Sch.-Insp. HEINRICH VOKOLEK: Begabungsproblem und
Vererbungslehre.

Freitag, den 14. Juni 1929

    Priv. Doz. RUDOLF CARNAP: Scheinprobleme der Philosophie
(von Seele und Gott).

 Sämtliche Vorträge finden im großen Hörsaal des Mathematischen Institutes der
Wiener Universität. Wien, IX. Strudelhofgasse Nr. 4, Erdgeschoß statt.

BEGINN 19 UHR

REGIEBEITRAG 50 GROSCHEN — FÜR MITGLIEDER EINTRITT FREI

Beitrittserklärungen werden im Sekretariate des
Vereines und bei den Vortragsabenden angenommen.

DER VORSTAND.

FIG. 2.—Flyer for lectures held by the Verein Ernst Mach, 1929. Carnap Papers in
the Archives of Scientific Philosophy, University of Pittsburgh Libraries, University of
Pittsburgh, document 029-30-01.

in Austria. Working with Oskar Strnad and Oskar Wlach, he had
produced, even before World War I, some of the most striking modern
residences in Austria. In 1927, the German Werkbund had invited him
to contribute to the Stuttgart Exhibition under the overall direction of
Mies van der Rohe. That modernist housing development was to
demonstrate the future of a whole neighborhood designed in the new
style, and included an astonishing collection of progressive architects,
among whom were Le Corbusier and Gropius. Philip Johnson later
called the enterprise, known as the Weissenhof development, "'the
most important group of buildings in the history of modern architec-

FIG. 3.—Josef Frank, Terrace Restaurant, 1925. From *Josef Frank, 1885–1967*, ed. Johannes Spalt (Vienna, 1981).

ture.'"[23] In addition to his strictly architectural accomplishments, Frank soon became the central theorist of the Austrian Werkbund. In that capacity he tried to navigate between left and right, between a naively progressivist (and, in his view, affected) functionalism of the Germans and his own countrymen's penchant for ornamentation, regionalism, and nationalism. It was a path he found difficult, at times even hopeless, to pursue.[24]

If Frank was at the center of the new architecture, he was not far from the vortex of the new scientific philosophy. For several years he designed architecture for Neurath's museum for picture statistics, a place where facts about the material conditions of the different classes could be presented in clear displays of billboards and graphs. It was a project that Neurath had held to be absolutely essential as a means of educating the masses; by its reliance on images rather than language, the picture museum would bridge the gap between nationalities. Neurath never lost faith that "just through its neutrality, and its independence of separate languages, visual education is superior to word education. *Words divide, pictures unite*" (*ES*, p. 217).[25] As with his commitment to the simplified universal jargon of "Basic English," his focus on the protocol sentences, and his apolitical politics, Neurath's pictures were intended as clear, universal building blocks on which all else could be built. Its international character, its constructivist dimension, and its visual simplicity all would have been appealing to the Bauhäusler when Neurath presented his work in 1929 (*HM*, p. 177).[26] Out of simple pictorial elements such as a machine, a worker, or coal, one could construct standardized representations of the distribution of industry, housing, and other aspects of material life. The ISOTYPE system (as it was called) was essentially a linguistic and pictorial form of transparent construction (fig. 4).[27]

In a letter, Neurath referred approvingly to Frank's design of the museum and its exhibitions: "The museum overflows with the old *Sachlichkeit.* Completely geometrical. Everywhere tables of commensurable quantities, and the whole put together with open space surrounding the tables."[28] Neurath's reference to the artistic movement of the *Neue*

23. Hans M. Wingler, *The Bauhaus: Weimar, Dessau, Berlin, Chicago,* trans. Wolfgang Jabs and Basil Gilbert, ed. Joseph Stein (1969; Cambridge, Mass., 1978), p. 534.

24. Friedrich Achleitner, "Wiener Architektur der Zwischenkriegszeit: Kontinuität, Irritation und Resignation," in *Das geistige Leben Wiens in der Zwischenkriegszeit,* ed. Peter Heintel et al. (Vienna, 1981), pp. 290–91.

25. See also Kinross, "Otto Neurath et la communication visuelle," p. 273, and Paul Neurath, "Souvenirs des débuts des statistiques illustrées et de l'isotype."

26. Neurath's lecture was titled "Picture Statistics and the Present."

27. For more on Neurath's picture language and his vision of the social uses of statistics, see Kinross, "Otto Neurath et la communication visuelle," and *Arbeiterbildung in der Zwischenkriegszeit.*

28. Neurath to Roh, n.d. [probably 1924], Roh Collection.

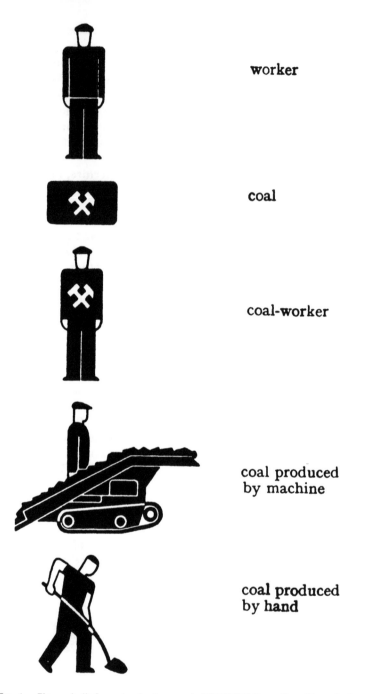

worker

coal

coal-worker

coal produced
by machine

coal produced
by hand

FIG. 4.—Figures built from simple elements in ISOTYPE. From Otto Neurath, *International Picture Language: The First Rules of ISOTYPE* (London, 1936).

Sachlichkeit or "new objectivity," a coldly clinical realist style, was apparently just one of many. According to Feigl, both Neurath and Carnap regularly referred to the Vienna Circle's logical positivism as "an expression of the *neue Sachlichkeit.*"[29]

3. Hausgewordene Logik

For the Vienna Circle, no philosophical work stood for this new objectivity as well as Wittgenstein's *Tractatus Logico-Philosophicus* of 1921. Indeed, it is nearly impossible to exaggerate the effect of the *Tractatus* on the Vienna Circle, where it was read out loud sentence by sentence twice, beginning in the Circle's Thursday meetings of 1926–27.[30] A trace of this devotion remains in the archives, where one finds a list of affirmations, a kind of positivist catechism, indicating, proposition by proposition, how each member of the Circle would vote on particular assertions before reading the *Tractatus* and after. For example: "The meaning of a sentence is through its method of verification."[31] Though Wittgenstein himself resisted being assimilated into the positivist camp, his commitment to a building up from verifiable propositions was similar enough to the aspirations of the Vienna Circle for the positivists to have seen Wittgenstein as a prophet of their philosophical modernism.

Wittgenstein's success with the *Tractatus* did not rescue him from his turbulent inner life. During World War I he voluntarily left philosophy to become a frontline soldier. War wounds left him hospitalized; when they healed, he turned first to gardening and then to teaching school in a remote mountain village. Just as the instruction of the young began to pale as a career, one of Wittgenstein's sisters, Margarethe Stonborough, commissioned a long-time family friend (and student of the architect Alfred Loos), Paul Engelmann, to design a large house on the Kundmanngasse in Vienna.

"You are me!" Loos once said to Wittgenstein.[32] Their common search for elimination of the superfluous, and their commitment to

29. Feigl, "The Wiener Kreis in America," p. 637.
30. Haller, "New Light on the Vienna Circle," p. 27. For more on the relation between the Vienna Circle and Wittgenstein, see the essays in *Wittgenstein, der Wiener Kreis und der kritische Rationalismus.* Allan Janik and Stephen Toulmin's well-known *Wittgenstein's Vienna* (New York, 1973) ties Wittgenstein's thought to controversies about language originating in the gap that arose between the political structure of the declining Austro-Hungarian empire and the new social and economic realities of turn-of-the-century Vienna.
31. Notes by Rose Rand, "Entwicklung der Thesen des 'Wiener Kreises,'" Nov. 1932–Mar. 1933, CP, PASP, document 081-07-01.
32. Paul Engelmann, *Letters from Ludwig Wittgenstein, with a Memoir,* trans. L. Furtmüller, ed. B. F. McGuinness (1967; New York, 1968), p. 127.

basic forms out of which the more complex would be derived, brought them to mutually sympathetic renditions of modernity. Loos's prewar volume, *Ornament and Crime*, had already laid the groundwork for his economic, moral, and aesthetic crusade against the decorative: "'I have discovered and given to the world the following notion: the evolution of civilization is synonymous with the elimination of ornament from the utilitarian object.'"[33] After the war, Engelmann found Loos's buildings and Wittgenstein's *Tractatus* opposing parallel targets: Loos aimed at the arts-and-crafts movement with its claims to higher spirituality while Wittgenstein laid his sights on the metaphysical system-builders of philosophy.[34] The *Tractatus* ends on a cautionary note, calling for philosophy to act as a kind of conscience against going beyond what can be said (such as the verifiable propositions of natural science). In sympathy with Loos's *Ornament and Crime*, Wittgenstein had written a kind of *Metaphysics and Crime*, with philosophy acting as the police.

At the same time, the philosophical and architectural projects shared an ideology of modernism: among other things, the *Tractatus* is a testimonial to the possibility of *building up* from simples into larger wholes. Its doctrine about logical propositions is precisely that from the elementary truth tables of simple propositions all others can be formed. "Mechanics determines one form of description of the world by saying that all propositions used in the description of the world must be obtained in a given way from a given set of propositions—the axioms of mechanics. It thus supplies the bricks for building the edifice of science, and it says, 'Any building that you want to erect, whatever it may be, must somehow be constructed with these bricks, and with these alone.'"[35] The architectural metaphor, I suggest, is not accidental. Whether he uses the verbs *bilden*, *bauen*, or the nouns *Konstruktion* or *Bau*, the Wittgenstein of the *Tractatus* is after an image of language, logic, and the world that starts at the basics and works up from there using logic alone. When complete, the structure would be without superfluity. At the time such a constructivist reading was encouraged all the more by Bertrand Russell's introduction to the first edition of the work, in which the British philosopher identified Wittgenstein's starting points as atomic sentences and continued the building metaphor through the idea of "molecular propositions" made from them.[36]

Wittgenstein's long-standing friendship with Engelmann and early admiration for Loos's architecture gave him personal and aesthetic

---

33. Benedetto Gravagnuolo, *Adolf Loos, Theory and Works*, trans. C. H. Evans (New York, 1982), p. 67.

34. Engelmann, *Letters from Ludwig Wittgenstein*, p. 128.

35. Ludwig Wittgenstein, *Tractatus Logico-Philosophicus*, trans. D. F. Pears and McGuinness (London and Henley, 1976), prop. 6.341, p. 68.

36. Bertrand Russell, introduction to ibid., p. xiv.

reasons to be enthusiastic about the Engelmann designs. Basic block elements characterized Loos's exterior designs, and in this respect his student Englemann followed the master in his early (1926) drawings of the basic structure of the Wittgenstein house. If Neurath, Carnap, and Feigl came to view the form of life surrounding modern art as an inspiration for their philosophy, Wittgenstein went further. Riveted by the Engelmann sketches, Wittgenstein began to reformulate every interior detail of the house, from windows to radiators. The emphasis on "industrial" design elements that expose the inner workings is apparent not only in the larger commitment to the hard-edged, but in details such as the transparent glass panels that leave the pulleys and counterweights of the elevator visible, or in the industrial-style columns with their recessed heads.[37] In its commitment to clean lines, simple elements, exposed functional elements, and empty spaces, the house at Kundmanngasse took the style of Loos's exteriors and brought them inside. At least one architectural historian suggests that Wittgenstein's designs may have inspired Loos to include interiors in his drive against ornamentation.[38]

In the end, the most striking contemporary characterization of the link between Wittgenstein's philosophy and his architecture came from one of Wittgenstein's other sisters, Hermine Wittgenstein. Shocked by the cold formality of the building, its absence of ornament and comforting decoration, she dubbed it "hausgewordene Logik" ["logic become house"]—an entirely appropriate appellation capturing the spirit of construction from simples that characterized both sides of the equation (*ALW*, p. 32).[39]

By the end of 1926, Wittgenstein and Engelmann are listed together as "Architekten" on the building permit, and in 1928, Witt-

---

37. Bernhard Leitner, *The Architecture of Ludwig Wittgenstein: A Documentation, with Excerpts from the Family Recollections by Hermine Wittgenstein*, ed. Dennis Young (New York, 1976), pp. 82–83, 102–3; hereafter abbreviated *ALW*.

38. See Gravagnuolo, *Adolf Loos, Theory and Works*, p. 82.

39. The later Wittgenstein famously rejected the linear constructivism associated with his *Tractatus*. Commencing with the *Blue Book* (dictated in 1933–34), Wittgenstein introduced a raft of concepts that opposed the notion of a building up from primitive elements. In the *Blue Book*, for example, he explicitly denounced the notion that there is an essence of an object that can be characterized by necessary and sufficient conditions. Instead, similarity or conceptual unification occurs through family resemblance, where no single specifiable property is present in all cases. Even later, in the *Philosophical Investigations*, Wittgenstein clarified the notion of a language-game in order to make clear that he meant it to designate more than verbal behavior: "Here the term 'language-game' is meant to bring into prominence the fact that the *speaking* of language is part of an activity, or of a form of life [*Lebensform*]" (Wittgenstein, *Philosophical Investigations*, 3d ed., trans. G. E. M. Anscombe [New York, 1958], § 23). But even in these later works Wittgenstein's links with the wider Austrian culture show through: the notion of a form of life was, as I have stressed above, a commonplace in post-World War I Austria, and nowhere

genstein, by himself, signed at least one plan as "Architekt" (*ALW*, pp. 9, 10). So engrossed was Wittgenstein in his architectural endeavors that when members of the Vienna Circle (including Schlick, Carnap, Feigl, and Friedrich Waismann) made their pilgrimages to him in the

FIG. 5.—Paul Engelmann, sketch for the Stonborough House at Kundmanngasse. From *Ludwig Wittgenstein Sein Leben in Bildern und Texten* (Frankfurt, 1983).

as prominently as in the world of architecture, as it reached out to transform personal and public life. Of course, in Wittgenstein's hands, *Lebensform* was divorced from the left-political discourse that Gropius, Meyer, Neurath, and others had bestowed on it. In no sense did the architects and philosophers I am discussing here intend the term to convey a sense of the epistemic relativism that the term acquired after Wittgenstein.

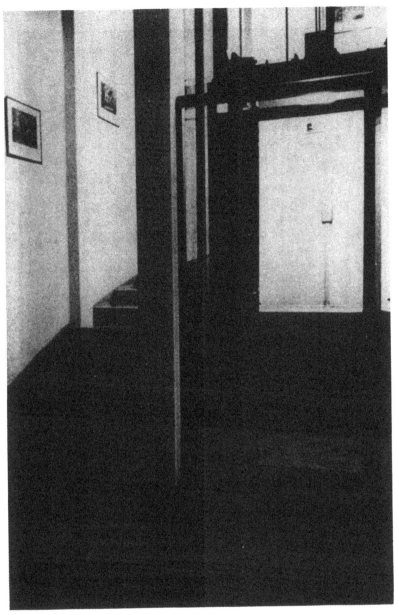

FIG. 6.—Elevator in Stonborough House, 1926–28. Note the exposed workings. From Bernhard Leitner, *The Architecture of Ludwig Wittgenstein: A Documentation, with Excerpts from the Family Recollections by Hermine Wittgenstein,* ed. Dennis Young (New York, 1976).

FIG. 7.—Stonborough House, exterior, 1928. From Leitner, *The Architecture of Ludwig Wittgenstein.*

late 1920s, he left them with the distinct impression that he was done with philosophy.[40]

4. The Architecture of the Aufbau

For Carnap, by 1929, architecture would have stood out as an exemplar of modern culture. The new building style would have been a subject of concern not only in his meetings with Wittgenstein but in his conversations with Neurath and in the Verein's own 1929 lecture series, in which Carnap took part. That series began with Josef Frank's discussion linking the modern worldview with modern architecture. Carnap's contribution was "Pseudo Problems of Philosophy: God and the Soul." That summer, perhaps inspired by the lecture series, the Verein Ernst Mach decided to celebrate their leader's (Moritz Schlick's) decision to return from Stanford to Vienna and to decline a tempting offer from Bonn. Their project was to write a manifesto incorporating their earlier proclamations and invitations, and to present it to their returning hero. As he finished his contribution to the draft, Carnap wrote to Neurath: "You see, I couldn't decide to pass this draft—with or without blessings—even to you; this Opus that I formulated by the sweat of my brow and in the same sweat typed. No, the bitter obligation and sweet right of last formulation have remained with me. But at the very last minute before publication you can still make corrections!"[41]

In its final form, the group's *Wissenschaftliche Weltauffassung* resembled far more the polemical manifestos of art, architecture, and politics than the staid volumes of philosophy. Even the style of writing, with its declamations and call to action, paralleled the daring pronouncements of the Italian futurists or the Russian constructivists far more than the dense philosophical works of the British Hegelians or the German neo-Kantians. Its stated ambition was grand: "The Vienna Circle does not confine itself to collective work as a closed group. It is also trying to make contact with the living movements of the present, so far as they are well disposed toward the scientific world-conception and turn away from metaphysics and theology" (*ES*, p. 305).

According to the manifesto, all was to be grounded on the simplest elements of observation and then built up from them: "*First* [the scientific world-conception] is *empiricist and positivist:* there is knowledge only from experience, which rests on what is immediately given. This sets the limits for the content of legitimate science. *Second*, the scientific

---

40. Feigl, "The Wiener Kreis in America," p. 638. For more on the relation of Wittgenstein to the Vienna Circle, see *Wittgenstein, der Wiener Kreis und der kritische Rationalismus.*
    41. Carnap to Neurath, 26 July 1929, CP, PASP, document 029-15-14.

world-conception is marked by application of a certain method, namely *logical analysis.*" Through this analysis the goal is to reach a unified science by "constituting" all scientific theories out of the elementary bits of perception. From the elementary aspects of the individual psyche it would rise to "a layer above" containing physical objects; these would then "constitute" other minds, and finally the objects of social science. With this building-up method, the constructional form [*Aufbauform*] of unified science would become clear (*ES*, p. 309).

The commitment to "removing the metaphysical and theological debris of millennia" was a distinctly modernist, and political, endeavor, as Carnap and his colleagues made explicit when they situated their dispute with traditional philosophy as issuing from "fierce social and economic struggles." The manifesto declaimed: "one group of combatants, holding fast to traditional social forms, cultivates traditional attitudes of metaphysics and theology whose content has long since been superseded; while the other group . . . faces modern times, rejects these views and takes its stand on the ground of empirical science." As the Bauhäusler did on every possible occasion, Neurath, Carnap, and the others used the manifesto to tie their mission to the image of industrial machinery, to the "modern process of production, which is becoming ever more rigorously mechanised and leaves ever less room for metaphysical ideas" (*ES*, p. 317). The modernism both groups had in mind would not stop at the traditional boundaries of science or art; they would reform fundamental aspects of daily life. Again the Vienna Circle manifesto: "We witness the spirit of the scientific world-conception penetrating in growing measure the forms of personal and public life, in education, upbringing, architecture, and the shaping of economic and social life according to rational principles" (*ES*, pp. 317–18).

Not surprisingly, since Carnap helped to draft it, the goals set out by the *Wissenschaftliche Weltauffassung* were closely tied to the goals of his just-completed masterwork, *Der Logische Aufbau der Welt*, usually translated as *The Logical Structure of the World*, but perhaps better rendered as *The Logical Construction of the World* since Carnap uses the terms *Struktur* and *Strukturform* in other, distinct ways (*ES*, p. 309).[42] Indeed, Carnap was enormously impressed with Bertrand Russell's

---

42. In the original German edition: "In die wissenschaftliche Beschreibung kann nur die *Struktur* (Ordnungsform) der Objeckte eingehen, nicht ihr 'Wesen'" (*Wissenschaftliche Weltauffassung der Wiener Kreis* [Vienna, 1929], p. 20). Carnap's own *Der Logische Aufbau der Welt: Scheinprobleme in der Philosophie* (1928; Hamburg, 1961) has a separate section on "Die Strukturbeschreibung" (pp. 14–15); hereafter abbreviated *A*. The *Logical Structure of the World: Pseudoproblems in Philosophy*, trans. Rolf A. George (Berkeley and Los Angeles, 1969) is the standard translation of Carnap's *Aufbau*; hereafter abbreviated *LS*. Where possible I have used this translation, though occasionally I have modified the translation on certain crucial points.

foundational view of objects as a logical construction of simple sense perceptions. As the epigram for the *Aufbau*, Carnap quoted (in English) from Russell's 1914 book, *Our Knowledge of the External World:* "The supreme maxim in scientific philosophizing is this: Wherever possible, logical constructions are to be substituted for inferred entities" (*A*, p. 1; *LS*, p. 5). In his idiosyncratic shorthand, Carnap has inscribed a comment near the end of chapter three, where Russell argues that a simplified construction, reconciling physics and psychology, is probably possible, but that he did "not yet know to what lengths this diminution in our initial assumptions" could be carried. The remark reads: "This deepening and diminution of the initial assumptions is my task!"[43]

In the *Aufbau*, Carnap tried to realize the constructional program announced in the *Wissenschaftliche Weltauffassung* and promised in the margins of Russell's *Our Knowledge of the External World:*

> Unlike other conceptual systems, a constructional system undertakes more than the division of concepts into various kinds . . . it attempts a step-by-step derivation or "construction" of all concepts from certain fundamental concepts, so that a genealogy of concepts results in which each one has its definite place. It is the main thesis of construction theory that all concepts can in this way be derived from a few fundamental concepts, and it is in this respect that it differs from most other ontologies. [*A*, p. 1; *LS*, p. 5]

Even Carnap's imagery is strongly architectural: the system has its *Grundbegriff, Grundgegenstand, Grundelemente, Grundwissenshaft,* and all the levels that build on them. Indeed, in summing up the task facing the scientific philosopher, Carnap insists that it is "no longer the task of the individual to erect the whole structure [*Gebäude*] of philosophy in one bold stroke." Elsewhere he adds that the philosopher's task is one of a "long, planned construction [*Aufbau*] of knowledge upon knowledge"; "a careful stone-by-stone erection of a sturdy edifice [*Bau*] upon which future generations can build" (*A*, p. xix; *LS*, pp. xvi–xvii).

It may be possible to interpret some of the above remarks as metaphorical, as such foundationalism was a long-standing theme in German philosophy. But in the preface to the *Aufbau*, Carnap makes the link to architecture literal and relaxes his otherwise technically encumbered language:

> We do not deceive ourselves about the fact that movements in metaphysical philosophy and religion which are critical of such [a

---

43. Carnap, personal copy of Russell's *Our Knowledge of the External World as a Field for Scientific Method in Philosophy* (Chicago and London, 1914), p. 97, in CP, PASP. My thanks to Richard Creath for the location of Carnap's copy of Russell's book.

scientific] orientation have again become very influential of late. Whence then our confidence that our call for clarity, for a science that is free from metaphysics, will be heard? It stems from the knowledge or, to put it somewhat more carefully, from the belief that these opposing powers belong to the past. We feel that there is an inner kinship between the attitude on which our philosophical work is founded and the intellectual attitude which presently manifests itself in entirely different walks of life; we feel this orientation in artistic movements, especially in architecture, and in movements which strive for a meaningful form of human life [*Gestaltung des menschlichen Lebens*], of personal and collective life, of education, and of external organization in general. We feel all around us the same basic orientation, the same style of thinking and doing. . . . Our work is carried on by the faith that the future belongs to this attitude. [*A*, p. xx; *LS*, pp. xvii–xviii]

Again, Carnap is after more than a contribution to philosophy; he is trying to participate in the creation of a "form of life" of which the *Aufbau*, the scientific world-conception, and modern architecture are all a part. Carnap finished the preface to the *Aufbau* in May 1928, and the book appeared later that year. With the new outward push of the Verein Ernst Mach in 1929, the architects and artists whom Carnap hoped would welcome the work did so, and Carnap accepted Meyer's invitation to Dessau.

5. Carnap in Dessau

Carnap arrived in Dessau on Tuesday, 15 October 1929 and was plunged immediately into a discussion about whether one should pursue only the aesthetic properties of materials. For the Bauhäusler this was a pressing issue, and the split between the "functional" and the aesthetic divided the faculty. Meyer led the charge against the aesthetic because it was metaphysical, that is, it included purely compositional content over and above what was technically demanded. After lecturing on "Science and Life," Carnap met with Ludwig Hilberseimer, Meyer's crucial appointment to the architectural department. Hilberseimer and his colleagues insisted that not only the artists' theories but also their objects (such as the Bauhaus lamps) still contained metaphysics, and that these needed to be purged.[44] In fact, the Bauhaus lamps provide an exemplary illustration of the tensions between conflicting impulses within the movement.

Moholy-Nagy, responding to Gropius's plea for the metal workshop to become a laboratory for industrial production, supported

44. Carnap diary, 15 Oct. 1929, CP, PASP, document RC 025-73-03.

FIG. 8.—Karl Jucker and Wilhelm Wagenfeld, glass lamp, 1923–24. Bauhaus-Archiv, Museum für Gestaltung, Berlin. Photo: Markus Hawlik.

inquiries into new lighting fixtures. Indeed Moholy-Nagy's own paintings inspired Wilhelm Wagenfeld and Karl Jucker in their creation of the best known of the Bauhaus lamps, one they advanced in 1924. It was a sleek design incorporating basic geometrical elements: a hemispherical opalescent glass shade, a transparent cylindrical glass stem, and a disk-shaped glass base, and visible inner wiring (fig. 8). As was so often the case, however, what struck the Bauhäusler as the very quintessence of industrial practicality was viewed quite differently from the factory floor. "'Retailers and manufacturers laughed at our efforts,'" lamented Wagenfeld. "'These designs which looked as though they could be made inexpensively by machine techniques were, in fact, extremely costly craft designs'" (*BR*, p. 112). By the time of Carnap's visit in 1929, this conflict between artisanal reality and industrial aspirations had evidently broken to the surface, for it is the residual craft component that Hilberseimer derided as "metaphysical." By coordinating their causes and language, Hilberseimer and Carnap located a common foe in the ornamental and nonfunctional, be it in decorative art or metaphysical philosophy.

On Wednesday, 16 October, Carnap gave his lecture "The Logical Construction of the World," beginning with the logical positivists' rallying cry: "There is only one Science ('Unified Science'), not separate subjects. . . . for all knowledge stems from one source of knowledge: experience—the unmediated content of experience such as red, hard, toothache, and joy. These make up the 'given.'" In summary, he educed four theses: (1) there are no things outside of the experiential— no realism about things; (2) there are no forces over and beyond relative motions—no metaphysics of force; (3) there is no psychology of the other that is not grounded in an individual's own experience—no psychorealism; (4) there are no social objects such as the state or the *Volk.* On this last point—and this would have been well received by Meyer's faction of the Dessau Bauhaus—he insisted that the Marxist conception of history *was* allowable because it was based on the empirically determinable. Carnap's basic slogan: Exclude metaphysics and limit utterances to those about the given. For example: dispose of the idea of God. And feelings attributed to others as well as *Verstehen* in history *are not knowledge.* For over an hour the architects and painters of the Bauhaus vigorously discussed the lecture, until Carnap retired at one that morning.[45]

Over the intervening years, much of what Carnap was opposing has lost its direct political significance. But in 1929 Carnap's four theses bore a manifest coherence in their opposition to powerful right-wing forces that sought to unify these ideas of *Volk,* metaphysics, the state, and God. The journal of the German Philosophical Society, *Blätter für Deutsche Philosophie,* is replete with examples. Consider, for example, the volume for 1929/30, which included lead articles such as "*Volk* as the Bearer of Education," "The Historical-Metaphysical Sense of Germanity [*Deutschtums*] and Its Surrounding World," and favorable book reviews of *The Logic of the Soul, The Doctrine of the State as Organism,* and *Godliness in the Character of the "Volk."*[46] The avowedly politicized, religious, and nationalistic character of such polemics helped bind together, by their opposition, the left-wing modernists of the Vienna Circle and the Dessau Bauhaus. Both were committed to a rationalism, secularism, and internationalism that they hoped to secure by a logical and empirical construction. In the days that followed, Carnap lectured on the four-dimensional world and on the misuses of language. Following his main interest—the elimination of all that did not flow from the simple unifying elements of experience—Carnap argued in one discussion that the Bauhäusler still had not rid themselves of metaphysics in their theoretical work. His example was that the proposition "black or white is heavy" could not be interpreted directly; its only significance came through psychological association.[47]

45. Transcript of Carnap's lecture, "Der logische Aufbau der Welt," 17 Oct. 1929, document dated 10 Oct. 1929, CP, PASP, document 110-07-45.

46. See *Blätter für Deutsche Philosophie* 3 (1929/30).

47. Carnap diary, Saturday, 19 Oct. 1929.

Fig. 9.—Geometrical paper form from Josef Albers's preliminary course, 1927. Bauhaus-Archiv, Museum für Gestaltung, Berlin.

On Sunday, Alfred Arndt took Carnap to the Bauhaus exhibition, where the philosopher was particularly impressed by the fundamental researches of the preliminary course: geometrical surface theory and forms made out of paper and wire screens (fig. 9).[48] Carnap's fascination with these etherial geometrical forms was perfectly understandable: ever since his doctoral dissertation in Jena on "Space,"[49] he had pursued his interest in geometry; moreover, the subject of geometry, as axiomatized and revived by the mathematician David Hilbert, provided a model

48. Carnap diary, Sunday, 20 Oct. 1929.

49. Carnap, "Der Raum. Ein Beitrag zur Wissenschaftslehre," *Kant-Studien*, Erg. Heft no. 56 (1922): 1–87.

for the construction process he had in mind for all of philosophy. At the preliminary course exhibition, Carnap met Kandinsky for the first time; it was, of course, no surprise that Kandinsky was there, as he was one of the leaders of the constructivist curriculum.

Not only would Carnap have found the subject matter of these geometrical explorations interesting, but the sentiment would surely have been returned. Carnap's thesis on space and his *Aufbau* were cited, for example, when the Bauhäusler wrote on space.[50] Carnap and Kandinsky shared the basic faith in a building up from the elementary. In the book that grew out of his preliminary course, Kandinsky called his artistic goal "practical" science.[51]

> The work in the Bauhaus is synthesis.
> The synthetic method naturally embraces the analytical one. The interrelation of these two methods is inevitable.
> The instruction in the fundamental elements of form must also be built on this basis.
> The general problem of form must be divided into two parts:
> 1. Form in its narrower sense—plane and space.
> 2. Form in its broader sense—color and the relation to form in its narrower sense.
> In both cases the work has to begin with the simplest shapes and systematically progress to more complicated ones. Hence, in the first part of the investigation of form the plane is reduced to three fundamental elements—triangle, square, and circle—and space is reduced to the resulting fundamental space elements—pyramid, cube, and sphere.[52]

The analysis into parts and reconstruction from geometry and color directly paralleled the project of Carnap's *Aufbau*. In the place of color and geometry, Carnap and his Vienna Circle had protocol sentences (expressing primitive sense experiences) and combinations of these protocol sentences using logic. Carnap's *Stufenform* [ascension forms] built up the complexities of all scientific terms out of these elements just the way Kandinsky's elementary geometrical forms made up the human figure. In both Bauhaus and Aufbau, construction from the intelligible simples eliminated the metaphysics of the unnecessary, the merely decorative.

Despite Kandinsky's attempt to make a "practical" science of color and form, he and others often referred to the "temperature" or the

50. László Moholy-Nagy, *The New Vision: Fundamentals of Design, Painting, Sculpture, Architecture*, trans. Daphne M. Hoffmann (New York, 1938), p. 162.

51. Wassily Kandinsky, *Point and Line to Plane*, trans. Howard Dearstyne and Hilla Rebay (New York, 1947), p. 20.

52. Kandinsky, "Staatliches Bauhaus Weimar 1919–1923" (Weimar-Munich, 1923), p. 26; reprinted in Wingler, *Bauhaus*, p. 74.

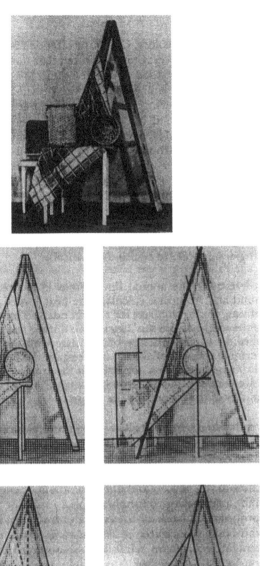

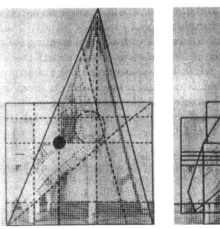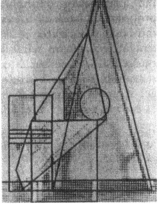

FIG. 10.—Wassily Kandinsky, analysis of still life, 1929–30. Bauhaus-
Archiv, Museum für Gestaltung, Berlin.

"weight" of particular colors. Obviously offended by the "metaphysical" quality of such utterances, Carnap insisted that such propositions could only properly be understood as psychological. Jost Schmidt, one of the most versatile sculptors and painters at the Dessau Bauhaus, gave such a view a sympathetic hearing. But though Schmidt "was clear" on these issues, Carnap recorded his impatience to see Meyer himself. On Monday, 21 October Meyer returned, and he and Carnap met. Despite the fact that it was Gropius who had appointed Meyer head of the Dessau Bauhaus, Meyer was determined to break with the old guard. To Carnap he commented that, in the old Bauhaus of Gropius, one found the expression of an individual-sentimental attitude.[53] Nothing of the sort would be appropriate under Meyer's leadership, as he had made clear in his article "Building" for the journal *bauhaus* the previous year. Instead of sentiment, historicity, or nationality, the basic elements of housing design were to be fixed empirically:

> we determine the annual fluctuations in the temperature of the ground and from that calculate the heat loss of the floor and the resulting depth required for the foundation blocks. . . . we calculate the angle of the sun's incidence during the course of the year according to the latitude of the site. with that information we determine the size of the shadow cast by the house on the garden and the amount of sun admitted by the window into the bedroom. . . . we compare the heat conductivity of the outside walls with the humidity of the air outside the house.

As Meyer insisted, the logical-empirical construction was inseparably associated with its internationality: "this constructive world of forms [*konstruktive Formenwelt*] knows no native country. it is the expression of an international attitude in architecture."[54] Meyer sought to render architecture in the neutral and universal idiom of engineering; Carnap pursued the analogous goal for philosophy.

Appropriately, the diary breaks off with Carnap meeting Reichenbach, who had just arrived in Dessau to give the artists their next installment of lectures on scientific philosophy. In the months that followed, Neurath came back to deliver two more lectures at the Bauhaus, and Philipp Frank offered a series of three presentations on the impact of modern physics on ideas of space and time (*HM*, p. 178).

53. Carnap diary, Sunday, 20 Oct. and Monday, 21 Oct. 1929.
54. Meyer, "Building," *bauhaus* [Dessau] 2, no. 4 (1928); reprinted in Wingler, *Bauhaus*, pp. 154, 153.

6. Neutrality and Nazism

Building up concepts and uniting the sciences out of simple propositions remained central to the logical positivists, so central, in fact, that different interpretations of their significance contributed to a division between Neurath and Schlick and, for a short period, strained relations between Neurath and Carnap. It seems that Neurath thought his colleague had infringed on his priority in the invention of these neutral bricks of knowledge.

If Neurath's faith in the neutral, scientific underpinnings served him in his quest for political unity in the 1919 revolution and its aftermath, political undercurrents again pressed on Neurath in the tumultuous years of the early thirties. And again Neurath responded with a commitment to technocratic Marxism that was of a piece with his more abstract philosophy. In October 1932, Neurath planned a trip to Moscow to discuss, inter alia, his plans for a branch of his picture museum. To Carnap he explained:

> In the middle of October, I have to travel to Moscow and am not too pleased about having to deal with my ideology. Over there I'm a technical specialist and abstain from all arguments which only seem to lead to differences. If today something's a no, tomorrow it's a yes, as soon as there is change in the party line. I realize all that. But I accept the consequence of this ideological abstinence and concentrate on *the technical.*

Continuing in the same letter, without any apparent break, Neurath then switches to philosophy and speaks about the importance and priority of his work on the neutral protocol sentences that lie at the basis of the unified sciences.[55] So committed was Neurath to the idea of the technical, and so against the ideological, that at the outset of the Vienna Circle he was dead set against even mentioning "philosophy" when speaking of the new enterprise. "The word 'philosophy,'" Neurath wrote Reichenbach, "is above all laden with associated meanings of 'system,' 'basic statements about the world,' 'values,' etc." Even in talk about "positive philosophy," or "exact philosophy," Neurath saw danger. "The scientific stands in the center for us, the indeterminate on the periphery! With the philosophers it is backwards . . . !"[56] In both philosophy and politics Neurath had faith that a rigorous technical

55. Neurath to Carnap, 1 Oct. 1932, CP, PASP, document 029-12-29.
56. Neurath to Reichenbach, 22 July 1929, CP, PASP, document 029-15-15.

analysis would solve problems that had resolutely resisted solution when infused with values and worldviews.[57]

The interpenetration of political and wider societal concerns in Neurath's reasoning is much more apparent in his letters than in his published philosophical tracts, as one sees in a letter of October 1932. At one point in the letter, his philosophical worry was about the relation of parts and wholes, and whether the basic statements of the constructional system can be Gestalt wholes rather than the constituent bits of perception: "I am amazed at how we can bring Schlick's Gestalt qualities into harmony with the philosophy of totalities. Give me a complex and I'll make a whole out of it—that's the slogan." Without missing a beat, Neurath moves from the wholes of philosophy to the cracks in society. "The tear is running . . . It'd be to throw up, if one didn't have to laugh. And behind it all stands Hitler. . . . Here comes God and Religion to the front and ancestral truths and the German *Volk,* and what you need to stab a jewish socialist with a knife between the ribs. . . . Oh Carnap! Oh World!"[58]

March 1933, Neurath to Carnap: "On Friday the Circle [met to discuss] Protocol sentences. Schlick was out of line [*ungehörig*]; started already arrogantly by saying that the thing didn't interest him, etc. Waissman in his own way [also objected]. They want instantaneous experience with 'now' and 'here'; they challenged the right to determine these [experiences] by means of coordinates" (this was necessary for Neurath's physical protocol sentences, which were to be interpersonal, not individual).[59] On this point Carnap would have concurred with Neurath, for after the *Aufbau* he increasingly came to view the starting points of the construction to be matters of convention. And given this conventional freedom, it was Neurath's firm view that the choice should be dictated by the practical advantage to the community. This demanded a language of physical effects not individual perceptions.[60] Given this division and his colleagues' relapse into what he considered crass idealism, Neurath commented that he thought the Vienna Circle would be misrepresented by Schlick at the forthcoming conference on the Unity of Science. Schlick, Neurath feared, would stand for the Vienna Circle the "way the third [Reich] claims alone to represent the nation."[61] As a result Neurath was now ready to renounce the validity of the Vienna Circle's right wing—including Wittgenstein,

57. Neurath very deliberately used "Wissenschaftliche Weltauffassung" and not "Wissenschaftliche Weltanschauung" precisely because of the system-building implications of the latter.

58. Neurath to Carnap, 9 Oct. 1932, CP, PASP, document 029-12-24.

59. Neurath to Carnap, 13 Mar. 1933, CP, PASP, document 029-11-20.

60. Carnap, "Intellectual Autobiography," pp. 51–52.

61. Neurath to Carnap, 13 Mar. 1933, CP, PASP, document 029-11-20.

whom he judged hopelessly metaphysical. The perimeter of the circle shrank as Neurath grasped for a cohort of sympathetic souls. A few months earlier he told Carnap: "I want to belong to a *Gemeinschaft* consisting of [Philipp] Frank, Hahn, Carnap, Neurath and few younger people who are all driven by the unity of science."[62]

By June 1933 Neurath's letters become increasingly despairing. "Sad times. But I'm looking around to see whether we can't find possibilities in the west. Carnap, Frank, Hahn, Neurath that should be the eternal quartet, for Schlick and his followers are slipping away into idealistic doubletalk."[63] Even Neurath's greatest commitment, his faith in the power of unity (personal, social, political, and philosophical), began to wane. "Up until now," he writes to Carnap, "I have had the inclination to emphasize the positive side and to leave criticism to the side in order to further community [*Gemeinschaft*]. But, I now feel—and I very much regret it—that I did not emphasize the Marxist deficiencies. . . . One sees *how weak* the foundation was in its components. We must build [*aufbauen*] anew, for this factual work is necessary and the many-sided refusal to accept Marxist superficiality. Youth is ready . . . to rebuild."[64] Defeated, the left was splintering, and the Vienna Circle itself began to fall apart along political lines.

With both Marxists and positivists on the run, the German Philosophical Society celebrated the Nazis' election to power. Their meeting of October 1933 opened with the collective singing of the "Deutschland Lied" and "The Horst Wessel Song." Now, the Nazi representative proclaimed, philosophy would be applicable to the people and fulfill the spiritual needs of the *Volk*. Hitler telegraphed a laudatory greeting, part of which read: "May the forces of true German philosophy contribute to the building and strengthening of the German worldview."[65] The philosophers complied with talks on *Deutschtum, Volk*, Soul, and Spirit.

Less than a year later, when the Vienna Circle confronted the right-wing philosophers at the International Philosophy Congress in Prague, a clash was inevitable. The principal nationalistic philosophy journal reported excitedly that the Congress had revealed philosophy to be at a turning point, as "a certain *Volk*" took its place in the development of the World Spirit. One of the heroes of nationalist philosophy, Hans Driesch, presented a plenary lecture, arguing for vitalism and

62. Neurath to Carnap, 22 Oct. 1932, CP, PASP, document 029-12-19. Even in the *Tractatus*, Wittgenstein left some qualified room for the mystical. Though such statements were sharply delineated from the verifiable, any such discussion was manifestly too much for Neurath.

63. Neurath to Carnap, 18 June 1933, CP, PASP, document 029-11-14.

64. Neurath to Carnap, 6 Apr. 1933, CP, PASP, document 029-11-18.

65. "Bericht über die 12. Tagung der Deutschen Philosophischen Gesellschaft zu Magdeburg vom 2. bis 5. Oktober 1933," *Blätter für Deutsche Philosophie* 8 (1934/35): 65–70.

guarding a place for metaphysics. Here the Vienna Circle jumped into the fray with what its enemies characterized as a "vehement and well organized attack," in which the Circle decried metaphysics as meaningless. Viewed from the right, the positivists "stood in the way" of the metaphysical concept of the world that was to underwrite the German worldview. Reichenbach blasted Driesch's organicism as "mystical," while Carnap denied that Driesch's organicism was sufficiently lawlike to make it scientific. Schlick remained silent, but the next day he presented an entire lecture, "On the Concept of the Totality," in which he claimed that while the distinction between totalities and aggregates might be linguistic or pragmatic, it was not a substantive distinction: there was no whole over and beyond the sum of parts.[66]

For both sides, the debate over the totality concept [*Ganzheitbegriff*] was crucial. Carnap, Reichenbach, and Schlick denied the idea of a transcendent reality to the *Deutschtum, Nation,* or *Volk* and so threatened to undermine central tenets of right-wing ideology. According to Schlick, in sociological as well as physical or biological systems, one could build up higher levels of organization from an adequate understanding of constituent individuals. There simply was nothing further left to add about the "totality" or "whole." To the Nazis and their allies, individuals had to be more than isolated entities; they were members of "higher totalities" whose full existence and whose cultural and spiritual acts could be understood only insofar as they were embedded in a larger inheritance, including their genetic material.[67] Similarly with the Bauhaus: the Nazi press cited internationalist tendencies and attacked Bauhaus art as a "calculating construction" that sought to abstract pure color and form from the world. Such an enterprise reduced man to a "geometrical animal" and was utterly incapable of capturing the "German essence."[68] On many counts, then, the Vienna Circle and Dessau Bauhaus's vision of transparent construction was anathema to the Nazi movement; it cut any transcendent national purpose from the state, from architecture, and from nature.

Violence superseded argument. Neurath wrote desperately to Carnap: "Of the atrocities of devastation let's not speak. *All* of my friends are either sitting still, or fired, or arrested or in flight. . . . A young friend is probably in the worst camp, others disappeared. Desperation. Misery. Bert Brecht came over, Brentano and others. Everyone's getting out. We're collecting money. And yet: working on. One knows where one stands and where one falls."[69] Elsewhere Neu-

---

66. Joh. Gauter, "Der VIII. Internationale Philosophenkongress 1934 (vom 2.–7. September in Prag)," *Blätter für Deutsche Philosophie* 8 (1934/35): 437–48, esp. pp. 437–40.

67. Ibid., pp. 440–41.

68. Peter Hahn, *Bauhaus Berlin: Auflösung 1932, Schliessung Berlin 1933, Bauhäusler und Drittes Reich* (Berlin, 1985), p. 124.

69. Neurath to Carnap, 6 Apr. 1933, CP, PASP, document 029-11-18.

rath recalled painfully, "If I think back, how many of those I knew have been killed? Rathenau, Landauer, and so on. . . . The four apocalyptic riders are in full form."[70] The same riders were now moving against the Circle's allies in the Dessau Bauhaus. For under Meyer the left-wing politics of the Bauhaus had sharpened, encountering in the process increasing trouble with the press and the town authorities. A group of students (about ten percent of the student body) formed a communist cell. The press reported, to a well-prepared opposition, that the students had even sung Russian revolutionary songs at a Carnival party in 1930. Under pressure both from outside the Bauhaus and from within, Meyer resigned later that year; his fall was catalyzed in part by charges that he had donated money to striking miners in the name of the school (*B*, pp. 190–91).

In his letter of protest and resignation to the mayor of Dessau, Meyer reiterated his accomplishments, among which was the extraordinary series of visitors he had drawn to the Bauhaus during his tenure. First on the list was Neurath; prominently displayed were the names of Carnap and Feigl. Moreover, support from the logical positivists was indicated not merely by their past presence; Neurath and Josef Frank authored a ringing denunciation of Meyer's removal from the directorship of the Bauhaus. In a clear reference to the participation of the Vienna Circle, they reminded their readers that Meyer had not only brought technical subjects to the Bauhaus, he had imported the more general scientific world-conception [*wissenschaftliche Weltauffassung*]. "In Meyer's view," Frank and Neurath emphasized, "only people with a fundamental understanding of societal phenomena and science would become architects." Such a scientific orientation was at one with politics: "He was an advance guard in the great struggle over the new form of life [*Lebensordnung*] of socialism. He was truly a thorn in the side of the reactionaries."[71] Having already forged bonds based on a common internationalist and constructivist sense of modernity, the two movements were now brought even closer together by the common terms of their persecution under the Nazis.

Mies van der Rohe, as the new director of the Bauhaus, swung the school away from the Marxist, the sociological, and the functional toward the formal, the elegant, and the aesthetic. It was a last, desperate attempt to preserve the Bauhaus under the Nazis, whom van der Rohe hoped might eventually soften their stand against the school. But the Bauhaus teachers were resigning in droves, the Nazi party took control of the Dessau city parliament, and the Bauhaus now came under direct fire for their international style: the flat geometrical roof, for example, was obviously not right for the north. According to the

70. Neurath to Carnap, 13 Mar. 1933, CP, PASP, document 029-11-20.
71. Josef Frank and Neurath, "Hannes Meyer," *Der Klassenkampf: Sozialistische Politik und Wirtschaft* 3 (1930): 574–75.

Nazis, the modernist challenge to traditional roof design was a throw-back to the "oriental" and "Jewish" "subtropical" regions (*B*, p. 195). After the Dessau Bauhaus was closed in October 1932, van der Rohe kept the institution alive for a few more months in Berlin.[72] Pius Pahl, one of the Bauhaus students at the time, remembered that "'the end came on 11 April 1933 during the first days of the summer term. Early in the morning police arrived with trucks and closed the Bauhaus. Bauhaus members without proper identification (and who had this?) were loaded on the trucks and taken away.'" The Bauhaus was dis-solved on 10 August 1933 (*B*, p. 196).

"Hard times," Neurath wrote Carnap in March 1933, "really hard times. And what will become of physicalism [Neurath's doctrine of building up from simple elements of experience]? When will we be able to go from our foundation [*Unterbau*] to the superstructure [*Überbau*]? When?"[73]

7. The American Incarnation

If the promised superstructure was ever made, it was in the Bauhaus's fourth and last incarnation in the United States. Feigl left for America in 1930; Carnap followed suit, leaving Prague in 1936 and settling at the University of Chicago. Hans Hahn died in 1934, and in 1936 Schlick—already very alienated from Neurath on political, philo-sophical, and personal grounds—was murdered by a deranged student. For some years Neurath remained in The Hague, Holland, where his work on the International Foundation for Visual Education continued; and from there he continued to participate in work on the *Encyclopedia of Unified Science* and ISOTYPE. The rather unstable life he had constructed fell apart on 10 May 1940, when the fighting in Holland came within audible distance. On 13 May he could see the sky red over Rotterdam, and he, along with his collaborator and later wife Marie Reidemeister, joined a desperate group of refugees in a lifeboat headed for England.[74]

But the modernist endeavor, a joint enterprise of the old Vienna Circle and the old Bauhaus, had already begun to reassemble with the cautious blessing of the University of Chicago under the direction of Moholy-Nagy. Moholy-Nagy had come from Hungary to Berlin after

---

72. On the last months of the Bauhaus under Mies van der Rohe, see Dearstyne, "Bauhaus Berlin—Bauhaus Finis," *Inside the Bauhaus*, ed. David Spaeth (New York, 1986), pp. 239–55, and Spaeth, *Mies van der Rohe* (New York, 1985).

73. Neurath to Carnap, 13 Mar. 1933, CP, PASP, document 029-11-20.

74. On Neurath's escape from the continent, see Marie Neurath's account in *ES*, pp. 68–73.

World War I to absorb Germany's "'highly developed technology'" and began incorporating gears, wheels, and machinery into his art.[75] Soon, however, his interest in machines began to merge with a fascination with light and photography, and his belief that the different artistic media were all part of the same unity became a refrain in his work. In 1921 he came into contact with Hilberseimer and Gropius; two years later, Gropius invited Moholy-Nagy to the Weimar Bauhaus to teach, edit, and write. He remained there until his split with Meyer drove him out.

As events turned worse in Germany, Moholy-Nagy, the artist of all trades, left the Bauhaus for Amsterdam, then went from Amsterdam to London and then to Chicago. As part of his scientific vision of art, Moholy-Nagy recruited at least four members of the positivists' Unity of Science movement. Carnap himself occasionally lectured there, but it was Carnap's colleague and devotee of logical positivism, Charles Morris, who maintained the closest affiliation with the New Bauhaus.[76] Morris, a philosopher at the University of Chicago, had acted as an American clearinghouse for contact between Americans and the Vienna Circle throughout the 1930s. Indeed, Morris played a leading role in getting Carnap to the philosophy department at the University of Chicago. In addition, for several years Morris had been the most active American in the ambitious series of conferences the Circle ran on the Unity of Science, and was a coeditor (with Carnap and Neurath) of the successor journal to the Vienna Circle's *Erkenntnis:* the *International Encyclopedia for the Unity of Science.* To the New Bauhaus Morris also brought two scientist members of the Unity of Science movement, Carl Eckart (from physics) and Ralph Gerard (from biology).

In the 1937 Prospectus for the New Bauhaus, Morris recalled: "Moholy-Nagy knew of the interest of Rudolf Carnap and myself in the unity of science movement. He once remarked to us that his interest went a stage farther: his concern was with the unity of life."[77] Now

75. Lloyd C. Engelbrecht, "The Association of Arts and Industries: Background and Origins of the Bauhaus Movement in Chicago" (Ph.D. diss., University of Chicago, 1973), pp. 246–47.

76. Ibid., p. 286. For more on the establishment of the New Bauhaus, see also William H. Jordy, "The Aftermath of the Bauhaus in America: Gropius, Mies, and Breuer," in *The Intellectual Migration*, pp. 485–543; *50 Jahre New Bauhaus: Bauhaus-Nachfolge in Chicago* (Berlin, 1987); and James Sloan Allen, "Marketing Modernism: Moholy-Nagy and the Bauhaus in America," *The Romance of Commerce and Culture: Capitalism, Modernism, and the Chicago-Aspen Crusade for Cultural Reform* (Chicago, 1983), pp. 35–75.

77. Morris, Prospectus for the New Bauhaus (hereafter abbreviated P), American School of Design, founded by the Association of Arts and Industries, p. 10, accession record 70-65 F65 in the Institute of Design Collection, The University Library, Special Collections Department, The University of Illinois at Chicago (hereafter ID/UIC). Morris to Lloyd Englebrecht, 3 June 1968, ID/UIC.

Morris was ready to further this expanded sense of unity, embedding it within a new nationalistic framework to facilitate its reception in an often xenophobic late-1930s America. "The general program [of the New Bauhaus] accords with the deepest American insights and needs—the dovetailing of Bauhaus plans with Dewey's *Art as Experience*."[78]

But whatever its similarity with American pragmatism, Morris continued the Vienna Circle's preoccupation with the reduction of all utterances to protocol sentences. In the New Bauhaus prospectus he insisted that "we need desperately a simplified and purified language in which to talk about art . . . in the same simple and direct way in which we talk about the world in scientific terms. For the purposes of intellectual understanding art must be talked about in the language of scientific philosophy and not in the language of art" (P, p. 10). By expanding the program of the *Wissenschaftliche Weltauffassung* to include art itself, Morris, in a sense, had made the project of the Aufbau and of the Bauhaus one and the same. Both now would find a common ground and unity in the foundations of the protocol sentences.

Even the two movements' ambition of producing a "new form of life" found resonance in Morris's hope that the "mentality of the scientists" would be incorporated into that of the artist. Given such a scientific artistic formation, Morris surmised, "Presumably no future Keats will arise from the New Bauhaus to drink a toast to the confusion of Newton for having destroyed the beauty of the rainbow." Rather it will be "the same man who seeks knowledge and a significant life, and it is the same world that is known and found significant. Art as the presentation of the significant and science as the quest for reliable knowledge are mutually supporting. Each supplies material for the other and each humanly enriches the other" (P, p. 10).

But there is another sense in which the two movements were "mutually supporting," as Morris dubbed their relation. Each legitimated the other. For the Bauhäusler, the Vienna Circle stood for the solid ground of science, the power of technology and the machine age. As such it gave their artistic movement a credence beyond that of taste or style. For the logical positivists, their association with the larger world of modern art certified them as progressive, and identified them with the future in a world in which their philosophical prospects were dim and their ties with traditional philosophy weak.

As the curriculum of the New Bauhaus began to take shape with Moholy-Nagy's backing, Morris fashioned his ideal of a unified course on art and science into a fundamental part of the New Bauhaus education program. In a course summary, prepared at the end of the first school year (1937–38), Morris reiterated his goals: "The treatment of

78. Morris, "The Intellectual Program of the New Bauhaus," unpublished typescript, 1937, folder 87, ID/UIC.

science was based on the study of the interrelationship of the terms of the various sciences; the aim was to show the unity of science by showing how all the terms of the sciences can be stated progressively on the basis of a few terms drawn from the everyday language. . . . We are now discussing the question as to how far art can be regarded as a language."[79]

It is in this context that Morris's published writing of the late 1930s should be understood. Above all, he hoped his course would, in his words, "give the verbal correlate of what, as I understand it, the Bauhaus is attempting to accomplish in practice."[80]

8. Conclusion: The Construction of Modernism

Morris's course at the New Bauhaus serves well as a summary of my principal thesis: the modernist construction of form out of elemental geometric shapes and colors is a correlate of the verbal development of theories out of logic and elementary bits of perception. Both artist and philosopher fastened on the simple and the functional; both sought to unify disparate domains through a common foundation. But what linked logical positivism and the Bauhaus went beyond mere structural parallels. The two movements drew on a common set of scientistic and machine-centered images; both called for their domains to be brought into accord with "modern methods of production." They were bound together through personal and familial relations, through Feigl's, Philipp Frank's, Reichenbach's, Carnap's, and Neurath's visits to the Dessau Bauhaus, through Josef Frank's collaboration with Neurath and his contribution to the Circle's lecture series, and through a complex process of mutual legitimation: the Vienna Circle bestowed an aura of scientificity on the Bauhaus and the Bauhaus conferred an image of progressivism and postwar reform on the Vienna Circle. If the Bauhäusler and logical positivists needed an external force to drive them even closer together, the anthroposophists and mystics did so in the beginning while the Nazis and nationalists served that purpose in the later period. In their common persecution, both movements rapidly became joined as heroes of internationalism and antifascism. By the time Moholy-Nagy set up the New Bauhaus in Chicago, it was overdetermined that he make common cause with the logical positivists in the creation of a scientifically and artistically modern form of life: logical positivism was in the form of life espoused by the Bauhaus, and the Bauhaus rationalization of the objects around us played a part in the form of life advocated by the logical positivists. Both were attempts to

79. Morris, "Intellectual Integration," unpublished typescript, n.d., F73-199; 1–2, ID/UIC.
80. Ibid.

117

interiorize an image of the machine world they saw on the outside, one through language, logic, and thought, the other through color, geometry, and architecture. Personal and collective forms of life would be reformed by the same means.

This process of interiorization took many forms, but above all the Bauhäusler and Vienna positivists of the late 1920s espoused a neutral stance modeled on their image of technology. Theirs was to be an *apolitical politics* (even when it was Marxist) predicated on organization, planning, and analysis. Here was ground on which Neurath could find common cause with the leaders of the Dessau Bauhaus. Similarly, Meyer and many of his colleagues pressed for an *unaesthetic aesthetics*, a move away from the decorative, historical, spiritual, or nationalistic toward the world of knowledge predicated only on a scientific orientation. Finally, the logical positivists urged a doctrine of an *unphilosophical philosophy*, a conception of the world of knowledge that would be predicated only on science. This triad of philosophy, politics, and aesthetics was grounded in a building up from clear, technical, first principles. Together these elements were supposed to form a joint enterprise; they were to be moments of the same drive toward a "modern" way of life, freed from ideology and grounded on a vision of the machine age, if not its reality.

If the left wings of the Dessau Bauhaus and the Vienna Circle made common cause in their espousal of a certain image of the machine and modernity, it does not mean that any commitment to machines and things technical was leftist, nor does it follow that the right-wing opposition was necessarily against technology. Quite the contrary. As Jeffrey Herf has argued eloquently in *Reactionary Modernism*, there were all manner of technology-embracing philosophies that glorified new means of transport, killing, and communication while denying reason an essential role in the conduct of individuals and society. What distinguished Carnap, Neurath, Meyer, Schmidt, and the other figures discussed here from right-wing technologists is the cultural significance they accorded technology. For the right, technology was part of a glorification of work, power, and domination. As one writer put it, technology was defined as the "'mobilization of the world through the *Gestalt* of the worker,'" where "'in the *Gestalt* lies the whole, which encompasses more than the sum of its parts.'" This whole meant that the symbols of technology—the hydroelectric dam, tanks, motorcycles— were to be considered as an inseparable part of a new authoritarian world order in which the technical was inseparable from the intentions and desires of the worker-soldier.[81] Though right and left shared a

81. Jeffrey Herf, *Reactionary Modernism: Technology, Culture, and Politics in Weimar and the Third Reich* (Cambridge, 1984), pp. 101–8.

picture of modernity embodied in technology, nothing could be further from a transparent Bauhaus lamp or the quasi-axiomatic image of philosophy that Carnap presented in his *Aufbau,* in which every action had its visible purpose and function. Technology, like modernism more generally, was coveted ideological ground.

Looking back at this modernist ambition from the present, a time in which modernism is being reexamined, we can no longer take for granted claims of neutrality. It is clear that many of the Bauhaus products were infused with a style that was not only independent of pure function but often impeded function. Similarly, with each passing year in the late 1930s, the belief that a purely technical approach to social problems could avoid politics also began to falter. As fascists, communists, and Christian Democrats fought it out in the 1930s, there was no demilitarized zone left for social, artistic, or philosophical neutrality. Even the realm of philosophy held no privileged position of neutrality. In the United States and England, where logical positivism came to rest, metaphysics (and antimetaphysics) no longer carried the political weight they had in the German-speaking world of the 1920s and 1930s. Torn from those roots, the ideal of a philosophy without metaphysics seemed ever more elusive as the years passed.

For a brief time, however, Carnap's ideal of "a single life" of artistic and scientific dimensions seemed possible. It was the dream of a world where a rational engineer could fashion not only the basis of philosophy and architecture but of the way of life that went with them. By coupling the Vienna Circle to this larger cultural effort, the "modern" in "modern philosophy" gains a sense deeper than merely a discontent with what came earlier. In subsequent years, some scholars have identified logical positivism as the arch foe of a progressive and holistic postmodernism. Others have defended the older logical positivism as a last vestige of Enlightenment thought against an obscurantist right. My own view is that the search for new directions in the philosophy of science must be integrated with a cultural and historical reassessment, not only of the logical positivists but of the antipositivist philosophical movements in the mid-1960s. I suspect that such a reappraisal would situate the antipositivists as solidly within, not outside, the particular modernist tradition of their "opponents." An alternative to modernism in the history and philosophy of science would, it seems to me, need to confront science as a variegated set of scientific practices that mesh together without a privileged foundational level either in observation or in theoretical assumptions. The characterization would be avowedly contextual: each strand of instrumental, theoretical, or experimental practice would be embedded in a wider cultural world. In such a picture the strength of the scientific enterprise would come not by building up from a privileged "foundation," but by the intercalated

FRIEDRICH STADLER

OTTO NEURATH—MORITZ SCHLICK: ON THE PHILOSOPHICAL AND POLITICAL ANTAGONISMS IN THE VIENNA CIRCLE*

In drawing the intellectual and personal profiles of Moritz Schlick and Otto Neurath, the two dominant adversary thinkers of the Vienna Circle, I present the outline of the physiognomies of two individuals as well as the inner physiognomy of a 'scientific community'. The example of these two strong personalities in this heterogeneous circle of scientists furnishes us with the minimum base of doctrinal consensus required to present some aspects of the pluralistic dynamic of theories which characterises this community; it shows the essentials of a 'psycho-sociogram' typical of all scientific communities; and it also provides evidence of the mixture of personal, world-view-related political, and philosophical-scientific elements.

My result relativises the myth (based on the combative programmatic pamphlet *The Scientific World-Conception. The Vienna Circle* (Carnap, Hahn, Neurath 1929)) that the Vienna Circle represents a closed anti-metaphysical school of philosophy, and serves to criticise the ahistorical perspective which describes neo-positivism as a traditional philosophical movement—similar to neo-Kantians and neo-Wagnerians—that was possessed from its inception by a tendency towards self-dissolution and showed disintegrative scientific communication amongst its only contingent forms of organisation (*Verein Ernst Mach, Erkenntnis*).

The following consideration of internal and external factors can sketch the specifically collective aspect of the Vienna Circle, which points beyond its description as a loose scientific community, but only in suggestive contours. The extreme positions in this grouping are here drawn very pointedly, partly because of the remarkable conclusion that there *did* exist what was demonstrated to the outside despite all of the personal and doctrinal differences—namely, the consciousness of representing a common movement which, from the beginning to the end of the Circle, prefaced all divisions by the credo of being an open experimental forum for discussion of the scientific world-conception.

T. E. Uebel (ed.), Rediscovering the Forgotten Vienna Circle, 159–168.
© 1991 *Kluwer Academic Publishers. Printed in the Netherlands.*

In his *Allgemeine Erkenntnislehre* (*General Theory of Knowledge*, 1918, 2nd. ed. 1925) up to the mid-1920's, Moritz Schlick[1] owed allegiance to non-positivist critical realism—as it has been advocated since by Herbert Feigl, Viktor Kraft, also by Karl R. Popper—but from 1925/6 he presented, with Friedrich Waismann, the strongly Wittgensteinian wing of the Circle with his 'Turning Point in Philosophy' (1930a). It is not the case, however, that one should speak of a congruence of the positions of Schlick and Wittgenstein, as can be seen from the example of their conceptions of philosophy later in the 1930's and their divergent conceptions of ethics and aesthetics. Schlick presented a (methodological) phenomenalism like that of Carnap in his *The Logical Structure of the World*. This occurred only *after* his critique of metaphysical school-philosophy at the height of the second scientific revolution at the beginning of the 20th century; there he summarised the results of the natural sciences within the framework of a programmatic conception of philosophy representing a system of most general principles. In this development, philosophy became a therapeutic activity of linguistic analysis of statements of the individual sciences, an activity designed to overcome metaphysics and to clarify the meaning of statements with the all too well-known and misunderstood dictum that the meaning of a statement lies in its method of verification. So we must conclude somewhat sketchily that Schlick changed, via the 'linguistic turn' à la Wittgenstein, from his early realistic position to a loosened and liberalised standpoint towards the end of his life. The latter found expression in his little noted essay 'The Vienna School and Traditional Philosophy' (1937) in which he distanced himself from the concept of Neurath's radical physicalism as a strictly anti-metaphysical, 'non-philosophical' unified science (later, encyclopedism).

Analogously we can similarly distinguish stages in the development of Neurath[2]—though with greater continuity of content and aim—from the programmatic pamphlet (co-)written by him (Carnap, Hahn, Neurath 1929)—which, as is known, was criticised by Schlick together with Wittgenstein as simplistic and propagandistic—up to the historically referenced self-portrait of logical empiricism in his essay on *The Development of the Vienna Circle and the Future of Logical Empiricism* (1936a). In the latter, the program of 'unified science' becomes the widened encyclopedic program, and the Vienna Circle or the Unity of Science is classified as a movement—provoking amongst other things Schlick's just mentioned repositioning of the Vienna Circle. From the relativising of the meaning criterion by the principle of tolerance (introduced by Karl Menger) on up to his death, Schlick still accorded a positive and constructive role to philosophy—a role which extended to ethics—whereas throughout his life Neurath was convinced of the dissolution of philosophy as an autonomous discipline into the comprehensive empiricist system of encyclopedism. In this, he unintentionally remained indirectly 'philosophical' or philosophically relevant, as can be seen by considering his work (Rutte 1982a [ch. 6], Haller 1982a [ch. 9]).

Neurath recognised the doctrinal divergences in the Circle and described them already at an early stage, still before the controversies around the 'problem of the basis'; nevertheless, he fortified its organisation and stressed in terms of content what was common to the neo-positivist movement against the outside. With certain reservations this harmonising tendency also holds of Schlick, for his internal criticism was objectively blunted to a large degree by his public-institutional role and function, as in his activities in the Verein Ernst Mach, and in particular his struggle against its dissolution in February 1934 (Stadler 1982b, pt. 2). One year earlier Neurath stated somewhat too optimistically and euphorically (and from today's view, wrongly):

But while Schlick came from 'realism', Frank as well as Hahn and Neurath were already 'free from philosophy' ... Wittgenstein provided the 'Circle,' which has been meeting around Schlick during the last few years, with a powerful stimulus and a fertile ground for discussion by forcing it to take issue with many problems. But it is possible to accept Wittgenstein's theory of truth and truth-functions and his extremely fruitful approach to the radical analysis of language and nevertheless reject decisively and without reservation his attempt to legitimize at least provisionally some form of idealistic, even mystical metaphysics in an indirect way, via preliminary elucidations. ...

Neurath concludes with some Utopianism that he has

taken the radical position within the framework of the 'Vienna Circle' that one cannot make even preliminary remarks in anything other than a physicalistic language. Any discussion of whether "language or the world is more complex" must be dismissed in advance—or, in short, any confrontation between 'world' and 'language,' such as the one that Wittgenstein suggests. To formulate this as strictly as possible, "*sentences can be compared only with sentences*" ... It seems that this uncompromising attitude to the basic metaphysical sentences of Wittgenstein's *Tractatus* is becoming more and more prevalent, and that it will be possible to come to some kind of agreement about the thesis of 'protocol sentences' as the basis of *logical empiricism*, which is opposed to idealistic half-measures and others like them. (Neurath 1933a, fn. 2 (pp. 274/5).)

It is known that this antagonism was sharpened in 1934 in the debate between Neurath's coherence and Schlick's correspondence theory (cf. Hempel 1981), and that Neurath's hope for a universally accepted 'materialistic' (read: physicalistic) solution of the problem of the empirical foundation of scientific knowledge and truth remained unfulfilled. His vision of harmony had failed and the basis problem of empiricism remained, together with the concepts of physicalism and unified science, one reason for the thorough-going pluralism in the Vienna Circle up until Schlick's death.

Neurath's consequential and unyielding position becomes intelligible only in a greater historical and developmental context, which shows the stunning continuity of his thought and action (Stadler 1982d). Through the influence of his father, his intellectual socialisation was formed by liberalist, social-reformist, economical and visual elements which raised the central question of a just social order; this was to be reached by 'social engineering' through planned economy, socialisation, social reform, picture language, and the scientific world-conception. Working in this sprit, Neurath could not but aim for a conception of enlightenment which comprehended theory and practice. This conception manifested itself variously in his organising, in his scientific and in his popular-educational work in the Vienna Circle, the Verein Ernst Mach, the Social and Economic Museum, and abroad in the Unity of Science movement: we can trace a developmental line from models of scientific philosophy, 'scientific world-conception', to encyclopedism in signs and language.

Neurath criticises the opposition to the scientific enlightenment by means of his functionalist concept of anti-metaphysics (i.e. according to varying analyses of society), a concept which found its positive doctrinal and methodological transformation in the 'unified science' of the early 1930's. It seems possible to demonstrate a systematic connection between his rational-empirical historicism as an epistemological principle and his holistic coherence-theoretical relativism in the theory of science. In one of his last self-portraits Neurath described for us how strongly he was impressed until the end of his life by the idea of collective work and planning in society and science for the purposes of humanisation and democratisation:

I also stress planning for freedom. I mean by planning for freedom orchestration of variety. Others may prefer other programs. But none can use Logical Empiricism to ground a totalitarian argument. It conceals not a loophole for dogmatism. Pluralism is the backbone of my thought. Metaphysical attitudes lead very often to totalitarianism, but I do not know of any consistent logical empiricist who came thereby to a totalitarian view. (Quoted after Kallen 1946, p. 533.)

The intellectual biography of the grand-bourgeois Moritz Schlick, who, on a superficial view, was also indebted to 'this-worldly' anti-metaphysical enlightenment thinking, was rather different. Coming from an aristocratic background, trained in the natural sciences, this favoured student of Max Planck and friend of Einstein was pre-destined to criticising the speculative-idealistical school philosophy, becoming one of the first who, with Einstein's agreement, approached relativity theory philosophically and who first presented scientific philosophy in his *General Theory of Knowledge* at the end of the World War I. In contrast to Neurath, Schlick always remained within a domain of scientific discourse which he understood to be autonomous. Philosophical and world-view/political differences could not but result from this contrast.

Though I cannot go further into the the Vienna Circle's problem of empirical foundations, I may, however, give a cursory pointer towards the difference between Neurath's Machian, syntactically oriented 'relativism' and Schlick's semantically oriented 'absolutism'. This opposition, characterised in these terms by Neurath himself, also involved the definition and function of anti-metaphysics in the elimination of idealistic speculation from the domain of science which was intended by both. These antagonisms find open and occasionally polemical expression in the correspondence of Schlick and Neurath. Neurath classified Schlick's 'On the Foundation of Knowledge' (Schlick 1934; the opposing position: Neurath 1934) as "philosophical absolutism", "rather worrying" and "sadly mystical" and mocked the "Wittgensteinery" of the "Schlick-sect" in their criticism of ethics[3], whereas Schlick complained, with allusion to Neurath, about the dogmatism of some older members, about the "anti-intellectual character" of some publications, about Neurath's terminology of a "right" and "left wing" of the Vienna Circle, and about the "unscientific and unserious" character of Neurath's book *Empirical Sociology*: "If one declares triumphantly on nearly every page that one can do without God and the angels, then this is very boring for comrades-in-arms, appears dogmatical to opponents, and is ridiculous for both."[4]

Quite apart from the fact that here the well-known personal animosities of the finely attuned, aesthetic Schlick against the combative, vital Neurath found expression, there also was another difference which cannot be discounted, namely, how the envisaged aim of a 'consistent' or 'logicising empiricism' was pursued. For Neurath, the levelling of philosophy and science was to help in achieving a physicalistic (behaviourist, materialistic) unified science and a corresponding encyclopedism, whereas for Schlick a scientific philosophy could coexist in parallel with science. What is decisive for Neurath in this context is the holism and fallibilism of the theory of science and a pragmatically and sociologically understood conventionalism which points to the social conditions of scientific communication and signals the necessity for a synthesis of internalist and externalist approaches—long before the works of Kuhn and Feyerabend. The community of scientists must itself therefore become a scientific problem which cannot but escape the (absolutist) 'pseudo-rationalism' (Neurath 1935b).

Here we touch on their world-view *cum* political differences. Neurath demanded the theoretical and practical unity of thought and action and—like Philipp Frank and Edgar Zilsel—joined the history of science with the sociology of knowledge and of science, whereas Schlick endorsed an autonomist conception of philosophy and science which, complemented by 'questions of life and meaning', amounts to a scientific ethics and a traditional philosophy of culture and society, and correspondingly, to the liberal-bourgeois understanding of politics in the sense of 'lowly' party and day-by-day politics.

A shortened social analysis and political abstinence determined Schlick's aesthetic-moral perspective on socio-political conditions, as can be seen from his posthumously published small book *Natur und Kultur* (Nature and Culture, 1952) which was written against German National Socialism. What becomes transparent in Schlick's self-understanding, in his vocabulary and in his practical proposals for solutions, is a laudable humanism, pacifism and cosmopolitanism, anchored—unlike Neurath's social eudaeumonism—in an individual-hedonist picture of society. The enlightenment postulate of cultural development and the increased sensibility of customs and consciousness thus lead Schlick—with a rigorous criticism against Spengler—to an abstract eudaemonism and libertarianism characterised by an anti-state liberalism which operates with ideal-type constructions like 'state' and 'economy' without references to empirical-causal results. The state merely has the task of the 'protection of life', even if the latter may lack the 'morality' necessary for a happy life. This idealistic understanding of politics by a liberal intellectual in the wake of Europe's descent into fascism could not but lead to the kind of helpless anti-fascism that is characteristic of his proclaimed moral-voluntarist conception of the state. Starting from this world-view with the motto "As little state as possible!" (1962, p. 42) Schlick inevitably arrived at a plea for a political '*Laissez fair*': "Liberalism is the only political form of thought which is adapted to the modern form of life determined by communication and technology." (Ibid., p. 47.)

126

Against this, Neurath urged the application of the socialist idea of planning on all levels of theory and practice—from war economics, total socialisation, the settlement movement and trade union movement, all the way to the idea of planning for freedom in a humanist world society in which a democratic science was to function according to the form of non-capitalist and collective production process.

This background makes intelligible Schlick's vain struggle against the dissolution of the Verein Ernst Mach, his surprising declaration of loyalty to the Austria of Dollfuss and his critique of the repression of academia in an étatist, authoritarian fashion (Stadler 1979a [ch. 5]), as well as his pointed remarks and confessions in his correspondence. This background also makes intelligible the disappointment of the emigré Neurath who expressed his regret over the role "for Dollfuss against unified science" taken by Schlick (who stylised the Dollfuss strategy as a bulwark against National Socialism).[5]

There also is no more surprise in Neurath's sharpening of the dualism of empiricism and metaphysics (the latter with reference to Plato and Hitler) and in his praise of the time of the Circle as the best period of co-operation despite these disappointments.[6] In the Social and Economic Museum, the Verein Ernst Mach, the Vienna Circle, and as the organiser of unified science in exile, the enlightenment worker, encyclopedist and social reformer Neurath was unable to separate science and politics (given his more concrete political consciousness and the diction of the workers movement), so that the 'publicity work' (viewed sceptically by Schlick) became a necessary correlate to the 'academic playing field'.

And although according to his 'Lost Wanderers of Descartes' (Neurath 1913b) it was the case that, due to the dualism of fact and value, political actions could not be derived as norms from the scientific world-conception, Neurath nevertheless aimed at a concerted action of scientific work and social practice: with monism as the means of empiricist communication and pluralism as an attitude to the development of hypotheses, this aim amounted to a uniform rational-empiricist program.

Schlick's aristocratic context in the area of academic communication and in his private circle—against propagandistic positions with the bourgeois values of value-free science, the Socratic ethos of clarity, cleanliness and love of truth—collided with Neurath's impulsive, spontaneous spirit and organisational vehemence in the context of the Viennese adult education movement. As promoter, innovator and fiery speaker at the 'base' between proletariat and *petit bourgeoisie*, the polyhistor Neurath demonstrated the practical cosmopolitanism of his wide interests in literature, art, politics and architecture through his inter-disciplinariy orientation and his concrete utopias, his popularisations by concrete example, situated between encyclopedism and social reform, like ISOTYPE, 'universal slang', and 'unity of science'.

If in conclusion we must take note of a number of personal and doctrinal differences between Schlick's and Neurath's paradigmatic profiles (which are exemplary for the pluralism in the Vienna Circle), then we must also evaluate the objective dimension of the Vienna Circle—over and above, that is, the commonalities which from their personal perspectives made up for the disagreements. It is in this objective dimension that the lasting influence of the neo-positivist movement lies, after all.

Commonalities show up in their opposition to Plato, Spengler, Kant, in their this-worldly empiricist ethical and scientific attitude informed by criticism of language, pluralism of theories, monism of cognition and method as based on the natural world-conception, common public work in the Verein Ernst Mach despite their internal differences, dualism of facts and values, their nominalism, hypothetico-deductive theory construction, and in the world-view-related and political embedding of rationalism and empiricism within the cultural syndrome of the 'late enlightenment' (Stadler 1981). Both thinkers were in their own way dominant figures in the Vienna Circle and in the Verein Ernst Mach, representing the maximum of differences and a still cohesive minimum of a common consciousness and public self-understanding in their publications and institutions. The latter served to translate their basic values and aims—like science as providing a basic attitude (empiricism, logic, and criticism of language)—into the framework of an implicit program of enlightenment.

Schlick made it possible for the Vienna Circle to gain an academic platform under difficult conditions in the University of Vienna, whereas Neurath's cooperation shows not only in his original scientific work, but also in his work as organiser and populariser in Austria and abroad, in particular for the International Congresses for the Unity of Science. Schlick's personal achievement was the natural scientific and logico-linguistic founding of a scientific philosophy and world-conception based on the natural sciences, against the 'chaos of philosophical systems'. And even though his early tragic death prevented a comprehensive conclusion to his work, his scientific ethos and spirit, his personal authenticity and his professional competence bestowed upon Schlick the central role of *doyen* in the Vienna Circle and allowed him to promote the international reputation of the 'Vienna School' as its ambassador abroad.

Of course, despite the 'Turning Point in Philosophy' which he initiated, Schlick remained a genuine 'philosopher' (see Schlick 1937). In his last retrospective of the neo-positivist movement he drew—against Neurath's 'unified science'—the harmonious picture of an anti-metaphysical, but essentially philosophical movement which employs anti-metaphysics only as a methodological principle in the elimination of philosophical errors. (It is noteworthy here, however, that despite all of his internal criticism Schlick did not renounce the collective concept of a philosophical-scientific movement.) The distinction between 'questions of fact' and 'question of meaning' allowed him to determine the different domains of science and philosophy: the philosopher seeks to clarify the meaning of our statements; the scientist attempts to decide their truth. In the end, Socrates is presented as the true father of philosophy.

Schlick postulated ethics as an equally genuine philosophical task, namely, the clarification of ethical concepts as a more important task than the solution of theoretical problems. With everyday and common sense philosophy understood in that way—oriented contrary to the scientifically narrowed conception of logic—Schlick indirectly meets again with Neurath's encyclopedism, which argued against metaphysics as a relevant political ideology between the green and brown fascisms, but without the help of a genuine philosophy besides or above the individual sciences.

For Schlick, there was no abstract frame of reference like that given by a pluralistic encyclopedia as an open research program, but there was the wholly concrete perspective of democratic science in the socio-political climate of contemporaneous English society. Due to his subjectivist self-understanding as an apolitical thinker, the strictly anti-Nazi 'stay-at-home' Schlick had to sublimate his moral position in a philosophy of culture and society and express it as a timeless philosophical-literary reproach. His dichotomy of 'experience' (*Erleben*) and 'cognition' (*Erkennen*) demanded a high price in comparison to pragmatic-empirical historicism and relativism, namely, that of projecting political forces into a domain that appeared to render them rationally decidable, the domain of philosophical and scientific discourse—with the consequences of neutralising practical action and indifference against everyday politics.

Yet if we consider the revaluation of the Vienna Circle which has begun in the last few years and was stimulated by the Schlick-Neurath centenary—a historical-critical inventory which followed in the wake of the emotional and mythologised 'positivism dispute' and the static criticism of the 'received view'—then Neurath's prognosis, made in 1938 to Herbert Feigl, happily seems to come to true: "Commonalities will remain; differences will pass as temporary phenomena."[7]

Universität Wien

NOTES

[*]   First published as 'Otto Neurath-Moritz Schlick: Zum philosophischen und weltanschaulich-politischen Antagonismus im Wiener Kreis', in *Schlick und Neurath. Ein Symposion*, Hrsg. R. Haller, *Grazer Philosophische Studien* 16/7 (1982) 451-463, © 1982, Rodopi, Amsterdam. Translated with kind permission of Editions Rodopi and the author by T. E. Uebel.

[1]   On Schlick's life and work see: Schlick 1979a, 1979b; Feigl 1937; Waismann 1938; Zilsel 1937; Rutte 1976, 1977b; Juhos 1963, Parthey/Vogel 1969; Wheeler 1969; Stadler 1982e.

[2]   On Neurath's life and work see: Neurath 1973, 1979, 1981, Hegselmann 1979a, Fleck 1979, Nehmet 1981, Stadler 1982a.

[3]   Neurath's letters to Carnap, 7 May 1934, 14 May 1934, 17 June 1934, 13 December 1934, 18 January 1935, 22 September 1935. All in the Vienna Circle Archive, University of Amsterdam. (The publication of the Neurath-Carnap correspondence is in preparation.)

[4]   Schlick to D. Rynin, 4 November 1933, to W. Köhler 13 March 1934, to R. Carnap 20 January 1935, to P. Frank ca. 1930. Vienna Circle Archive, Amsterdam.

[5]   Neurath to Carnap , 18 July 1934, Vienna Circle Archive, Amsterdam.

[6]   Neurath to Carnap, 1 July 1936.

[7]   Neurath to Feigl, 1938, Vienna Circle Archive, Amsterdam.

BJHS, 1994, 27, 153–75

Planning science: Otto Neurath and the *International Encyclopedia of Unified Science*

GEORGE A. REISCH*

INTRODUCTION

In the spring of 1937, the University of Chicago Press mailed hundreds of subscription forms for its latest enterprise – a projected series of twenty short monographs by various philosophers and scientists. Together the monographs were to form the first section of the *International Encyclopedia of Unified Science*. Included in each mailing was an introductory prospectus which began:

> Recent years have witnessed a striking growth of interest in the scientific enterprise as a whole and especially in the unity of science. The concern throughout the world for the logic of science, the history of science, and the sociology of science reveals a comprehensive international movement interested in considering science as a whole in terms of the scientific temper itself. A science of science is appearing. The extreme specialization within science demands as its corrective an interest in the scientific edifice in its entirety. This is especially necessary if science is to satisfy its inherent urge for the systematization of its results and methods and if science is to perform adequately its educational role in the modern world. Science is gradually rousing itself for the performance of its total task.[1]

Charles Morris, a philosopher at the University of Chicago, wrote the prospectus announcing this 'science of science'. He and Rudolf Carnap, then also at Chicago, were the *Encyclopedia*'s associate editors while Otto Neurath was Editor-in-Chief. Neurath outlined the prospectus for Morris and would probably have liked to write it himself. The *Encyclopedia* was his brainchild and he had been vigorously organizing the project for several years. Yet no sooner had Neurath rallied the University of Chicago Press to undertake this ambitious project, than he was off to Mexico extending further this 'comprehensive international movement' through another of his enterprises – ISOTYPE, his International System of Typographic Picture Education. For Neurath, science's total task included both its development as an integrated whole, as 'unified science', and also the spread of an international, empirical and scientific sensibility. The *Encyclopedia* would help to develop unified science, while his ISOTYPE exhibits would promote this

* Department of Humanities, Illinois Institute of Technology, Chicago, IL 60616-3793, USA.

I would like to thank Loren Butler, Jordi Cat, Hasok Chang, the IIT philosophy workshop, Robert Richards, Howard Stein, Thomas Uebel and two anonymous referees for suggesting ways to improve earlier drafts of this paper.

1 C. Morris, 'Foundations of the Unity of Science', prospectus, University of Chicago Library, Department of Special Collections, 'University of Chicago Press Papers', Box 347, Folder 2.

131

Figure 1. Examples of ISOTYPE's visual dictionary and grammar. The symbol for 'shoe' composed with that for 'factory' yields the symbol for 'shoe factory', and so on. From, O. Neurath, 'Visual education', in Neurath, *Empiricism and Sociology* (ed. M. Neurath and R. S. Cohen; tr. P. Foulkes and M. Neurath), Boston, 1973, 225.

Wissenschaftliche Weltauffassung: his charts and diagrams depicted economic and demographic facts important to Marxist sensibilities like Neurath's and they conveyed this information in a culturally independent way, ideally to even non-literate people. Ultimately, Neurath hoped, unified science would be a tool for a scientifically minded world society to plan and manage its operation and development. Morris probably thought of Neurath in Mexico as he concluded the prospectus, asking all those who support the 'spread of the scientific habit of mind'[2] to offer their aid, that is their money, to the new *Encyclopedia*.

2 Morris, *op. cit.* (1).

Today, the *Encyclopedia* is perhaps less well known than Neurath. Along with Carnap and Moritz Schlick, he was a central member of the Vienna Circle of philosophers, in which he cemented his reputation in the 1920s and 1930s as an arch-foe of metaphysics. He had an imposing physical presence as well. Karl Popper recalled 'a big, tall, exuberant man with flashing eyes, a big red beard, and a loud voice... who would not look behind him or, when rushing ahead, care very much about whom his big stride might knock down'.[3] Nor, it would seem, did he ever stop working. He was an economist by training yet, by the time of his death in 1945, had published essays in logic, philosophy, history of science, social and economic planning, and education theory. He was president of the Central Planning Office for Bavaria's short-lived socialist government in 1919;[4] founded museums in Vienna for social and economic information and for housing and town planning; produced and promoted his ISOTYPE[5] (see Figure 1); and in the 1930s he organized a series of International Congresses for the Unity of Science as well as the *International Encyclopedia of Unified Science.*

Because of all this, Neurath is remembered as an activist and 'indefatigable organizer'.[6] As the current 'rediscovery'[7] of Neurath shows, however, his physical girth and

3 K. Popper, 'Memories of Otto Neurath', in O. Neurath, *Empiricism and Sociology* (ed. M. Neurath and R. S. Cohen; tr. P. Foulkes and M. Neurath), Boston, 1973, 52.

4 O. Neurath, 'Experiences of socialization in Bavaria' in Neurath, op. cit. (3), 18–28.

5 Neurath was not a graphic artist. He articulated the grammar, syntax and goals of ISOTYPE and had others design appropriate figures. For more on ISOTYPE, see O. Neurath, *Modern Man in the Making*, New York, 1939; *International Picture Language*, Psyche Miniatures no. 83, London, 1936; chapter 7 in Neurath, op. cit. (3), 214–48; K. Müller, 'Neurath's theory of pictorial-statistical representation', in *Rediscovering the Forgotten Vienna Circle* (ed. T. Uebel), Boston, 1991, 223–51.

6 J. Joergensen, 'The development of logical empiricism', *International Encyclopedia of Unified Science*, Chicago, 1950, ii (no. 9), 43.

7 Neurath's 'rediscovery' began in 1973 with the translation and publication of some of his essays in Neurath, op. cit. (3). A collection of his complete methodological and philosophical writings was published in 1981, some of which soon appeared in English (*Gesammelte Philosophische und Methodologische Schriften* (ed. R. Haller and H. Rutte), 2 vols., Vienna, 1981; *Philosophical Papers: 1913–1946* (tr. and ed. R. S. Cohen and M. Neurath), Boston, 1983 (herein after *Philosophical Papers*)). Austrian historians and philosophers have contributed most to the explication of Neurath's ideas. Some of these essays have been recently translated and collected by Thomas Uebel (op. cit. (5)). Uebel describes how Neurath's reputation as 'the original – confused – neo-positivist caveman' is being replaced with that of 'a serious thinker, one whose belated recognition is bound to transform the picture of the past of analytical philosophy' (5, 3; see also 10–14).

For an extended analysis of Neurath's epistemological views and their development, see T. Uebel, *Overcoming Logical Positivism from Within*, Atlanta, 1992.

There are no published studies on the *Encyclopedia* in particular. Zolo discusses it at some length in his study on Neurath and identifies its cooperative, democratic ideal (D. Zolo, *Reflexive Epistemology: The Philosophical Legacy of Otto Neurath* (tr. D. McKie), Boston, 1989, especially ch. 5). Nemeth compares Neurath's unified science to his vision of a planned, non-monetary economy in kind but stops short of identifying the logos of economic planning with that of unified science and the *Encyclopedia* (E. Nemeth, 'The unity of planned economy and the unity of science', in Uebel, op. cit. (5), 275–83). While some come close, none of these studies sufficiently emphasize and articulate the role planning plays in Neurath's conceptions. Friedrich Stadler, for instance, notes that Neurath

urged the application of the socialist idea of planning on all levels of theory and practice... all the way to the idea of planning for freedom in a humanist world society in which a democratic science was to function according to the form of non-capitalist and collective production processes ['Otto Neurath – Moritz Schlick: on the philosophical and political antagonisms in the Vienna Circle', in Uebel, op. cit. (5), 159–68, on 165].

enthusiasm were matched by an extraordinarily broad set of social, epistemological and historical beliefs and goals. His many writings give the clear impression of an intellectual socialist reformer, one eager to purge philosophy and the sciences of metaphysics and to democratize and disseminate scientific knowledge. In the details of this ambitious agenda for *Aufklärung*, however, Neurath's thought becomes murky. If, as Popper tells us, he walked quickly, without regard for whom he might knock down, he wrote the same way, often without regard for those unable to catch his many subtle allusions and conclusions, the premises of which hide between the lines. Since many of his ideas appear today as unique and original anticipations of Quinean and Kuhnian ideas, the task of clarifying them is only more urgent.[8]

This paper concerns three aspects of Neurath's work. First, it aims to reconstruct Neurath's conception of unified science and its *Encyclopedia*, in particular, their epistemological underpinnings and his evolving view of how the sciences were to become unified. Secondly, my account bears on the more difficult question of how Neurath's various political and scientific interests and activities cohere. Although he saw unified science as a tool for socialization and a widespread, scientific world-view as an antidote against political obscurantism, there is a closer connection between Neurath's concepts of social reform, on the one hand, and scientific reform, on the other: both rely on *planning*. For Neurath, planning – of social orders, economies, or unified science – was a simple two-step strategy for problem solving: first, find or construct all possible ways in which a goal might be reached; secondly, choose one of these options and pursue it. The technique is perhaps too simple to deserve mention, but in Neurath's day it was a part of a modern and rational approach to life, especially when scientific tools were brought to the task of exploring these possibilities. Not only would unified science be a tool most suitable for social and economic planning, this unification of the sciences *itself* would be democratically planned through the forum of the *Encyclopedia*. Finally, I shall document how Neurath wrestled – successfully, I think – with the criticism that unified science and its *Encyclopedia* were essentially undemocratic and totalitarian.

More abstractly, Uebel characterizes Neurath's project as one of 'controllable rationality' and 'conceptual responsibility by collective management' (op. cit. (5), 9, 10). To understand more concretely Neurath's vision of these 'collective production processes' and the mechanisms of this 'collective management' is, I suggest, to understand the *Encyclopedia* as the medium of planning in science. I developed this thesis following Marx Wartofsky's discussion of the intimate connection between Neurath's non-foundationalism and his vision of the unified scientist as 'a servant of the political and social decision making process'. 'What is at issue', in Wartofsky's discussion of the Vienna Circle, 'is the ways in which the "internal" ideational-philosophical content [of the Vienna Circle's scientific philosophy] bears upon the "external" social content and role of the movement' ('Positivism and politics: the Vienna Circle as a social movement', in *Schlick und Neurath – Ein Symposion* (ed. R. Haller), Amsterdam, 1982, 79–101, on 100, 93). This relation, I suggest, is nearly identity: for Neurath, unified science was scientific planning much like the social and economic planning that would become *de jure* in the new enlightenment.

8 Neurath appears to have anticipated, for example, certain aspects of (1) the non-foundational, holistic view of theories popularized in the 1950s and 1960s by Quine and Kuhn; (2) a synthetic historical and philosophical approach to understanding theories and scientific change associated usually with Kuhn; and (3) a philosophical naturalism aligned with contemporary naturalistic and evolutionary epistemology. Uebel treats Neurath's anticipation of (3) in T. Uebel, 'Neurath's programme for naturalistic epistemology', *Studies in History and Philosophy of Science* (1991), 22, 623–46. See also note 29, where I list several important ways in which Neurath's views were unlike Kuhn's.

UNIFIED SCIENCE AND THE *ENCYCLOPEDIA*

The Vienna Circle announced itself to the world in its famous pamphlet of 1929, *Wissenschaftliche Weltauffassung: Der Wiener Kreis*.[9] With Carnap and mathematician Hans Hahn, Neurath described how the anti-metaphysical sensibilities of the Circle pointed toward this goal of unified science: 'The endeavour is to link and harmonise the achievements of individual investigators in their various fields of science.' In his characteristically dense prose, Neurath hurriedly described what was involved:

> From this aim follows the emphasis on *collective efforts*, and also the emphasis on what can be grasped intersubjectively; from this springs the search for a neutral system of formulae, for a symbolism freed from the slag of historical languages; and also the search for a total system of concepts. Neatness and clarity are striven for, and dark distances and unfathomable depths rejected. In science there are no 'depths'; there is surface everywhere; all experience forms a complex network, which cannot always be surveyed and can often be grasped only in parts. Everything is accessible to man; and man is the measure of all things.[10]

Scientific cooperation, axiomatic systems, empiricist epistemology and Protagoras all figured in Neurath's conception. Most of these ideas find place in a two-dimensional conception of unified science. First, there was a horizontal component which linked the sciences to each other. Since 'all experience forms a complex network', Neurath reasoned that the sciences should also be connected. '"Making predictions" is what all of science is about', Neurath explained, and,

> it must be possible to link the laws of all sciences with each other to make *one* definite prediction. One can only know whether a certain house will burn down if one can take into account how the building components behave, how the human groups behave who push on to fight the fire. The various scientific disciplines together make up the 'unified science'. It is the task of scientific work to create unified science with all its laws.[11]

Because the sciences must work together in this way, they must belong to one structure, a 'physics in its largest aspect', Neurath explained, 'a tissue of laws expressing space-time linkages – let us call it: *Physicalism*'.[12]

'Physicalism' introduces the other, vertical axis of unified science. Since the sciences would be linked together by utilizing a common language, a 'physicalist' language, they would each gain a certain continuity with physics. As everyday language, purged of metaphysical, unempirical terms, this language was not that of physics itself. But, since all physicalist statements would 'contain references to the spatio-temporal order, the order that we know from physics'[13] they would be transformable into, or replaceable with, the language of physics. In the spirit of Carnap's *Der Logische Aufbau Der Welt*[14] – in which

---

9 O. Neurath, 'The scientific conception of the world: the Vienna Circle', in Neurath, op. cit. (3), 299–319. Here, Marie Neurath notes that Otto wrote this paper and later edited it with Carnap and Hans Hahn (p. 318).

10 Neurath, 'The scientific conception of the world', in Neurath, op. cit. (3), 306.

11 O. Neurath, 'Physicalism' in *Philosophical Papers*, op. cit. (7), 52–7, on 53–4. Neurath's favourite illustration of this point is a forest fire. See 'Sociology and physicalism', in ibid., 58–90, on 59; 'Individual sciences, unified science, pseudorationalism', in ibid., 132–8, on 132–3.

12 O. Neurath, 'Physicalism: the philosophy of the Vienna Circle', in *Philosophical Papers*, op. cit. (7), 48–51, on 49.

13 Neurath, 'Physicalism', in *Philosophical Papers*, op. cit. (7), 54.

14 R. Carnap, *The Logical Structure of the World* (tr. R. George), Berkeley, 1969.

Carnap attempted to construct such a 'neutral system of formulae' in which all scientific concepts would find place – Neurath explained that 'I see blue', for instance, could be explicated in terms of statements about physiology and electromagnetic oscillations.[15] In general, Neurath felt that unified science would allow any two scientists to speak to each other, and especially to the physicists.

In the early 1930s, however, Neurath's vision of unified science began to change as he increasingly emphasized this horizontal component of unity. In an essay of 1931, for instance, he alluded to Carnap's work less as an example of vertical unity and more as a model of horizontal consistency and 'harmony' among the sciences. He proposed an 'empirical sociology' as a 'linking of history, political economy and other social sciences', explaining that

> we must create a system of concepts which provides the possibility of deriving concepts from each other, like a pyramid (cf. Carnap, *Konstitutionssystem*)... Indeed, our linking of history, political economy and other social sciences presupposes that a group of concepts may be articulated that are in harmony with each other.[16]

In later essays, Neurath further distanced unified science from this image of a pyramid.[17] The sciences would be coordinated and unified, but none, he felt, should be considered more fundamental than others. Although individual sciences might evolve toward hierarchical axiomatization, and the task of logically reducing some sciences to others was worth while, unified science as a whole, Neurath felt, should be conceived not as a pyramid but as a 'mosaic'.[18]

Among the reasons for this shift was Neurath's long-standing critique of epistemological foundationalism. I shall show later how this image of a pyramid of sciences, where those above are dependent on, or less fundamental than, those below, was out of step with this distinctive component of Neurath's thought. Equally important, however, was the support Neurath found for producing an actual encyclopedia of the sciences. If, as he had suggested earlier, the task of arranging the sciences in such a hierarchical pyramid required some 'system of concepts' or a 'neutral system of formulae', none had appeared.[19] For an

15 Neurath, 'Physicalism', in *Philosophical Papers*, op. cit. (7), 55; see also 'Sociology and physicalism', in ibid., op. cit. (7), 63–4, 72. These comments suggest the spirit – and not the letter – of Carnap's *Aufbau* since they depict the terms of all sciences as reducible to (in some unspecified sense) the language of physics instead of constructible out of (not the language of physics, but) 'elementary experiences', as Carnap called them. Although Carnap allowed that physicalist statements could be used as a basis in such constructions (op. cit. (14), §62), he came to agree with Neurath that the physicalist basis should be preferred. R. Carnap, 'Intellectual autobiography', in *The Library of Living Philosophers* vol. 11, *The Philosophy of Rudolf Carnap* (ed. P. A. Schilpp), LaSalle, Ill. 1963, 3–84, on 50–2.

16 O. Neurath, 'Empirical sociology', in Neurath, op. cit. (3), 319–421, on 390.

17 O. Neurath, 'Departmentalization of unified science', in *Philosophical Papers*, op. cit. (7), 200–5, on 203–5; see also Neurath, 'The social sciences and unified science', in ibid., 209–12, on 211.

18 Neurath, 'Departmentalization of unified science', in *Philosophical Papers*, op. cit. (7), 204; see also O. Neurath, 'Unified science as encyclopedic integration', *International Encyclopedia of Unified Science*, Chicago, i (no. 1), 1–27, on 3. On the possibility of achieving local axiomatizations within the larger unified science, see O. Neurath, 'Encyclopedia as "model"', in *Philosophical Papers*, 145–58, on 145, 148–9; Neurath, 'The new encyclopedia of scientific empiricism', in ibid., 189–99, on 194.

19 Zolo, op. cit. (7), 83; C. Hempel, 'Logical positivism and the social sciences', in *The Legacy of Logical Positivism* (ed. P. Achinstein and S. Barker), Baltimore, 1969, 163–94, on 172–3.

activist like Neurath, however, there was no impediment to developing a more 'horizontal' unified science and an encyclopedia to help organize and disseminate the results.

It was launched in 1934 at the Eighth International Congress of Philosophy at Prague. He proposed that the idea be considered formally at the First International Congress on the Unity of Science, to be held the next year in Paris. It was quite well received, both at the Congress and across the Atlantic at the University of Chicago Press. With assistant editors Morris and (by 1936) Carnap at the University of Chicago, the university's Press was eager to meet Neurath and detail plans to publish the *Encyclopedia*. They agreed to undertake the project as an experiment. If an introductory series of monographs, *Foundations of the Unity of Science*, could support itself, they would allow further volumes to be announced. With suggestions from Carnap and Morris, Neurath then drew up a list of twenty individuals variously representing the Vienna Circle, the Berlin group of scientific philosophers centering around Hans Reichenbach, Bertrand Russell and English scientists, and finally American philosophers including, most notably, John Dewey. Each would treat an aspect of logic, mathematics, or one of the sciences with an eye toward how they could be unified.[20]

The experiment was a success. By 1939, after a campaign of advertising and 'propaganda' – as Neurath called it – over twice the required 250 subscriptions had been received.[21] Sales were brisk and additional volumes were soon announced. If the introductory monographs were to address the sciences at hand and the tools for their unification, the new volumes were to reflect life in the trenches of unified science. Containing ten monographs each, volumes III–VIII would treat: general problems encountered in unifying the sciences; the structural role of logic and mathematics in unified science; physics; biology and psychology; the social and 'humanistic' sciences; and the history of the 'scientific attitude' that the unity of science movement represented. Neurath's plans for the *Encyclopedia*, however, soon grew much larger. As Morris recalled later, he envisioned another section of eight volumes which would explore 'the actual state of systematization within the special sciences' and another ten volumes aiming to incorporate education, engineering, medicine and law into unified science. Finally, each

20 Between 1938 and 1970, the following monographs appeared. Volume I began with 'Encyclopedia and unified science' which combined articles and comments by Neurath, Niels Bohr, John Dewey, Bertrand Russell, Rudolf Carnap and Charles Morris. Subsequent monographs are: Morris, 'Foundations of the theory of signs'; Carnap, 'Foundations of logic and mathematics'; Leonard Bloomfield, 'Linguistic aspects of science'; Victor Lenzen, 'Procedures of empirical science'; Ernest Nagel, 'Principles of the theory of probability'; Philipp Frank, 'Foundations of physics'; E. Finlay-Freundlich, 'Cosmology'; Felix Mainx, 'Foundations of biology'; and Egon Brunswik, 'The conceptual framework of psychology'.

Volume II consists of: Neurath, 'Foundations of the social sciences'; Thomas Kuhn, 'The structure of scientific revolutions'; Abraham Edel, 'Science and the structure of ethics'; John Dewey, 'Theory of valuation'; Joseph H. Woodger, 'The technique of theory construction'; Gerhard Tintner, 'Methodology of mathematical economics and econometrics'; Carl G. Hempel, 'Fundamentals of concept formation in empirical science'; Giorgio de Santillana and Edgar Zilsel, 'The development of rationalism and empiricism'; Joergen Joergensen, 'The development of logical empiricism'; and Herbert Feigl and Charles Morris, 'Bibliography and index'.

21 Neurath's relative, Waldemar Kaempffert, was a science editor for the *New York Times* and wrote several articles about the *Encyclopedia*: 'Toward bridging the gaps between the sciences', *New York Times Book Review*, 7 August 1937, 2; 'Sciences to be unified through a common language', *New York Times*, 14 February 1938. Announcements also can be found in various philosophical journals published during these years.

volume in this last section would be accompanied by a volume of ISOTYPE diagrams together forming a 'Visual Thesaurus'.[22] Clearly, Neurath saw the *Encyclopedia* as a forum not only for unifying the sciences, but for extending their range of application in society and, through ISOTYPE, illustrating their power in a vivid, accessible language.

This use of illustrations and attention to applied science, Neurath thought, would allow the new *Encyclopedia* to continue the aims of the French *Encyclopédie*. He closed his lecture at the Paris congress by saluting '*Unité de la science et fraternité entre les nouveaux encyclopédistes*'.[23] Neurath wrote about other encyclopedias and dictionaries as well,[24] but the new one would exceed all of them: while the *Encyclopédie* was an exception, he felt, most mistakenly relied on certain overarching metaphysical, non-scientific schemes to arrange their contents. The new encyclopedia would avoid this pitfall. Most importantly, however, it had access to the new logic of Russell and Whitehead. Logical analysis would be used 'in the new service of unified science' to illuminate the horizontal and vertical relations that exist among the sciences, highlight the gaps between them, and perhaps expose 'unsuspected possibilities'[25] for further unification. In this way, the very spirit of the new encyclopedia was unique: it looked forward, toward facilitating progress in unifying the sciences, and not only to past or present achievements. It would be 'a living intellectual force growing out of a living need, and not a mausoleum or a herbarium'.[26]

As the *Encyclopedia* was becoming a reality, Neurath's essays continued to introduce perspectives and tools needed to unify the mosaic of the sciences. He wrote less about a 'physicalist' language of unified science and spoke increasingly of a 'universal jargon' or 'universal slang'.[27] Like physicalist terminology, this language would be natural, purged of metaphysical or unempirical terms, and peppered with technical terms from all the sciences. It would serve, however, to connect the sciences horizontally,[28] and not to secure any special connection with the language of physics. Attending this shift, the broader perspective of 'physicalism' gave way to a new programme which Neurath called, appropriately enough, 'encyclopedism'. Much like Kuhn's 'paradigms' or Foucault's 'epistemes', Neurath's 'encyclopedias' denoted 'the totality of scientific matter' in different historical eras:

22 C. Morris, 'On the history of the *International Encyclopedia of Unified Science*', *Synthese* (1960), 12, 517–21, on 518, 519–20.

23 O. Neurath, 'Individual sciences, unified science, pseudorationalism', in Neurath, op. cit. (7), 132–8, on 138; on the *Encyclopédie*, see O. Neurath, 'Unified science and its encyclopedia', in *Philosophical Papers*, op. cit. (7), 172–82, on 179; Neurath, 'Unified science as encyclopedic integration', op. cit. (18), 7.

24 Neurath gives his most extended treatment of the encyclopedic works of Diderot, Hegel, Comte, Leibniz and Spencer in 'Unified science as encyclopedic integration', op. cit. (18), 2, 7–8. He mentions also Comenius' *Orbis pictus* and internationalist Paul Otlet's *La Cité mondiale*, for example, in 'An international encyclopedia of unified science', in *Philosophical Papers*, op. cit. (7), 139–44, on 143.

25 Neurath, 'An international encyclopedia of unified science', in *Philosophical Papers*, op. cit. (7), 139–44, on 139–40.

26 Neurath, 'Unified science and its encyclopedia', in *Philosophical Papers*, op. cit. (7), 172–82, on 181.

27 Neurath, 'Encyclopedia as "model"', in *Philosophical Papers*, op. cit. (7), 145–58, on 155; see also Neurath, 'Unified science and its encyclopedia', in ibid., 175, 178; Neurath, 'The new encyclopedia of Scientific Empiricism', in ibid., 191.

28 Neurath, 'Universal jargon and terminology', in *Philosophical Papers*, op. cit. (7), 213–29, on 214; see also Neurath, 'Unified science and its encyclopedia', in ibid., 180; O. Neurath, 'Orchestration of the sciences by the encyclopedism of logical empiricism', in ibid., 230–42, on 235.

we are dealing with encyclopedias each of which is a model of science, and one of which is applied at a definite period. The march of science progresses from encyclopedias to encyclopedias. It is this conception that we call *encyclopedism.*[29]

The actual *Encyclopedia*, therefore, would present the 'encyclopedia' of science currently in use and also encourage research toward further unifying and extending the application of the sciences. If Neurath's physicalism of the early 1930s saw unified science as a static 'pyramid' of the sciences, under encyclopedism it denoted both a 'mosaic' of sciences and an ongoing process – a process of examining their logical relations, creating and refining the universal jargon, establishing 'cross-connections',[30] and continually incorporating new scientific developments into the network represented by the *Encyclopedia.*

INTERNATIONALISM

Although his prospectus allied the *Encyclopedia* with a 'comprehensive international movement', Morris did not detail how it shared elements with many other movements popular at the time. Several years before, in 1929, Neurath was more explicit. He had explained that logical empiricism and the quest for unified science implied certain 'attitudes towards questions of life' that found expression in a variety of modern movements:

> Endeavours toward a new organization of economic and social relations, toward the unification of mankind, toward a reform of school and education, all show an inner link with the scientific world-conception.[31]

Some of these links are evident in the *Encyclopedia.*[32] The goal of a 'universal jargon' for the sciences and Neurath's plan to incorporate ISOTYPE echoed the goals of Esperantists

29 Neurath, 'Encyclopedia as "model"', in *Philosophical Papers*, op. cit. (7), 146. There is a tendency among Neurath scholars to inflate the degree to which Neurath seems to have anticipated now entrenched, Kuhnian ideas. Heiner Rutte, for example, implies the anticipation is nearly complete in 'The philosopher Otto Neurath', in Uebel, op. cit. (5), 92 (see also Zolo, op. cit. (7), 90, 92). Neurath, however, did not suggest what are (or at least were) perhaps the most controversial aspects of Kuhn's account of scientific revolutions: (1) Observations are theory-laden. The precise features of sensations and perceptions were not germane to Neurath's account of protocols, much less any *gestalt*-like or theory-influenced character they might possess (see Neurath, 'Protocol statements', in *Philosophical Papers*, op. cit. (7), 91–9). (2) Theories as world-views – one meaning of 'paradigms' – are monolithic and incommensurable. Neurath's 'encyclopedias' share the holistic nature of Kuhn's paradigms whereby each is globally affected by local alterations. For Neurath, however, the relative ease of making such changes (to mitigate inter-theoretic inconsistencies, for instance) concerns the normative heart of his programme. Kuhn, on the other hand, feels that such innovation will often be impeded by epistemic and even perceptual obstacles. Unlike Kuhn's, Neurath's holism sees theoretical wholes as plastic and easily manipulable, as suggested by his proposals to construct unified science by first resolving the sciences into their smallest theoretical parts. (See Neurath, 'Departmentalization of unified science', in ibid., 200–5; Neurath, 'The social sciences and unified science', in ibid., 200–5.) On one occasion, at least, Neurath further denied that (3) there were separate historical encyclopedias that contradict or compete with each other in history (Neurath, 'Encyclopedia as "model"', in ibid., 157). For an opposed statement, however, see Neurath, 'Individual sciences, unified science, pseudorationalism', in Neurath, op. cit. (7), 137–8.

30 Neurath, 'Encyclopedia as "model"', in *Philosophical Papers*, op. cit. (7), 155; Neurath, 'Unified science and its encyclopedia', in ibid., 175; Neurath, 'The new encyclopedia of scientific empiricism', in ibid., 191, 194.

31 Neurath, 'The scientific conception of the world', in Neurath, op. cit. (3), 304–5.

32 For an account of Neurath's, Carnap's, Morris' and other philosophers' relations with architectural modernism, see P. Galison, 'Aufbau/Bauhaus: logical positivism and architectural modernism', *Critical Inquiry* (1990), Summer.

who expected that an international language might foster increased communication and trust among nations.[33] These hopes for international scientific terminology were also shared by the League of Nations' International Committee on Intellectual Cooperation.[34] Moreover, from its inception, the *Encyclopedia* was an international effort. Its authors and committee members were from Europe, England and the United States. Since cooperation and dialogue were paramount, Neurath planned to issue English, French and German editions. He expected the encyclopedists to

> remain in permanent contact with one another, so that the Encyclopedia may serve the building of bridges from one science to another, without, however, limiting personal expressions of opinion.[35]

In this way, they would be a microcosm of the international, scientifically minded world-community Neurath envisioned.

When the war interfered with this dialogue and greatly slowed the production of monographs – in part because Neurath himself, an Austrian national, was interned for several months on the Isle of Man after narrowly escaping the Nazi occupation of The Hague[36] – the editors variously described the *Encyclopedia* as a war-effort. Morris seemed to view the Sixth International Congress, held in Chicago in September 1941, as a symbolic antidote to the schisms the war created. 'The Organizing Committee', he wrote in its promotional flyer, 'feels that the present world condition enhances rather than restricts the need for the vigorous continuation of the unity of science movement'.[37] In 1943, when the University of Chicago Press decided to suspend the project in the face of rising costs and greatly diminished sales, Neurath explained to them the *Encyclopedia*'s importance for the war. He deftly praised what had been the project's 'business as usual'[38] spirit and excused the unavoidable delays in securing monographs. This 'real international enterprise', he explained, 'has to be written partly by refugees'.[39] To them, he seemed to suggest, suspending the project might appear as a concession: 'it would be like defaitism[*sic*] to

33 For Esperanto inventor L. L. Zamenhof's beliefs to this effect, see P. Forster, *The Esperanto Movement*, New York, 1982, 96.

34 The International Committee on Intellectual Cooperation was staffed at one time or another by Einstein, Marie Curie, Henri Bergson and many other intellectuals and internationalists. From 1922 until the Second World War, the Committee concerned itself with issues such as universal languages, exchange scholar programmes, scientific property rights, and bibliography and abstract standardization in scientific publications. Proposals to initiate the standardization and unification of scientific terminology in meteorology, political theory, economics, archaeology and anatomy were also discussed. As did Neurath, the Committee sometimes referred to its work as 'coordination' of the sciences and shared his hopes for advancing science by refining scientific language. The Committee, furthermore, aimed to enlist intellectual cooperation in the League's campaign against war. By facilitating scientific communication and progress, and also by promoting science education, the Committee understood itself to be laying necessary foundations for world peace. See League of Nations, International Committee on Intellectual Cooperation. *Minutes*. Geneva: League of Nations Publications, 1922–30.

35 Neurath, 'Unified science and its encyclopedia', in *Philosophical Papers*, op. cit. (7), 178.

36 M. Neurath, 'Memories of Otto Neurath', in Neurath, op. cit. (3), 68–71.

37 Promotional flyer, University of Chicago Library, Department of Special Collections, 'University of Chicago Press Papers', Box 346, Folder 3. Similar perceptions of the Fifth International Congress, held at Harvard in 1940, can be found in M. Singer and A. Kaplan, 'Unifying science in a disunified world', *The Scientific Monthly* (1941), 52, 79–80.

38 Neurath to McNeill, 24 May 1943, University of Chicago Library, Department of Special Collections, 'University of Chicago Press Papers', Box 346, Folder 3.

39 Neurath to McNeill, 9 September 1943, op. cit. (38).

suspend now anything'.[40] Neurath's arguments were successful and the *Encyclopedia* was never suspended. 'During wartime science and logical analysis cannot rest', he wrote to Morris. 'We have to prepare future peace life, particularly in Europe.'[41]

Ideally, one feature of this 'peace life' would be widespread participation in unified science. Especially in Europe, however, Neurath continually saw the unempirical methods of German historicism as a threat to both unified science and its internationalism. The belief that *Verstehen* and empathy could probe the 'dark distances and unfathomable depths' of history and culture opposed Neurath's empiricism.[42] 'There is surface everywhere'[43] and this empirical surface of the world – described by the universal jargon – was the proper subject of all the sciences. Each would remain tethered 'in one plane'[44] of empirical statements in which theories and hypotheses could be understood and unified by an international community of scientists.

This international empiricism, however, was not simply an end in itself. Neurath believed that once methodological overspecialization was minimized, unified science would nourish scientific growth. In this respect, he once invoked the interdisciplinary sensibilities of Marxian studies as a paradigm.[45] In 1931, Neurath explained that 'bourgeois' research strove simply to extend knowledge in all directions,

> proceeding here in an anthroposophic way, there mathematicising, here psychologising, there pursuing the idea of fate, here in a technical manner, there in an occultist one.

Under Marxism, however, 'what otherwise remains isolated fact, is set...into a comprehensive view which stimulates some to apply their minds to more complex connections'.[46] Accepting a common empirical method and recognizing the general interconnectedness of social and natural processes, scientists within unified science would likewise be able to explore such 'complex connections'. They would create threads binding their specialties to the rest of unified science and thereby bolster the whole fabric. Ideally,

40 Neurath to McNeill, 24 May 1943, op. cit. (38).

41 Neurath to Morris, 7 January 1942. University of Chicago Library, Department of Special Collections, 'Unity of Science Movement Papers', Box 2, Folder 14.

42 Neurath, 'The scientific conception of the world', in Neurath, op. cit. (3), 306. Neurath discusses the 'metaphysical countercurrents' in sociology stemming from Dilthey and others in Neurath, 'Empirical sociology', ibid., 353–8.

43 O. Neurath, 'Personal life and class struggle', in Neurath, op. cit. (3), 249–98, on 306.

44 Neurath, 'Departmentalization of unified science', in *Philosophical Papers*, op. cit. (7), 204.

45 The role Marxism plays in Neurath's agenda for unified science and *Aufklärung* has yet to receive comprehensive treatment. Here I shall point out only Neurath's beliefs that (1) the contents of a culture are in some way causally shaped by realities of economy and production (see, for instance, Neurath, 'Empirical sociology', in Neurath, op. cit. (3), 324), or they are at least consistent with a society's '*Lebenspraxis*' (see Neurath, 'Personal life and class struggle', in ibid., 293; Neurath, 'Encyclopedia as "model"', in *Philosophical Papers*, op. cit. (7), 157); and (2) the reality of class struggle informs public and intellectual debate (see Neurath, 'Personal life and class struggle', in Neurath, op. cit. (3)). If these beliefs render Neurath a 'marxist', however, he denied for epistemic reasons – the same that underlie his conception of unified science (treated below) – that long-term theoretical predictions of society or economy could be successful (see, for instance, Neurath, ibid., 293; O. Neurath, 'Foundations of the social sciences', *International Encyclopedia of Unified Science*, Chicago, 1944, ii (no. 1), 1–51, 28–30). In this way, Neurath's beliefs confound Karl Popper's characterization of him as a 'historicist' sociologist whose 'aim is to make forecasts, preferably large-scale forecasts' (K. Popper, *The Poverty of Historicism*, New York, 1957, 103 (no. 1), 39).

46 Neurath, 'Personal life and class struggle', in Neurath, op. cit. (3), 292.

unified science would exhibit the methodological consistency Neurath admired about Marxist studies, in which it was no longer 'left to chance whether a man thinks about some linguistic formations in Chinese or about a medieval legal text, about African beetles or about wind conditions at the North Pole'.[47] Under unified science, thought would follow (or create) wide-ranging and constructive interdisciplinary paths.

'We live', Neurath wrote, 'in a period of conscious shaping of life'[48] and the projected *Encyclopedia* played several roles in his hopes for an international, socialistic, rational world order. First, it would educate. 'We want to educate people to become critical member[s] of a democratic community',[49] he explained to his publisher. Like his ISOTYPE projects, it would popularize the scientific world-view in which a planned, socialist world seemed within reach. This enlightenment would also combat political obscurantism, often couched in metaphysical and spiritual language.[50] Furthermore, with increasing socialization, an educated, scientifically minded populace would be equipped to use the sciences, to weigh and choose among the possibilities they offered to 'ameliorate personal and social life'.[51] Finally, by coordinating the sciences, the *Encyclopedia* would help strengthen them. Since multiple sciences were involved in understanding even the simplest of events, the most powerful science would be, in one sense, all of them together – unified science.

EPISTEMOLOGY AND PLANNING

While the *Encyclopedia* reflected these internationalist and socialist currents, it is equally important to understand the *Encyclopedia* in terms of Neurath's distinctive views about epistemology. For his social goals, on the one hand, and hopes for unified science, on the other, were more than complementary. An enlightened, scientifically minded society would benefit from unified science and it, in turn, would be nourished by the widespread, international participation of enlightened citizens and unified scientists. Yet, these social and scientific goals were best achieved through a common technique – planning – which is justified, even demanded, by Neurath's epistemology of *non-foundational holism*.

Like other logical empiricists, Neurath broadly understood science as the activity of producing empirically accurate statements about the world. Yet, early in his career, Neurath introduced an idea which complicated this interpretation: statements are one kind

47 Neurath, 'Personal life and class struggle', in Neurath, op. cit. (3), 294–5.

48 O. Neurath, 'Utopia as a social engineer's construction', in Neurath, op. cit. (3), 150–5, on 150.

49 Neurath to McNeill, 9 September 1943, op. cit. (38). Here Neurath characterized both the *Encyclopedia* and his ISOTYPE projects as 'work dealing with education'. The latter, he said, 'deal[s] with wider circles of the public' while the *Encyclopedia* 'deal[s] with scientific analysis, but with education, too. We want to educate people.'

50 In an early essay on Descartes published in 1913, Neurath explains the prevalence of superstition, belief in prophecy, and 'the striking lack of criticism with which...election speeches of parliamentarians are received' as a symptom of non-scientific, 'pseudo-rationalist' thinking (Neurath, 'The lost wanderers of Descartes and the auxiliary motive (on the psychology of decision)' in *Philosophical Papers*, op. cit. (7), 1–12, on 8). Eckehart Köhler describes how metaphysics, along with irrationalism, obscurantism and spiritualism 'could not but appear to the Vienna Circle as the intellectual dumping ground of the Western mind' (E. Köhler, 'Metaphysics in the Vienna Circle', in Uebel, op. cit. (5), 131–42, on 138).

51 O. Neurath, 'Unified science as encyclopedic integration', *International Encyclopedia of Unified Science*, Chicago, (1938), i (no. 1), 1–27, on 1. As Marx Wartofsky put it, Neurath saw unified science as a 'servant of the political and social decision making process' (op. cit. (7), 100).

of thing, the world they describe another. Not too unlike the way young Wittgenstein infuriated Russell by refusing to accept that 'there was not a rhinoceros in the room'[52] in which they talked, Neurath infuriated some of his colleagues by insisting that there was an epistemological gulf between statements and the world.[53] The truth of a statement, for instance, could not be ascertained by comparing it with experience or the facts it reports because the comparison involved two distinct sorts of things. Science was indeed empirical, according to Neurath. However, it made contact not with experience itself but with statements about experience: '*statements are always compared with statements*, certainly not with some "reality", nor with "things"'.[54] He explained:

> If a statement is made, it is to be confronted with the totality of existing statements. If it agrees with them, it is joined to them; if it does not agree, it is called 'untrue' and rejected; or the existing complex of statements of science is modified so that the new statement can be incorporated; the latter decision is mostly taken with hesitation. *There can be no other concept of 'truth' for science.*[55]

The 'truth' of statements lay exclusively in their consistency or agreement with all other, accepted statements. This is the holistic component of Neurath's epistemology. Since there could be 'no *other* concept of "truth" for science', there was no epistemological foundation for knowledge, no stable, extra-linguistic framework against which to evaluate statements. This indicates Neurath's non-foundationalism.

Over twenty years before the *Encyclopedia* was born, Neurath explained why epistemic foundationalism, and Descartes' in particular was essentially flawed:

> Each attempt to create a world-picture by starting from a *tabula rasa* and making a series of statements which are recognised as definitively true, is necessarily full of trickeries. The phenomena that we encounter are so much interconnected that they cannot be described by a one-dimensional chain of statements. The correctness of each statement is related to that of all others.[56]

As Neurath saw it, Descartes was simply mistaken. No statement (*Cogito ergo sum*, for example) can be taken by itself as true and used as a foundation for the rest. Since 'it is absolutely impossible to formulate a single statement about the world without making tacit use at the same time of countless others',[57] the ambiguity and imprecision of other statements would infect any that one tried to isolate as 'true'.

---

52 B. Russell, cited in R. Monk, *Ludwig Wittgenstein: The Duty of Genius*, New York, 1990, 39.

53 Carl Hempel objected that Neurath's account could not discriminate between science and 'fairy-tale' (R. Haller, 'History and the system of science in Otto Neurath', in Uebel, op. cit. (5), 33–40, on 37). Moritz Schlick felt much the same way (M. Schlick, 'The foundations of knowledge', in *Logical Positivism* (ed. A. J. Ayer), New York, 1959, 209–27, on 215–16), and Edgar Zilsel felt that, according to Neurath, 'all symbolic edifices stay hanging in mid-air and no structure may be distinguished from totally arbitrary other structures as the structure of the experienced world' (R. Haller, 'The Neurath principle: its grounds and consequences', in Uebel, op. cit. (5), 117–29, on 127). Popper succinctly stated that 'Neurath unwittingly throws empiricism overboard' (J. A. Coffa, *The Semantic Tradition from Kant to Carnap: To the Vienna Station* (tr. and ed. L. Wessels), New York, 1991, 365).

54 Neurath, 'Physicalism', in *Philosophical Papers*, op. cit. (7), 53; see also Zolo, op. cit. (7), 32–4.

55 Neurath, 'Physicalism', in *Philosophical Papers*, op. cit. (7), 53; see also O. Neurath, 'Physicalism and knowledge', in ibid., 159–71, on 161. Note that 'existing statements' must be construed to mean something similar to 'statements existing and *accepted by* practising scientists'.

56 Neurath, 'Lost wanderers of Descartes', in *Philosophical Papers*, op. cit. (7), 3.

57 Neurath, 'Lost wanderers of Descartes', in *Philosophical Papers*, op. cit. (7), 3.

For Neurath, the Cartesian ideal of certain knowledge, resting on unshakable foundations, was a mistaken ideal of the Enlightenment and no longer pertinent. Epistemic certainty was unavailable and it would always be the case that decisions would have to be made without relying on 'insight'. There were, after all, no epistemic foundations for insight to probe and against which to adjudicate options. For Descartes, this was the unique circumstances of the lost hikers he described in his *Discourse*.[58] They picked a direction – randomly – and wisely maintained it, thereby avoiding circuitous or closed paths. Neurath, however, saw more in this metaphor than Descartes intended: it represented our *regular* epistemic condition and had important consequences for how we make decisions in the world. Just as the wanderers had to choose a path, increasing modernization would lead society to choose its social order, its economy, its city plans, and so on. Yet, because of this epistemological circumstance, none of the available options or plans could be demonstrably correct or optimal.

Nor could one find a surrogate for epistemic certainty in Neurath's holism. As Neurath saw it, Descartes' failure underscores the innumerable interconnections that obtain between any statement and all others. Consequently, one could never isolate a statement as 'true' by virtue of its consistency with all others for one could never examine all these interconnections. Furthermore, some of the trains of thought required to chart them would consume an entire lifetime[59] and as a result, he once noted, 'the powers of an entire generation of scholars are hardly sufficient to perceive all the consequences of a single theory'.[60] Finally, even if one could examine *all* statements and their interconnections, scientific progress would alter the encyclopedia of the day and, with it, these innumerable relations. A statement once 'true' could become 'false' or it could lose all connection with others and become metaphysical.

Epistemic certainty had no place in Neurath's epistemological beliefs nor in his conceptions of unified science and its *Encyclopedia*. He distinguished unified science from the image of a 'pyramid', as we have seen, because the image invited an epistemological misconception – that knowledge depended on some external, epistemic foundation just as a real pyramid rests on the earth.[61] Nor could the image of such a stable, grounded structure capture the dynamic, changing nature of the sciences.[62] It was against this epistemological backdrop that Neurath refined his conception of unified science as a mosaic, as a modular and variously connectible network of sciences.[63] Just as the meaning of any statement is determined by its holistic connections to all others, and just as the image or pattern in a mosaic depends on all its tiles, Neurath saw unified science as an

---

58 The discussion occurs at the beginning of Part 3 of the *Discourse*.

59 Neurath, 'Lost wanderers of Descartes', in *Philosophical Papers*, op. cit. (7), 3, 3–4.

60 Neurath, 'Encyclopedia as "model"', in *Philosophical Papers*, op. cit. (7), 157.

61 For Neurath, this would be one example of how 'we often still drag with us the traditional absolutistic terminology that allows reference to the "real world", the "ideal totality of statements" and other similar things', Neurath, 'Encyclopedia as "model"', in *Philosophical Papers*, op. cit. (7), 147.

62 See, for example, Neurath, 'Departmentalization of unified science', in *Philosophical Papers*, op. cit. (7), 204; O. Neurath, 'The social sciences and unified science', in ibid., 209–12, on 211; O. Neurath, 'The unity of science as a task', in ibid., 115–20, on 116.

63 Neurath's proposal that 'we should regard the Social Sciences as a collection of a great many scientific units *which can be combined in very different ways*' illustrates this mosaical image of the sciences. Neurath, 'The social sciences and unified science', in *Philosophical Papers*, op. cit. (7), 209–12, on 211.

interconnected network of terms and statements in which each had equal epistemic importance.

Even though philosophical certainty was beyond their reach, Neurath saw this mosaic as a powerful tool. The sciences embodied the best knowledge available and they could be used – never to secure certainty, but – to provide options for action. In this way, Neurath's epistemological views eventuate in the notion of *planning*, a technique of approaching problematic situations by examining all possible courses of action and their expected consequences. Then, one course would be chosen without presuming that it was the best, or only, way to reach the goals in question. Planning was a common notion among Neurath's colleagues and many intellectuals of his day.[64] For him, it was a technique that enabled and justified action in a world where epistemic certainty was unattainable.

Neurath frequently appealed to the sciences as flexible and creative tools for planning. In an essay of 1913, he discussed how wartime economies might be used as a stepping stone to planned, peacetime economies. What was needed, he felt, was a *general* theory of economics, one whose special cases would model 'all possible forms of economy in a quite general way'.[65] In this case, various wartime and peacetime economies could be compared in order to formulate plans for consciously transforming those of one kind into those of the other. In 'Utopia as a social engineer's construction', Neurath more explicitly distinguished the task of examining hypothetical, even merely imaginable, options – in this case, plans for utopias – from the act of choosing one to pursue. The term 'utopias', he explained, should refer to 'all orders of life which exist only in thought and image but not in reality'.[66] Social engineers could devise various schemes for utopias, but it was incorrect – 'pseudorationalistic'[67] was Neurath's epithet – to believe that they could further evaluate or rank these schemes on some scale of viability. Each coherent plan, in its own terms, would be correct:

64 In general, 'planning' minimally involves applying the sciences to problems of society, economy and other institutions. The term gained currency in the West in the early decades of this century and figured in most alternatives to *laissez-faire* social and economic philosophies. Neurath most frequently refers to Joseph Popper-Lynkeus as an economic planner who understood the scientific nature of planning and who contributed to the rise of a scientific view of the world in Vienna (Neurath, 'The scientific conception of the world' in Neurath, op. cit. (3), 303; see also Neurath, 'Utopia as a social engineer's construction', in ibid., 151–2; Neurath, 'Personal life and class struggle', in ibid., 262; Neurath, 'Empirical sociology', in ibid., 339). Carnap saw 'the development of an efficient organization of state and economy' as a pressing need, and saw much of his own philosophical work as 'language planning' (see Carnap, op. cit. (15), 84, on 67–71; for a comparison of Carnapian language planning and Kuhnian revolutions in science, see G. Reisch, 'Did Kuhn kill logical empiricism?', *Philosophy of Science* (1991), 58, 264–77, on 270–4). Herbert Feigl, active in the Vienna Circle with Neurath and Carnap, similarly felt that 'cooperative planning on the basis of the best and fullest knowledge available is the only path left to an awakened humanity that has embarked on the adventure of science and civilization' ('The scientific outlook: naturalism and humanism', in *Readings in the Philosophy of Science* (ed. H. Feigl and M. Brodbeck), New York, 1953, 8–18, on 18).

65 O. Neurath, 'The theory of war economy as a separate discipline', in Neurath, op. cit. (3), 125–30, on 125. As Jordi Cat pointed out to me, Neurath appealed to this difference between considering all possible economic orders and actual orders in his essay of 1910, '*Zur Theorie der Sozial Wissenschaften*' in O. Neurath, *Gesammelte Philosophische und Methodologische Schriften*, op. cit. (7), 23–46, on 29.

66 Neurath, 'Utopia as a social engineer's construction', in Neurath, op. cit. (3), 151.

67 See for example, Neurath, 'Lost wanderers of Descartes', in *Philosophical Papers*, op. cit. (7), 8; Neurath, 'The unity of science as a task', in ibid., 118; Neurath, 'Individual sciences, unified science, pseudorationalism', in Neurath, op. cit. (7), 136–7.

If for example we raise the question what characteristics must a social order have in order that no one shall suffer hunger, that there shall be no credit crises, no living standards that are determined by chance and privilege, then we may find several equally correct solutions under quite definite presuppositions.

But, since we have only 'inadequate insight' about these presuppositions – just as Descartes' insight failed to isolate any genuinely true and certain statements – scientific planning could only, but not unimportantly, secure 'several possibilities'[68] for action. A choice among them could never be justified in the light of epistemological foundations, but it could be fully informed about the options available.

PLANNING UNIFIED SCIENCE

Neurath also viewed the history of science in terms of this distinction between theoretical, abstract possibilities, on the one hand, and historical realities, on the other. In 1916, soon after explaining why 'insight' was often unavailable to inform decisions made in the world, he offered a method for historiography which likewise avoided relying on any rare and 'lucky thrust' of historical insight.[69] First, one identified in a particular field the 'elements' out of which any particular theory could be reconstructed. The optical theories of Huygens, Newton, Malebranche and others, he illustrated, could each be identified with one articulation of this field of *possible* optical theories. This first stage, he noted, relies on 'a purely logical point of view' which sees every possible theory having equal value; no element is afforded special weight. In the second stage, however, this tolerance is dropped and the historian considers the actual life of particular theories, the reasons some were preferred over others, and their relations to the rest of the sciences and world view of their day.[70] Just as economic planning began with the 'purely technical question'[71] of what economies were theoretically possible, and just as utopian planning described 'all orders of life which exist only in thought and image but not in reality',[72] this historiographic method began with a survey of abstract possibilities. One first glimpsed all the theories that *could* have been pursued, given the conceptual elements of the time. Then, the historian examined how and why particular theories actually emerged from this field of historical possibility. The method involved a kind of retrospective scientific planning: one would examine history against the background of its conceptually possible variations, as if history could have been consciously planned.

If Neurath saw the history of science as if it could have been planned, he felt that science's future *should* be planned. His epistemology invited planning in science for the same kinds of reason it did elsewhere: since scientific statements did not connect with any epistemic foundations nor any extra-linguistic reality, scientific progress was contingent,

68 Neurath, 'Utopia as a social engineer's construction', in Neurath, op. cit. (3), 152; see also O. Neurath, 'International planning for freedom', in ibid., 422–40, on 426.

69 O. Neurath, 'On the classification of systems of hypotheses (with special reference to optics)', in *Philosophical Papers*, op. cit. (7), 13–31, on 13. Here 'insight' is a translation of *Einblick*; in his criticism of Cartesian 'insight', Neurath used *Einsicht*.

70 Neurath, 'On the classification of systems of hypotheses', in *Philosophical Papers*, op. cit. (7), 15, 16, 23, 30.

71 Neurath, 'Personal life and class struggle', in Neurath, op. cit. (3), 262.

72 Neurath, 'Utopia as a social engineer's construction', in Neurath, op. cit. (3), 151.

malleable and open-ended. Like utopias, economies, or classical theories of optics, unified science could take many possible forms, none of which could be uniquely correct or optimal. Unified science, therefore, would confront, in its own way, the question planning addressed throughout Neurath's work: *which of the many possible systems of unified science should be pursued?*

If unified science was to be planned, this logic of planning invited four general questions: (1) If economic plans, for instance, are ideally developed in the light of general economic theories, what could be the source of plans for unified science? (2) If social and economic plans are democratically chosen by an informed public, to whom will the resulting choice of plans for unified science be submitted? (3) What will be the criteria by which a choice among plans will be made? (4) Through what institutions or mechanisms will this process operate? Since answers to each of these questions can be found within Neurath's thoughts about unified science and the *Encyclopedia*, he understood that project, I suggest, as the first steps toward planning science.[73]

First, any claim to a general theory of knowledge which would provide different possible articulations of unified science would strike Neurath as metaphysical and epistemologically pretentious. History showed that 'grandiose syntheses' attempted by even the brightest metaphysicians had 'no chance of success'. Instead of issuing from some speculative meta-theory of science, any possible plan for unified science would be built from the ground-up: 'no system from above, but systematization from below'.[74] Consequently, planning unified science was very different from other kinds of planning. This planning 'from below', it would seem, could never secure at any one time *multiple* and complete plans for unified science. For if, in the light of Neurath's non-foundational holism, 'the powers of an entire generation of scholars [would be] hardly sufficient to perceive all the consequences of a single theory',[75] not even one complete and fully articulated plan could be realized. Rather than a choice among different plans, however, there would be a choice of how to *begin* to connect the terms and statements of the extant sciences. Just as one could plan *an* economy or *a* utopia without presuming to design *the* economy or *the* utopia, the task was to begin constructing one unified science, without pretending to create the only or best one possible. In unified science, just as in social and economic planning, Neurath's epistemology guaranteed a plurality of possibilities and simultaneously denied any meta-linguistic vantage point from which to survey and rank those possibilities.

The next questions, about who will choose plans, and by what criteria, refer to this process of arranging and connecting the extant sciences. The growth of unified science would always require selections and decisions about where and how to build inter-theoretic bridges. In this way, the scientists constructing unified science would implicitly choose among the innumerable possible coordinations. Their efforts, however, would unfold in a larger democracy whose concerns would influence their choices. Consequently,

73 Neurath referred to the *Encyclopedia* as a 'planned collective work' and unified science as a 'planned synthesis' but never articulated in any systematic way precisely how the enterprise could or should be understood as a kind of planning (Neurath, 'Protocol statements', in *Philosophical Papers*, op. cit. (7), 99; Neurath, 'Individual sciences, unified science, pseudorationalism', in Neurath, op. cit. (7), 132).
74 Neurath, 'Encyclopedia as "model"', in *Philosophical Papers*, op. cit. (7), 148, 153.
75 Neurath, 'Encyclopedia as "model"', in *Philosophical Papers*, op. cit. (7), 157.

the integrity of this network – at any stage of unification – would be evidenced both by its internal consistency as a holistic, theoretical network and, at the same time, by its effectiveness as a tool for managing society, economy, creating technology and so on.[76] Even the interests of the general audience, Neurath explained, would help determine 'the various possible lines of development' the *Encyclopedia* could pursue.[77] Thus, scientific and pragmatic desiderata would together guide scientists to coordinate their specialties and create a unified science.

The final question brings us to the *Encyclopedia*, for it would structure and institutionalize this process of planning. The first step in constructing anything is to appraise the materials at hand and this, Neurath implied, was reason to begin the *Encyclopedia*. Since *a priori* plans for a unified science were unavailable, the first step was to present the sciences in one place:

> If we reject the rationalistic anticipation of *the* system of the sciences, if we reject the notion of a philosophical system which is to legislate for the sciences, what is the maximum coordination of the sciences which remains possible? The only answer that can be given for the time being is: *An Encyclopedia of the Sciences.*[78]

By bringing the sciences together, the *Encyclopedia* would allow gaps and inconsistencies to be identified more easily. These could be rectified in many different ways and immediately unified scientists would have to make choices.

Since these choices would affect and alter the individual sciences, it would be incorrect to see Neurath's *Encyclopedia* as merely a forum to devise novel conjunctions of independent scientific theories. That individual sciences could not generally remain autonomous becomes clearer when two of Neurath's discussions are viewed in the light of his non-foundational holism. In the first, he asked his readers to follow him in a 'thought-experiment':

> Let us imagine that we have to choose between several completed encyclopedias that are in contradiction with each other: very different factors could determine our choice. One can assume, for example, that one of the encyclopedias would strongly develop a certain limited domain while through its structure it would be less favourable to the development of other domains. The other encyclopedia would be characterised, on the contrary, in our hypothesis, by its equal and uniform

76 Cat, Chang and Cartwright describe unified science in terms of Neurath's goal of economic socialism and articulate parallels between Neurath's 'unity' and current post-modern notions of 'disunity' of science. See, J. Cat, H. Chang and N. Cartwright, 'Otto Neurath: unification as the way to socialism', in *Einheit der Wissenschaften*, New York, 1991, 91–110. For an account of Neurath's evolving conception of unified science, in the light of his political activities, see J. Cat, H. Chang and N. Cartwright, 'Otto Neurath: politics and unity of science', forthcoming. The present account seeks less to identify Neurath's unified science and his politics than to show their common reliance on the notion of planning. Another difference concerns the degree of systematicity Neurath envisioned in unified science. Cat, Chang and Cartwright rightly oppose Neurath's conception to 'a single overarching' science but misconstrue it, I think, as 'a collection [of statements] that lacks intrinsic systematic order' (op. cit., 95, 99). Neurath surely did not envision unified science as an explicitly systematic structure and the kind of 'systematization' (op. cit. (27), 153) involved in its realization he had in mind is not clear. Without attributing some kind of structure or order to his conception, however, it becomes difficult to understand, for instance, the unifying function Neurath ascribed to the universal jargon; his explication of 'truth' as agreement among statements; or, in general, that there is any sense in which scientific statements can exhibit more or less 'unity'.

77 Neurath, op. cit. (51), 25.

78 Neurath, 'Unified science and its encyclopedia', in *Philosophical Papers*, op. cit. (7), 176–7.

elaboration of all disciplines but would not give rise to the hope for the certain perfection of work that the first seemed to promise.[79]

This is a thought experiment because, as I have explained, Neurath expected no such multiplicity of encyclopedias, much less *completed* encyclopedias, to arise. It does, however, raise the question of how the extant sciences are to be coordinated and developed. In this example, either (1) all sciences could be developed or (2) effort could be focused on 'a certain limited domain' which, in turn, would hinder the growth of others. This compromise follows from Neurath's holism: since all accepted statements are so interconnected, developing the terms and theories in this limited domain may elsewhere force adjustments, or perhaps deletions, of terms on which other sciences depend. Years earlier, when outlining his programme for 'empirical sociology', Neurath described such an outcome. By encompassing economics, history and political economy, this new discipline, he explained, would be equipped to answer new and important questions. However, this synthesis would irrevocably change the nature and scope of these and other fields. In this case, he explained, religion, art, science and law would become aspects of 'the total social process' and it would therefore be 'hopeless to write an autonomous "history" of religion, art, mathematics and so on'.[80]

In short, Neurath expected that unified science and its *Encyclopedia* could revolutionize the means by which scientific progress occurred. The history of unified science would not be a 'depressing compilation that strings together science after science in the accidental form that history has given them'.[81] Instead, the trajectories of each science and the forms they take would be informed by the deliberate choices made by unified scientists. Neurath's enthusiasm, however, was based on this new *mode* of scientific progress and not on any precise fore-knowledge of future encyclopedias or reasons why unified scientists will pursue one instead of others. About these imagined encyclopedias, he explained,

> since we do not possess a special theory of the importance of these possibilities for the progress of science or life in general or since we cannot assign a rank to these two encyclopedias from the point of view of this theory, we must function as a 'touchstone' ourselves and decide for the one or the other.[82]

As it was for Descartes' lost wanderers, the paths unified science would take would be determined ultimately by conscious decisions made without any such 'special theory' or foundational epistemological 'insight'. These choices would be deliberate, however, and Neurathian encyclopedists would be aware of the far-reaching effects they could have.

From this point of view, the encyclopedic format appears quite suited to this process. When new possibilities for unification emerged, future editions would incorporate and disseminate these developments:

> A hundred gateways are open. Thus the young will grow up to become collaborators in this 'eternal' encyclopedia that can publish each monograph in new editions according to need and deal with more and more details of the logic of science in later 'layers'.[83]

79 Neurath, 'Encyclopedia as model', in *Philosophical Papers*, op. cit. (7), 157.
80 Neurath, 'Empirical sociology', in Neurath, op. cit. (3), 352.
81 Neurath, 'Encyclopedia as "model"', in *Philosophical Papers*, op. cit. (7), 148.
82 Neurath, 'Encyclopedia as "model"', in *Philosophical Papers*, op. cit. (7), 157.
83 Neurath, 'The new encyclopedia of scientific empiricism', in *Philosophical Papers*, op. cit. (7), 195.

The *Encyclopedia*, that is, was much like the ship in Neurath's now famous metaphor: since our statements lack epistemological foundation, he explained,

> our actual situation is as if we were on board ship on an open sea and were required to change the various parts of the ship during the voyage. We cannot find an absolute immutable basis for science; and our various discussions can only determine whether scientific statements are accepted by a more or less determinate number of scientists and other men. New ideas may be compared with those *historically* accepted by the sciences, but not with an unalterable standard of truth.[84]

As the boat could be repaired and altered at sea, so the sciences could be further coordinated and altered without any 'immutable' epistemological basis. Over time, just as many repairs and alterations could yield an entirely different boat, the sciences too could become transformed as future encyclopedists pursue 'new open questions'.[85] The whole business will 'go on in a way we cannot even anticipate today'.[86]

A TOTALITARIAN ENCYCLOPEDIA?

Critics of centralized planning often saw it as a threat to the intellectual freedom science requires. Karl Popper, one of Neurath's most vociferous critics, offered perhaps the most thorough, and epistemologically sensitive, of these arguments: large-scale social planning risked the freedom most conducive to scientific progress, namely the freedom to conjecture hypotheses and attempt their refutation.[87] According to Popper, he and Neurath 'disagreed deeply...on almost all matters'[88] and, indeed, Neurath's advocacy of large-scale, utopian planning placed him squarely in Popper's sights. However, Popper did not seem to detect in unified science or the *Encyclopedia* the kernel of planning that informed Neurath's conceptions of these enterprises.[89] If he had, the two might have found some common ground. After all, Neurath expected unified science to grow gradually and cautiously, in much the same way that Popper's 'piece-meal engineers' would adjust society 'step by step...always on the lookout for the unavoidable unwanted consequences of any reform'.[90] And, in its own way, Neurath's programme assumed just as much freedom as Popper's. It demanded an unrestricted production of ideas about how terms and statements of different sciences could be coordinated and made consistent. Since the sciences could be interconnected in so many ways, each differentially advantageous, in the light of scientific and societal goals, unified science needed to be planned as freely as those goals themselves were chosen.

Unlike Popper, Horace Kallen did see unified science and the *Encyclopedia* as a threat to scientific freedom and engaged Neurath in an exchange of articles just prior to Neurath's

---

84 Neurath, 'Unified science and its encyclopedia', in *Philosophical Papers*, op. cit. (7), 181. See also O. Neurath, 'Anti-Spengler', in Neurath, op. cit. (3), 158–213, on 199.

85 Neurath, 'Unified science and its encyclopedia', in *Philosophical Papers*, op. cit. (7), 173.

86 Neurath, 'Foundations of the social sciences', op. cit. (45), 47.

87 See, for example, Popper, op. cit. (45), 154–5.

88 Popper, op. cit. (3), 56. For Neurath's critique of Popper's falsificationism, see Neurath, 'Pseudorationalism of falsification', in *Philosophical Papers*, op. cit. (7), 121–31.

89 Instead, Popper saw unified science as a misconceived attempt to demarcate science and metaphysics through theories of meaning. See K. Popper, *Conjectures and Refutations: The Growth of Scientific Knowledge*, 3rd edn, London, 1969, 268–70.

90 Popper, op. cit. (45), 67.

death in 1945.[91] Kallen, a philosopher at the New School for Social Research, and undying advocate of cultural pluralism, felt that unified science was allied with social and intellectual trends whose 'practical and logical limits' were 'the *de facto* totalitarianisms of Russia, Italy, Germany, Spain and the Roman Catholic Church'.[92] Underlying these stinging charges was a conception of science which, like Popper's, saw any sort of formal organization in science as a threat to 'the personal, the contingent, [and] the accidental' character of scientific innovation. By emphasizing scientific cooperation, Kallen reasoned, the *Encyclopedia* would suppress individual creativity and could only progress in some predetermined direction, toward a unity like that of the *Catholic Encyclopedia* – a unity 'assumed and found, not made'.[93]

Neurath argued, as he often did, that unified science was anything but an attempt to unify the sciences by imposing some assumed, *a priori* 'system' on them.[94] This was the point of intersection he saw between his project and the French *Encylopédie*. Its editors, he felt, also distrusted grand, rationalistic systems and saw their encyclopedia as 'an alternative'.[95] Neurath might have made his point clearer to Kallan, however, by distinguishing his project from that of the *philosophes* in this crucial respect. For D'Alembert believed the encyclopedic task began with a meta-scientific perspective of just the sort Kallen seemed to fear, yet which Neurath explicitly denied – a perspective from which one might attempt to assume, find and impose some particular global unity on the sciences. As D'Alembert explained it, organizing an encyclopedia

> consists of collecting knowledge into the smallest area possible and of placing the philosopher at a vantage point, so to speak, high above this vast labyrinth, whence he can perceive the principal sciences and the arts simultaneously. From there he can see at a glance the objects of the speculations and the operations which can be made on these objects; he can discern the general branches of human knowledge, the points that separate or unite them; and sometimes he can glimpse the secrets that relate them to one another.[96]

As did Neurath, D'Alembert denied that there was one unique or optimal way to arrange the sciences. Like geographical terrain, the tree of knowledge supported many different maps and 'projections'.[97] He still assumed, however, that this high vantage point was available and desirable, for certain 'secrets' about the sciences and their relations might appear.

91 Kallen studied philosophy at Harvard under William James and, in 1919, joined the original faculty of the New School for Social Research, where he worked until his death in 1974. An ardent Zionist, social philosopher and activist, Kallen is best remembered for his criticisms of assimilationism and Americanization (see M. R. Konvitz, 'Horace Meyer Kallen (1882–1974): in praise of hyphenation and orchestration', in *The Legacy of Horace Kallen* (ed. M. R. Konvitz), New York, 1987, 15–35). Kallen and Neurath corresponded privately for a year as these exchanges were published (H. Kallen, 'Postscript: Otto Neurath, 1882–1945', *Philosophy and Phenomenological Research* (1946), 6, 529–33, on 531).

92 H. Kallen, 'The meaning of "unity" among the sciences, once more', *Philosophy and Phenomenological Research* (1946), 6, 493–6, on 493.

93 H. Kallen, 'Reply', *Philosophy and Phenomenological Research* (1946), 6, 515–26, on 519, 520.

94 Neurath, 'Orchestration of the sciences', in *Philosophical Papers*, op. cit. (7), 241. The same point occurs in Neurath, 'Departmentalization of unified science', in ibid., 203; Neurath, 'The unity of science as a task', in ibid., 116; Neurath, 'Individual sciences, unified science, pseudorationalism', in ibid., 137.

95 Neurath, op. cit. (51), 7.

96 J. D'Alembert, *Preliminary Discourse to the Encyclopedia* (tr. R. Schwab), Indianapolis, 1963, 47.

97 D'Alembert, op. cit. (96), 48.

This was the circumstance Kallen seemed to fear: some, or perhaps all, of Neurath's encyclopedists would reach for such a meta-scientific vantage, form their own conclusions about the 'secrets' of the sciences, and control the development of the *Encyclopedia* and unified science accordingly. He sensed this totalitarian threat not from D'Alembert, of course, but probably from Morris' descriptions of unified science. In the *Encyclopedia*'s prospectus, and elsewhere, he described it as the 'science of science'[98] and implied, at least, that the very genius of unified science was some such reflexive, meta-scientific perspective.

For Neurath, however, this notion of some higher, privileged view of the sciences was epistemologically misconceived: one could gain no insight by pretending to view the sciences from 'high above', for the meaning and 'truth' of their statements lay only in their mutual relations and not in some ulterior reality – the 'objects of the speculations', the 'secrets' that relate the sciences – which might seem to appear from a higher vantage. Try as one may, viewing the sciences from above is really just viewing from within. Furthermore, since these interconnections (again) are long and innumerable, an individual, even a generation, could never fully explore this labyrinth. In other words, the sweeping, penetrating, and – in spirit, at least – Cartesian vision of the sciences D'Alembert described was a phantom. As Neurath understood it, unified science could never be seen all at once, much less 'in a glance'.

These considerations lay behind Neurath's responses to Kallen. He explained, for instance, that there is 'no point outside ourselves from which we may examine everything, ourselves included',[99] just as he had in his earliest writings about unified science: 'all experience forms a complex network, which cannot always be surveyed and can often be grasped only in parts'.[100] Ultimately, he did not reduce Kallen's fears. There could be no guarantee, after all, that some unified scientist or mob would not try to gain control. Neurath's conception certainly assumed scientific cooperation but it also provided cooperation an epistemological justification: since the only genuine epistemic support the sciences enjoy is in the connections between their terms and statements, the greater the unity of the sciences, the greater their epistemic integrity. Yet such unity should be forged cooperatively by the practitioners and theorists who best know their sciences, for this community as a whole is most able to detect and adjust the ways these terms are used and understood.

Far from developing a blueprint from which Neurath or some other gang might legislate the future of science, the *Encyclopedia* would present the sciences and views of how they might, in a piecemeal fashion, be further unified. Participants would contribute their own tiles to a larger mosaic whose emerging pattern, if one was to emerge at all, could not be predicted in advance. In this way, Neurath's conception did not deny Kallen's (and Popper's) view that science can grow from singular flashes of creativity. It did, however, urge scientists to address the ways other sciences epistemologically nourish their own and the ways in which the sciences worked together in the world. Ideally, the sciences would

98 On Kallen's reading of Morris, see H. Kallen, 'The meanings of "unity" among the sciences', *Educational Administration and Supervision* (1940), **26**, 81–97, on 87–8.

99 O. Neurath, 'For the discussion: just annotations, not a reply', *Philosophy and Phenomenological Research* (1946), **6**, 526–8, on 527.

100 Neurath, op. cit. (9), 306.

become more powerful and adaptable for use by the same free society in which they were created.

CONCLUSION

Neurath was busy with his ISOTYPE work during the *Encyclopedia*'s birth and during its demise as well. In 1941, after his release from internment, he settled in Oxford and evidently spent more time with other projects than Morris would have liked. He was making films for the Ministry of Information; hoping to begin a 'book-of-the-month' programme featuring books about unified science; managing his other new publishing projects, the *Journal of Unified Science* and the *Library of Unified Science*; and, he wrote to Morris, he was intrigued about cooperating with a new journal that would treat international relations 'from a real scientific point of view'. Morris surely wondered how the *Encyclopedia*, flagging during the war, could regain its pace given Neurath's attention to so many other projects. On several occasions during their correspondence, Neurath seemed rather defensive: 'we do not forget our Unity-of-Science work. NOT EVEN ONE DAY'.[101]

Perhaps because Neurath was so busy, and certainly because of the continuing war, the *Encyclopedia* never again enjoyed the pace it had in the late 1930s. In December 1945, however, Neurath died of a stroke and the outstanding monographs that were promised to the *Encyclopedia* came in very slowly. The first two volumes were not completed until 1970 and additional volumes were never pursued. By that time, of course, the philosophical spirit that fuelled the project had also changed. Back in 1937, Neurath asked Federigo Enriques to write a monograph on the history of science. This task was passed to I. Bernard Cohen at Harvard, and eventually to Thomas Kuhn. The result, Kuhn's famous *The Structure of Scientific Revolutions*,[102] has since been credited with supplanting logical empiricism literally from within one of its most ambitious publications, the *Encyclopedia*.[103] For these and undoubtedly other reasons, the *Encyclopedia* did not have the endurance Neurath expected. It does, however, allow us to understand one coherent axis within the sometimes dizzying array of Neurath's thoughts and deeds: as the social and economic future of modern society would be planned, so would the future of science. Through the *Encyclopedia*, scientists, philosophers and a scientifically minded public would each have a role in fashioning unified science as an organon for modern life.

101 Neurath to Morris, 1 December 1941; Neurath to Morris, 7 January, 1942, op. cit. (41).

102 T. Kuhn, *The Structure of Scientific Revolutions*, Chicago, 1962.

103 For a different view of Kuhn's philosophical relationship to Carnap, see Reisch, op. cit. (64); for an account of how Neurath pre-figured the epistemological naturalism of Quine – whose work is also reputed to have refuted logical empiricism – see Uebel, op. cit. (8).

Carnap, Tarski and the Search for Truth

ALBERTO COFFA

INDIANA UNIVERSITY

In 1933 Tarski published his celebrated monograph on the semantic conception of truth. In December of that year Carnap sent to
the printer the manuscript of his *Logical Syntax of Language*[1] (henceforth
LSL). Although it is well known that Tarski's work had a great
impact on Carnap's post-syntacticist philosophy, the specific nature
of this influence is not so widely recognized. Our main purpose
here is to determine how far Carnap's syntax was from Tarski's
semantics in 1933. To this end we will first determine what the
former (section 1) and the latter (section 2) had to say about truth,
and then examine the difference between them (section 3).

1. SYNTAX AND TRUTH

The single most decisive stimulus for the train of thought that
culminates in LSL was Gödel's 1931 paper on the incompleteness
of formal theories of arithmetic. The relevance of Gödel's conclusions to Hilbert's program are widely acknowledged. But its relevance
to what one might call "Wittgenstein's program" was no less
decisive.

Wittgenstein had explained in the *Tractatus* that "the characteristic
mark of logical sentences is that one can recognize (*erkennen*) from
the symbol alone that they are true; and this fact in itself contains
the whole philosophy of logic," and he then proceeded to designate
"as one of the most important facts, that the truth and falsehood
of non-logical sentences can *not* be recognized from the sentence
alone." ([1921] 6.113)

Wittgenstein's point included two distinguishable doctrines: (i)
that logical propositions have their truth-value determined by the
character of language and (ii) that this determination is embodied

NOÛS 21 (1987): 547-572
© 1987 by Noûs Publications

155

in constructive procedures that allow someone who understands the given language to "recognize" the truth-values in question.

The division between the range of what is linguistically determinate and what is not would eventually become the heart of the positivist conception of the a priori. Following the constructivist fashion of the twenties, the positivists had interpreted that linguistic determinacy in terms of constructive procedures such as derivability from effective rules of inference. But Gödel's [1931] had shown that no such set of rules would suffice to determine that mathematical sentences widely recognized as true *are* true. Thus Gödel's work seemed to threaten the idea that a logicist understanding of mathematics could be placed within the framework of a Wittgensteinian conception of logic.

Now the constructivist had two choices: he could stick to his guns and say that constructive proof is still all there is to mathematical and logical truth, thus dismissing the idea of general validity as meaningless and denying the philosophical significance of Gödel's work; or he could declare himself refuted, altering his views on mathematical truth and consequence. The first was Wittgenstein's way; the second Carnap's.

It is ironic that Carnap's syntactical philosophy is sometimes thought to be refuted by Gödel's discoveries which, we are told, establish the need to go beyond syntax. In fact, Gödel's discoveries were the decisive factor in determining both the technical problems that Carnap faced and his solutions to them. Far from having been written in ignorance of Gödel's results, Carnap's LSL was inspired by an appreciation of the significance of Gödel's work that only a handful of logicians could match at the time. Few people realized as clearly as Carnap the extent to which Gödel's [1931] had reactivated an old philosophical problem: what precise sense can we make of the notion of truth involved in Gödel's two major theorems? It is well known that Tarski not only saw the problem but also solved it. It is less widely known that, next to him, Carnap deserves credit for having come closer than anyone else to a solution.[2]

The main problem behind the technical developments in LSL was to define two notions: mathematical truth and logical consequence. Carnap's *starting* point was the recognition that truth and theoremhood are different things, and that consequence and proof are equally far apart. "One of the chief tasks of the logical foundations of mathematics," Carnap observed,

> is to set up a formal criterion of validity, that is, to state the necessary and sufficient conditions which a sentence must fulfill in order to be valid (correct, true) in the sense understood in classical mathematics. (p. 98)

Up to now, he added, it had been generally thought that the notion *valid* or *true on logical grounds* could be explicated in terms of processes of derivation (p. 41). Consequence, for example, had been explicated—prominently in Tarski's writings—as a purely syntactical notion. However, in view of Gödel's results,

> for our particular task of construction a complete criterion of validity for mathematics, this procedure, which has hitherto been the only one attempted, is useless; we must endeavour to discover another way. (p. 100)

Let us notice in passing that Carnap was never tempted to say that perhaps some mathematical statements are neither true nor false. Even in his strictest constructivist period he gave no thought to the idea of dropping the principle of excluded middle—which, like most others, he did not distinguish from bivalence. Even though Carnap's extreme reluctance to endorse a concept of truth different from that of well grounded belief was defeated only after he learned of Tarski's work, he never seems to have seriously doubted that in the range of mathematics, proof was one thing and truth an entirely different one. When Gödel convinced him that proof could not even grasp extensionally the concept of mathematical truth his instinctive reaction was: something else must. The most interesting technical portions of LSL are devoted to the task of explicating this new notion of mathematical truth.

If urged to present his explicandum in the dreaded material mode Carnap would have said that the languages under consideration in LSL are interpreted as dealing with the natural numbers (or an isomophoric domain of "locations") and all classes that may be built up from them. Since the domain of discourse is well defined, each sentence in those languages should have a truth value. Given a closed logical sentence in any of Carnap's languages, whether we know it or not, whether we can prove it or not, that sentence is either true or false. Thus, the class of mathematical truths is well determined and the problem is to "define" it, i.e., to find necessary and sufficient conditions for it. The definition will have to "decide" every mathematical sentence one way or the other; it will have to force *every* such sentence either into the "true" category or into the "false" category, and not into both. Moreover, all mathematical axioms must end up in the "true" category, which must also be closed under the consequence relation. The name that Carnap chose to characterize this "true" class was 'analytic' (and for the "false" one, 'contradictory'); but one should not be misled by the use of that ambiguous word:

> in relation to [logical languages] certainly, 'true' and 'false' coin-

cide with 'analytic' and 'contradictory', respectively. (p. 216)

There is a certain lack of coherence in Carnap's treatment of analyticity and consequence in LSL, suggesting that there may have been two stages in the development of his ideas on the topic. Carnap's first and least interesting strategy for defining mathematical truth and consequence is the one applied for language I and also in the general characterization of "consequence-concepts" (c-concepts) in section 48. The idea may have been inspired by an incorrect (pre-Gödelian) diagnosis of the problem: the assumption that what is wrong with standard proof-theoretic characterizations of mathematical truth is that the rules they use are not strong enough. Given this diagnosis the correct strategy would be to define truth and consequence as syntactic generalizations of *theorem* and *deduction*, respectively. Thus, for language I Carnap defined 'analytic' as the consequences of an empty class of statements, where the consequences of a class of formulas were conceived as the super-theorems that could be derived from that class using logical axioms, the standard rules of deductive inference and some other rules whose actual application is normally beyond the scope of human capabilities. Carnap's preferred "indefinite" rule is infinite (transfinite) induction: from $F(n)$, for each ordinal number n, infer $(x)Fx$. Hilbert and Tarski had studied this rule and Gödel's recent work had brought it to prominence by emphasizing the fact that even though the rule is intuitively sound, it is not a derived rule in standard systems of arithmetic.[3]

The *second* strategy for defining truth and consequence in LSL appears in section 34, when analyticity is defined for language II. The essentially semantic nature of this new approach becomes clear when we notice that the centerpiece of the new characterization of truth is not a generalization of the notion of inference, but the radically new idea of *valuation*.

Carnap never explained the reason for this change of strategy, but one may conjecture that at some point he realized that the first technique had worked for the case of language I only because of the weakness of its expressive power. By Gödel's theorem any system of rules representable in language II would inevitably fail at the task of deciding all mathematical statements.

In [1933] Tarski had faced the same situation in connection with the idea of consequence. Up to that point Tarski had given a purely proof-theoretic analysis of what he called the "ordinary" notion of consequence (truth-preserving inference). But Gödel's work had also led him to question the adequacy of that method. In [1933] he discussed at length the rule of infinite induction and its role in improving proof-theoretic explications of consequence. But he recognized that this strategy will not work in general:

> the profound analysis of Gödel's investigations shows that whenever we have undertaken a sharpening of the rules of inference, the facts, for the sake of which the sharpening was felt to be necessary, still persist, although in a more complicated form, and in connection with sentences of a more complicated structure. The formalized concept of consequence will, in extension, never coincide with the ordinary one. . . (p. 295)

Whereas Tarski concluded from Gödel's results that the notion of consequence could not be formally explicated on the basis of the logical resources at hand or even of forseeable extensions thereof, Carnap proceeded to identify a technique which allowed him to define those concepts for a wide class of languages.[4] Dropping the approach that turned on strengthened inference rules, Carnap developed a new technique closely associated with Tarski's notion of satisfaction, and on this basis he introduced non-proof-theoretic characterizations of consequence and mathematical truth, both defined for a large class of languages.

Carnap did not seem to be aware of how large a shift had taken place from his account of c-concepts for Language I to the account in section 34. A combination of philosophical prejudices led him to present the semantic ideas contained in that section in such an absurdly contorted fashion that even Carnap could not recognize them for what they were. Yet, when we look under the thick crust of nominalist and verificationist dogma, what we find bears an interesting relation to Tarski's work.

Language II is, in effect, a simple theory of types and a "coordinate language", i.e., a language whose terms of level zero are number expressions and, in particular, whose individual names are all numerals. The language is defined in the following way:

> The logical primitives of II are the propositional connectives, $=$, ' (successor), quantifiers (both bounded and unbounded) for variables of all types, and the minimum operator K (both bounded and unbounded) — e.g. Kx(...x...) stands for 'the least x such that ...x...').

> The system of type-indexes T may be defined as follows: 1. $O \in T$; 2. If $(t_1, ..., t_n) \in T$, then $(_1, ..., t_n) \in T$; 3. If $t_1, ..., t_n \in T$, then $(t_1, ..., t_k : t_{k+1}, ..., t_n) \in T$ (Carnap allowed functions to have a sequence of values); 4. that's all.

> Variables may be introduced as follows: For every type index t we have denumerably many variables x_t, y_t, ... which are called "variables of type t". The variables of type O are also

called "numerical variables". Finally, there are denumerably many propositional variables p, q, r, ...

This is one primitive constant 'O' and this together with all expressions O', O'', O''', ... (called St; also written '1', '2', '3', ...) are called "the constant numerals", and are all of type O. There are no more constants of type O but there may be as many constants of type t (t ≠ O and t ∈ T) as one wishes. There are, consequently, infinitely many languages II depending on the choice of non-logical primitives. The rules of formation are the usual ones. The type of a complex expression is defined so that a function taking arguments of types $t_1, ..., t_k$ and values of types $t_{k+1}, ..., t_n$ will be of type $(t_1, ..., t_k : t_{k+1}, ..., t_n)$, and a predicate taking arguments of types $t_1, ..., t_n$ will be type $(t_1, ..., t_n)$.

The axioms of language II include the standard propositional and functional axioms and rules of inference, the Peano axioms, a version of the axiom of choice, the law of extensionality for predicates and functors of all types and the axioms for the minimum operator.

Carnap's strategy for defining 'analytic' was roughly this. First, he identified an effective procedure that associates with each closed sentence of II another sentence S* provably (in II) equivalent to S. S* is in prenex normal form, i.e., it consists of a string of zero or more quantifiers followed by a quantifier-free matrix with a different free variable for each of its prenexed quantifiers. In the limiting case of zero quantifiers and zero free variables, if the formula has no descriptive signs it will be either 'O = O' or 'O ≠ O' (Theorem 34b). The problem of defining *analytic* for a language II without descriptive primitives is thereby reduced to that of defining it for these prenex sentences.

Take the simplest prenexed formula '(x)Fx', where 'x' is of type O (therefore, a number variable). Carnap said that this formula is analytic when each element in the class of instances in II {'F(O)', 'F(1)', 'F(2)', ...} is analytic. This is unobjectionable under the assumption that the domain of discourse of II is (isomorphic to) the class of natural numbers—for we would then have exactly as many individual names in II as we need. But Carnap thought at first that he could apply the same (substitutional) interpretation of quantification to all types. For example, if 'F' is of type (O), '(F)M(F)' will be analytic when all appropriate substitutions of expressions of type (O) for F in 'M(F)' turn out to give analytic sentences. In 1932 Carnap showed his definition to Gödel and received, in response, yet another gentle push in the direction of semantics.

In a letter written on 11.9.1932 Gödel explained to Carnap that the idea would not do as it stood:

> Consider, for example, the formula (F). F(O) v -F(O); in order to determine whether it is analytic, we must determine whether all formulas of the form P(O) v -P(O) are analytic. To this end we must replace each constant predicate P by its definiens, but they may contain again bounded predicate variables and so forth, so that one falls into a regressus ad infinitum. This becomes clearest in that, in some cases, *the same* formula can always recur. If, for example, in the formula (F). F(O) v O = O, we substitute for F the constant predicate (F)F(x) and derive the normal form, we end up with the original formula.
>
> In my opinion this error can be avoided only if we recognize as the range of variability of the function-variables not the predicates of a determinate language but all classes and relations in general.
>
> This does not involve any Platonistic standpoint, for I assert that this definition for 'analytic' can be carried through only within a determinate language, that already has the concepts "class" and "relation". ([CRR] 102-43-05)[5]

Carnap acknowledged the point.[6]

One may reasonably conjecture that Gödel's objection is what led Carnap to search for a radically new approach to the characterization of 'analytic' and to place the notion of *valuation* at the heart of the correct definition of mathematical truth.[7] As we shall now see, Carnap's new approach was based on the idea that in order to define 'analytic' we must begin by correlating expressions with what he called their "values", presumably, what they are agreed to stand for, (dare one say it?) their "referents". The quantifers are no longer treated as incomplete symbols calling for syntactic analysis on a narrow syntactic basis.

The revised definition of valuation starts as before: expressions of type O may take only numerals in II as valuations; beyond type O, however, we no longer evaluate with expressions of the corresponding type but, rather, with classes of the appropriate type which are ultimately built from the numerals. By a possible valuation of an expression of type (O), Carnap announced, "we shall here understand a class (that is to say, a syntactical property) of accented expressions," (p. 107) i.e., of numerals; the possible valuations of type (O:o) will be the functions from numerals to numerals, and so on. *All* classes of numerals, *all* functions from numerals to numerals, and so on, regardless of whether they are definable in the system or not, are to be included among the possible valuations for the expressions of the relevant type.

In effect (Definition VR 1 of section 34c) we can see Carnap as introducing the concept of a *range of possible values Vt for an expression of type t* as follows: V_O is the class of numerals; $V(t_1, \ldots, t_n))$ is the power set of the Cartesian product $Vt_1 \times \ldots \times Vt_n$; $V(t_1, \ldots, t_{j-1} : t_j, \ldots, t_n)$ is the set of all functions from the Cartesian product of $Vt_1 \times \ldots \times Vt_{j-1}$ into $Vt_j \times \ldots \times Vt_n$. A *valuation* for a quantifier-free matrix with variables and non-logical primitives ("value-bearing signs") v_i of types t_i is an assignment to each v_i of an element of Vt_i .

The valuations of the primitive signs are extended in the natural manner to all defined terms. Consider, for example, a valuation v such that v('x') = '7' and v(' + ') is the standard correlation between pairs of numerals and the numeral representing their addition (we are assuming that ' + ' is not defined). Then v('x + 3') = '10'. More generally, we have (Definition VR 2 of 34c) 1. 'O' is the valuation of 'O'; 2. If St is the valuation of Z_1 , St' is the valuation of Z_1 , ; 3. If $v_1, \ldots, v_n$ are the valuations of the terms $A_1, \ldots, A_n$, respectively, and if v_{n+1} is the valuation of (the appropriate typed) Fij, then the valuation of $Fij(A_1, \ldots, A_n)$ is the object correlated with the n-tuple $(v_1, \ldots, v_n)$ by v_{n+1}.

Since the only well formed expressions that have not been given a type are the sentences, they are the only ones waiting for a valuation. This is done employing, in effect, the sentences 'O = O' and 'O ≠ O' as the "honest" syntacticist versions of Frege's Truth and Falsehood. (The choice of mathematically determined sentences as representatives of truth-values is a further confusing element in Carnap's strategy; for it reinforces the assumption that what is being done relates to mathematical truth rather than to plain truth.)

At this point we are ready to explain what it is for a matrix to be mathematically true or analytic relative to a given valuation of all of its value-bearing signs. The idea is simply to find the truth value of the matrix relative to that valuation by evaluating it "from inside out" as it were, starting from the propositional atoms in the matrix and moving outwards. Since no quantifiers are involved at this stage, the process can be concluded in finitely many steps. Theorem 34c.1 states that for every quantifier-free reduced matrix M and for every evaluation of its value-bearing signs v, the described process of evaluation (in conjunction with the reduction process) leads in a finite number of steps from M and v to either 'O = O' or 'O ≠ O' (i.e., the truth values of all matrixes are determined, for each appropriate v).

Let us postpone for a moment the details of this process. Once we are given the notion of 'analytic relative to a valuation' for quantifier-free matrixes we have only one more small step to go to reach our goal. Basically, all we have to do now is to extend that idea to the prenex formulas in the obvious fashion[8] (see Definitions DA2C and DA3). For example, if our formula is

(*)   (EG) (Ex) (y) Gxy

we say that (*) is analytic if there are two valuations v' and v''
of the types of 'G' and 'x' respectively, such that for every valuation w of the type of 'y', 'Gxy' is analytic relative to the valuations
v', v'' and w. '(x) (Ey) (x + 1 = y)' turns out to be analytic, for
example, because (we can prove in the syntactic metalanguage that)
no matter how we evaluate 'x' (e.g., v('x') = '3'), given the definition of ' + ', there will always be a valuation of 'y' (in this case
v('y') = '4' such that the matrix will turn out analytic.[9] Finally,
bearing in mind that an arbitrary formula F is equivalent to its
reduct F*, we may define: a closed formula in II is analytic if its
reduct F* is analytic.

The link between all this and the semantics that would emerge
from Tarski's work is clear enough. But the similarities should not
be exaggerated. The key difference is manifested at the heart of
the enterprise, the characterization of 'analytic in a valuation' for
quantifier-free matrixes. A paragraph ago we skipped over its details.
Let us now focus on them.

Suppose, for example, that S_1 is the matrix we are evaluating,
and suppose that 'F' is a predicate-symbol of type (O); then, Carnap said,

> If V_1 is a particular valuation for 'F' of this kind [i.e., a class of
> numerals], and if at any place in S_1 'F' occurs with [a numeral]
> St_1 as its argument (for example, in the partial sentence 'F(2)'), then
> this partial sentence is—as it were (*gewissermassen*)—true on account
> of V_1 if St_1 is an element of V_1, and otherwise false. . .The definition of 'analytic' will be so framed that S_1 will be called analytic
> if and only if every sentence is analytic which results from S_1 by
> means of evaluation on the basis of any valuation for 'F'. (p. 107)

Nowhere was Carnap closer to the semantic conception of truth
than at this point. At about the same time that Carnap was writing
these words, Tarski was publishing a paper where he explained,
in effect, that under the envisaged conditions, Carnap's S_1 was not
just "as it were, true", but true in V_1, period. How far Carnap
was from that relaxed attitude towards truth becomes apparent in
the passage deleted from the preceding quotation:

> Now, by the evaluation of S_1 on the basis of V_1 we understand a
> transformation of S_1 in which the partial sentence mentioned [i.e.,
> 'F(2)'] is replaced by 'O = O' if that St is an element of V_1 and
> otherwise by 'O ≠ O'." (p. 107)

To further illustrate the point, consider Carnap's treatment of the two basic types of atomic sentences:

> Let a partial sentence S_2 have the form $Pr_2(Ar_1)$; and let the valuations for Ar_1 and Pr_2 be B_1 and B_2 respectively. If B_1 is an element of B_2 then S_2 is replaced by 'O = O'; otherwise by 'O $\neq$ O'. Let a partial sentence S_2 have the form $A_1 = A_2$, but not 'O = O'; and let the valuations for A_1 and A_2 be B_1 and B_2 respectively. If B_1 and B_2 are identical, S_2 is replaced by 'O = O'; otherwise by 'O $\neq$ O' (p. 110)

What Carnap's rules of evaluation do is to determine the truth value of a quantifier-free matrix for a given valuation of its value-bearing signs. But Carnap would not allow himself to use this "realistic mode of speech;" he preferred to say that what his rules do is tell us how to transform one sentence into another, so that in the end all mathematical sentences of II will be transformed either into 'O = O' or O $\neq$ O', the former being precisely the analytic sentences.

Carnap was clearly making an enormous effort not to say what one would naturally say under the circumstances: that he had identified a process that determines the truth-values of all the arithmetical claims of II in the domain of the natural numbers. By embedding his ideas in a Procrustean nominalist mold Carnap denied himself the possibility of grasping their true nature and their link with the non-mathematical notion of truth.

2. MEANWHILE, BACK IN WARSAW

According to Tarski's account in [1936] we can distinguish three stages in the evolution of the ideas incorporated in his monograph on truth. The earliest version, submitted to the Warsaw Scientific Society in March 1931 had as its centerpiece the thesis that a truth-predicate can always be defined for languages of finite order. The version published in 1933 included a revision of the original text partly inspired by Gödel's [1931], and leading to the conclusion that the truth-predicate is not definable for languages of infinite order. Finally, the German version of the monograph, [1936], is an extension of the Polish text. In a new postcript Tarski withdrew his earlier conclusion about the indefinability of truth and argued that the truth-predicate can *always* be defined for languages of arbitrary order in metalanguages of appropriate complexity. We will first review the central ideas of Tarski's monograph; in the following section we shall examine their link with Carnap's theory of 'analytic'.

Tarski's problem was: Can we define truth? Carnap had posed a similar question; but Tarski had had the good idea of asking a prior question first: What is a definition?

In Tarski's early work on definition there were two different senses of definability which he had aimed to explicate.[10] On the one hand, there was the purely syntactical problem of determining for a theory T already including a predicate P in its language, whether P is definable in T, relative to (some of) the remaining primitives $t_1, \ldots, t_m$. Under Tarski's explication, the question became whether an expression of the form (S) $P(x_1, \ldots, x_n)$ = $F(x_1, \ldots, x_n; t_1, \ldots, t_m)$ is provable in T, where $F(x_1, \ldots, x_n; t_1, \ldots, t_m)$ is a formula containint the same free variables as P and no primitives other than $t_1, \ldots, t_m$. For extensional languages, the question whether a term is definable in this purely syntactic sense is the question whether it is dispensable, whether it can be *eliminated* from the underlying theory without loss of content. To this category of problem belongs for example, the question whether mass is definable in classical mechanics. Padoa's method was designed to deal with situations of this sort. (See, for example, Tarski's treatment of Padoa's method in [1935].) We shall refer to these as problems of proof-theoretic definability (P-definability).

An entirely different problem of definability (discussed, for example, in Tarski [1931]) is that in which we start with a well defined semantic object (usually an individual or a class) and we ask whether there is in a given language, a propositional function whose extension is the object in question. If there is, we may say that the propositional function *defines* that object. We shall call this a problem of model-theoretic definability (M-definability). Whereas in the proof-theoretic case we are, in effect, raising the problem of the eliminability of an expression from a theory, in the model-theoretic case we are, if anything, dealing with the possibility of introducing a new expression into the language (in our case, a truth-predicate). Better yet, the problem of model-theoretic definability does not concern itself with the introduction or elimination of expressions but, rather, with the question of whether the expressive power of a language suffices to capture a certain semantical object.

In order to pose properly a question of (P- or M-) definability, one must therefore begin by specifying the language (and, for P-definability, the theory) relative to which the question is to be understood. In the case of a predicate (like Tr) which is supposed to apply to sentences, there is a further complication, since here we must specify both the language which will provide the instances of the predicate and the one which will constitute the framework in which the question is posed. These two languages might conceivably coincide, but at least the latter must be formally characterized if the problem of definability is to have a precise meaning. Let us call ML the (meta-)language in which Tarski will pose the problem of the definability of Tr, and OL the (object-)language for which Tr is to be defined. Tarski's ML is, in effect, a syntactic metalanguage for OL extended to include a translation of OL *and nothing else*. In particular, Tarski wanted to exclude all semantic primitives (such as *truth* or *designation*) from his ML.[11] Besides logical and syntactic primitives, therefore, ML will include translations of all the non-logical primitives of OL. Tarski also assumed that we have an axiomatic system in ML, its axioms being a standard and sufficiently powerful logic, plus an axiomatic account of the syntax of OL and (if OL is also endowed with an axiomatic structure) additional axioms which translate all the axioms of OL.

Now, even though Tarski's ML comes with an axiomatization, the problem which he was posing was not one of P- but of M-Definability. Tarski was not assuming that the language ML already contains a truth-predicate Tr whose eliminability was under consideration, nor was he asking whether an equivalence of the sort (S) can be proved in M. He assumed instead a language without any semantic primitives, a purely (extended) syntactic language, and he also assumed that there is a well defined class of true sentences of OL. The question he raised was whether there is a propositional function ϕ in ML whose extension is the class in question. If there is one, we may trivially introduce the predicate 'Tr' in ML through the axiom ("definition") 'Tr(X) $= \phi$(X)'.

The propositional function ϕ must qualify, to begin with, as a *formally correct* definiens. In the case of P-definability Tarski had, in effect, explicated formal correctness as follows: a propositional function ϕ (in the language of T) is a formally correct definition of a predicate P in the theory T, relative to the primitives t_1, ..., t_n when (i) ϕ has the same free variables as P and all its primitives are taken from the list t_1, ..., t_n, and (ii) we can prove in T
$$\phi = P.$$
The condition of formal correctness for M-definability, on the other hand, cannot possibly fulfill requirement (ii) and is therefore circumscribed to condition (i). A formally correct M-definition, therefore, is any propositional function whose free variables pertain to the appropriate categories, and whose non-logical signs are the ones to which we intend to reduce the *definiendum*.

When a problem of P-definability is posed, the requirement of formal correctness fully determines both the problem and its solution, for either there is a theorem in T of the appropriate sort, or there is none, whether we know it or not. But when a problem of M-definability is posed, the formal requirement vastly underdetermines the problem. This becomes obvious as soon as we notice that in the case of the class of true sentences, for example, any proposi-

166

tional function of ML whose extension is a class of OL sentences fulfills the requirement of formal correctness as intended by Tarski. One must therefore add further conditions restricting the class of acceptable solutions, and these conditions should obviously take account of the semantic aspects of the problem. This is what led Tarski to introduce his second requirement of material adequacy.

To say that a definition of the class of true sentences of OL must be materially adequate is simply to say that if $\phi(x)$ is the propositional function which M-defines the intended class, one should be able to prove in ML, for each sentence X of OL, the formula of the form

$$\phi(X) = p,$$

where 'p' is to be replaced by a metalinquistic sentence "having the same meaning" (p. 187) as X. This is the heart of Tarski's celebrated convention (T); for if we introduce the sentential predicate 'Tr' in ML adding the axiom '$Tr(X) = \phi(X)$', we have the famous condition '$Tr(X) = p$' ('snow is white' is true iff snow is white).[12]

The point of convention (T) is that its satisfaction will guarantee that the truth-predicate 'Tr' will have the class of true sentences as its extension and thus qualify—in the semantic sense—as a definition of it. For if OL is fully interpreted (as it is in the languages considered by Tarski) then all sentences of the OL as well as their metalinguistic translations will be endowed with a truth value, however unknown it may be to us. If a predicate 'Tr*' which applies only to sentences of OL satisfies also convention (T), then for each sentence X of OL we will be able to prove in ML

$$TR^*(X) = p;$$

hence,

$$TR^* = Tr;$$

and on the assumption (obvious to Tarski) that metalinguistic theorems are true, we have that an OL sentence X will be an instance of 'Tr' precisely when its translation p (and therefore X itself) is true. Thus, for interpreted languages, the problem of finding a formally correct and materially adequate $\phi(X)$ is a *precise* version of the problem of whether we can find a predicate whose extension is the class of true sentences in OL.

We can now restate the goal of Tarski's paper: it is, in effect, to determine the extent to which one can hope to find for arbitrary OLs a materially adequate and formally correct definition of truth.

Tarski's first target (section 1) was ordinary language, and there his results were entirely negative. In virtue of Lesniewski's version of the liar paradox (which Carnap had also discussed in LSL) Tarski concluded that Tr cannot be defined *within* ordinary language.[13] The question whether it can be defined *for* it from some other language is not examined due to the imprecise nature of ordinary language.

Tarski next turned (section 2) to the examination of formalized languages, raising the question whether truth can be defined (not within but) for them, from appropriate metalanguages. The languages with which Tarski was concerned here were formalized but by no means "formal" in Hilbert's sense. Indeed, he was unequivocal in his conviction that for "formal" lánguages the problem of defining truth "is not even meaningful" (p. 166), as it surely is not if one needs to assume a given class of true sentences as a starting point.[14]

> We shall always ascribe quite concrete and, for us, intelligible meanings to the signs which occur in the languages we shall consider. (p. 27)

These languages are, moreover, the vehicles of deductive systems of intuitively true claims. Thus they are normally assumed to be endowed with an axiomatic structure which captures, however partially, a domain of truth. "The sentences which are distinguished as axioms", Tarski explained, "seem to us to be materially true. . ." (p. 167). The intuitive conditions imposed on the languages under investigation therefore include

> a strictly determinate and understandable meaning of the constants, the certainty of the axioms, [and] the reliability of the rules of inference. . . (p. 211)

Moreover, the reason for including a translation of the OL axioms into the metalanguage is that

> as soon as we regard certain expressions as intelligible, or believe in the truth of certain sentences, no obstacle exists for using them as the need arises. (p. 121)

The object of Tarski's investigations were therefore, (not languages but) interpreted axiomatic systems of a mathematical nature—since certainty would hardly be claimed for any other systems—which embody fully interpreted, true, indeed, certainly true mathematical claims.[15]

The general type of language with which Tarski was concerned has a type-theoretic structure, even though he preferred the philosophical foundation provided by Husserl's theory of categories to the one displayed in Whitehead and Russell's *Principia Mathematica* (p. 215). Every meaningful expression in these languages has an *order* which is characterized in essentially the same way as Carnap's *levels* (p. 218): the order of individual names and variables is 1; the order of a predicate or functor is greater by one than the highest level of its arguments. Roughly speaking, the order of a predicate-expression F measures how many times we have to collect elements already available into new classes in order to have objects of the type appropriate to interpret the symbols in F.

Carnap had used the hierarchy of levels to observe that II contains infinitely many "concentric language-regions II_1, II_2, ..., which form an infinite series" ([1934], p. 88).

> Not counting [predicate and function symbols], all the symbols already occur in II_1, and thus in every region. . . In II_1, [predicates and functions of level 1] occur both as constants and as free variables, but not as bound variables. Further, in a region II_n (n = 2, 3, ...) [predicate and function expressions] occur as constants and as free variables up to the level n, but as bound variables only up to the level n-1. (The line of demarcation between II_1 and the further regions corresponds approximately to that between Hilbert's elementary and higher functional calculus.) (p. 88)

Tarski defined a *language of infinite order* as "one which contains variables of arbitrarily high order" ([1936] p. 20); thus, all of Carnap's $II_{i's}$ are of finite order, but their union, II itself, is of infinite order.

In sections 3 and 4 Tarski concentrated on languages of finite order. The *pièce de resistance* of the 1931 version occurs in these sections, where Tarski showed—against Carnap's expectations—that there is a very general strategy for defining a materially adequate truth-predicate in the extended syntax of such languages.

As is well known, the definition is based on a notion of satisfaction which differs from Carnap's concept of valuation in nothing of philosophical significance except for the fact that its range is not restricted to linguistic entities even at the first stage of the process. The main difference with Carnap arose in connection with the other major technical insight: the idea of defining truth recursively in a fashion that parallels the recursive definition of a well formed sentence. For example, when Carnap asked whether 'P v Q' is analytic in a given valuation v, his answer involved an often complex reduction process that transforms the sentence into another one and, in the end, into 'O = O' or 'O ≠ O'. On the other hand, when Tarski asked essentially the same question, his answer was "if P is analytic in v or Q is analytic in v."[16]

With the definition of Tr at hand, and using the metalinguistic translation of the OL axioms one can establish that the OL axioms are true and (presumably) that the OL rules preserve truth. Thus, in general (for a qualification, see the footnote on p. 237 of [1936]) one can prove in ML that all OL theorems are true. Since one can also show that some OL sentences are not true (the denials of the true ones), we have a metalinguistic proof of the consistency of the OL theory. As Tarski indicated, "The proof carried out by means of this method does not, of course, add much to our knowledge, since it is based upon premises which are at least as strong as the assumptions of the science under investigation." (p. 237) Indeed, this method amounts to asking the axioms under investigation to determine whether they are consistent. We already knew that if they are inconsistent they will lie, allowing us to prove (among other things) their own consistency. We also know, by Gödel's results, that in standard OL's the axioms will entail their consistency *only if* they are inconsistent. What Tarski added is that in their metalinguistic rendering axioms do not necessarily lie, for consistent axioms, no less than inconsistent ones, will entail the consistency of their OL translations.

Section 5 turns to the analysis of languages of infinite order. As we know, Carnap had defined in LSL a predicate ('analytic') whose extension is the class of sentences of II true under the standard arithmetical interpretation. Yet in this section Tarski would argue that such a truth-predicate cannot be defined either *in* or *for* infinitary languages. Languages of infinite order include variables of arbitrarily high finite order which, Tarski feared, require for their semantic treatment the use of expressions of "infinite order" in the metalanguage. "Yet," Tarski added,

> neither the metalanguage which forms the basis of the present investigations, nor any other of the existing languages, contains such expressions. It is in fact not at all clear what intuitive meaning could be given to such expressions. (p. 244)

Such a language would conflict with Husserl's and Lesniewski's theory of semantic categories[17]; but

> it is scarcely possible to imagine a scientific language in which the sentences have a clear intuitive meaning but the structure of which cannot be brought into harmony with [the doctrine of semantic categories]. (p. 215)

The reason behind this is that

> language, which is a product of human activity, necessarily possesses a 'finitistic' character, and cannot serve as an adequate tool for the investigation of facts, or for the construction of concepts, of an eminently 'infinitistic' character. (p. 253)

According to Tarski, the 1931 version of his monograph included only "certain suppositions" to the effect that truth was indefinable for languages of order ω (p. 247). After submitting the paper to the Polish Academy, however, Tarski read Gödel's [1931] and saw

a way to develop a rigorous version of the intended point. The basic
new result is Theorem I of section 5, stating for infinitary languages
that

> In whatever way the symbol 'Tr', denoting a class of expressions,
> is defined in the metatheory, it will be possible to derive from it
> the negation of one of the sentences which were described in . . .
> the convention (T), (p. 247)

i.e., a sentence entailing 'Tr(X) = p', where 'p' is replaced by
a translation of X. An immediate consequence is that

> assuming that the class of all provable sentences in the metatheory
> is consistent, it is impossible to construct an adequate definition of
> truth in the sense of convention (T) on the basis of the metatheory.
> (p. 247)

This result should not, of course, be confused with what is now
known as "Tarski's theorem", the claim that the truth-predicate
for OL is not definable *within* OL. That claim is, to be sure, contained in the proof of Theorem I, but the main point of this Theorem
(which he chose to emphasize, for example, in [1932], an abstract
of the 1933 version for the Viennese Academy) was the thesis that
one *cannot* define truth for an OL *even in its ML* when OL is of
infinite order. The basic reason was that the theory of semantic
categories determines that no meaningful language is more powerful than the language of order ω. To ask for the definition of truth
for an infinitary language is therefore to ask for the definition of
truth in a language within itself and, by Tarski's theorem, no consistent language can do that.[18]

Such were the basic philosophical ideas contained in the first
published version of Tarski's monograph. The German translation,
[1936], included a new postscript that registers a radical change
of perspective. Tarski withdrew the main negative conclusion of section 5; he no longer believed in the theory of semantical categories.
By appealing to Cantor's theory of transfinite ordinals he showed
that the same procedure used to define truth for languages of finite
order can be extended to apply to the infinite case.[19] Truth is, after
all, definable for languages of arbitrary order. The earlier negative
result of section 5 remains now in the form of the impossibility of
defining truth within the OL under investigation; but if the ML
is essentially stronger than its OL, the definition is always possible
in that ML.

3. TARSKI'S TRUTH AND CARNAP'S TRUTH

The analogies between many of the technical ideas in LSL and Tarski's monograph are striking. We have seen how Carnap's distinc-

tion of levels paralleled Tarski's "orders", and how, like Tarski (for the case of 'truth') Carnap explained that 'analytic' for a language of finite level can be defined in a language of higher level (e.g., section 34d, p. 113). Carnap also argued that we can do no better, i.e., that we cannot expect to define 'analytic' for a language of level n *within* that language on pain of contradiction (Theorem 60c.1, a verison of Tarski's theorem). And we have seen that he proved this by means of an argument which, like Tarski's, combined Lesniewski's and Gödel's insights. Moreover, like Tarski, Carnap realized that one could use the metalinguistic definition of 'analytic' to prove the consistency of the object-language: he proved the 'analyticity' of the axioms of II and the fact that its rules of inference preserve analyticity, and he concluded from the existence of non-analytic statements in II that the language was consistent. Yet, like Tarski, he reflected on the modest epistemological significance of this consistency proof (Tarski [1936] pp. 236-7; Carnap [1934] p. 129). Last, but not least, there was the role played by Carnap's *valuations* and their close link with Tarski's satisfaction-functions. Little wonder that in [1936] Tarski referred to section 34 of LSL as containing "quite similar ideas" (p. 277) to the ones developed in his monograph.

In two respects Carnap went even beyond Tarski: (i) as we saw, his "tolerance" allowed him to overcome the restrictions of semantic categories and put him in a position to acknowledge a predicate that defined the class of arithmetical truths in II; (ii) these same freedoms allowed him to recognize that the formal explication of consequence was not, as Tarski thought at first, impossible.

I would not dream of comparing Tarski's and Carnap's logical skills; but on the subject of semantics around 1933, on technical aspects in particular, Carnap and Tarski were not that far apart. It is therefore interesting to ask why it was that Tarski, rather than Carnap, got there first.

We have already detected one of the reasons: Carnap's nominalism, his decision to replace talk about objects by talk about their names. Gödel's letter of 1932 had dealt a death blow to this approach; but Carnap hoped at first that he could abandon his reductionist nominalism for all levels but the lowest. Thus, even though no effort was made to reduce quantification at all levels to anything remotely related to discourse about language (Carnap had, in effect, become a Platonistic homophonist[20] for higher level mathematical discourse), Carnap incongruously insisted on interpreting individual expressions as designating numerals. This waning nominalism was far from being the only source of confusion, however.

Verificationism had played a major role in the early stages of Carnap's work on syntax; in the first draft of LSL he had displayed an "intolerant" preference for what later on became language I; and we are familiar with his early reluctance to endorse concepts which are not effective or, as he called them, "definite." But, as we have also seen, this penchant for definitness was not strong enough to lead him to dismiss the call of the indefinite concepts emerging from Gödel's results.

The evolution of Carnap's logical thought in the early 30's is one of increasing acceptance of indefinite concepts and thereby of specific semantic patterns, slowly allowing for their introduction within the boundaries of meaningful discourse. This admission was gradual, however, and at the time of publication of LSL, it had not been completed. LSL still displayed a lingering verificationist bias and, together with the syntactical nominalism, this may have been the main reason for Carnap's failure to develop his ideas into a semantic framework.

At several places in LSL Carnap found himself bordering on the semantic conception of truth. A typical passage is the following one, where he turned to consider a sentence with descriptive primitives:

Let [the predicate-expression] pr_i be descriptive; here a valuation of the same kind as for [the predicate-variable] p is possible. Here also, [the sentence] S_1, in which pr_i occurs, will be called analytic if the evaluation on the basis of any valuation for pr_i leads to an analytic sentence. In contradistinction to the case of a p, however, s_1 will here only be called contradictory if the evaluation on the basis of any valuation of pr_i leads to a contradictory sentence. For, in the case of a p, s_1 means "So and so is true for every property", and this is false even if it does not hold for even one instance. Here, in the case of the pr_1, however, s_1 means: "So and so is true for the particular property expressed by pr_1" where we have a [descriptive predicate-expression] and therefore an empirical and not logically determinable property; and this sentence is only contradictory—that is to say, false on logical grounds—if there exists no property for which s_1 is true. (pp. 107-8)

Carnap was looking, at a non-logical sentence and asking when is it analytic and when is it contradictory? He was *not* asking, when it is true? But an answer to that question is implicit in his reasoning. Since the sentence with a descriptive predicate says that it is true (sic) for a particular interpretation of that predicate, in order to show that this claim is contradictory it does not suffice to show that under a different interpretation the sentence comes out mathematically false. In the middle of this reasoning Carnap found himself arguing, quite against his "better" judgment, about a sentence "true for the particular property expressed by pr_i". What, exactly, could he have thought that he meant by that phrase? Had he sat down to work this out, he would have encountered the "triviality" that to say that s_1 is true (under those circumstances) is to say that the coordinates under consideration were elements of the class associated with pr_i by the interpretation.

But it was not to be. In section 60b of [1934] Carnap raised explicitly the question whether truth can be defined within the confines of his syntactical techniques, and he argued that it cannot.

The argument—one of the worst Carnap ever advanced—begins with a Lesniewski-type analysis intended to show that the assumption that a truth-predicate can be defined in the object-language leads to a contradiction. Carnap assumed that the syntax of OL is formulated within OL, and that it contains the predicates 'Tr', 'Fa', and 'K' (for "true sentence", "false sentence" and "not a sentence", respectively). In their customary usage, he said, truth and falsehood are supposed to satisfy the following assumptions: (1) Let N be the name (in OL) of an expression S in OL; then Tr(N) iff S; (2) For every expression E, 'Tr(E)', 'Fa(E)' and 'K(E)' are sentences; (3) Every expression which is a sentence is either true or false but not both. But these assumptions, Carnap explained, lead to a contradiction. The sentence 'Fa(S)' suffices to show this, but in order to avoid the confusing effect of direct self reference Carnap derived the contradiction from the assumptions that

(A₁)   Tr(A₂)

and

(A₂)   Fa(A₁).

From (2) we have that either Tr('Fa(A₁)') or Fa('Fa(A₁)'). From the first disjunct and (1) we derive Fa(A₁), and thereby Fa('Tr(A₂)'). From (3) and (1) this leads to not-Tr(A₂) and thus to not-Tr('Fa(A₁)'). An analogous reasoning leads us from the second disjunct to *its* denial.[21]

"This contradiction," Carnap explained,

> only arises when the predicates 'true' and 'false' referring to sentences in a language S are used in S itself. On the other hand, it is possible to proceed without incurring any contradiction by employing the predicates 'true (in S_1)' in a syntax of S_1 which is not formulated in S_1 itself but in another language S_2. (p. 216)

Up to this point Carnap had paralleled Tarski's reasoning in the considerations prior to his introduction of the concept of truth. He was now posing the question of definability for the metalanguage. Like Tarski, Carnap had no objection to including in the metalanguage a full translation of the object language.[22] But he was so convinced that this extra portion can have no philosophical significance that he didn't even think it was worth keeping in mind. (Tarski would soon make him change his mind on this.) It is here that Carnap made the single most damaging error in his treatment of semantic matters, as he offered an utterly confused argument for the impossibility of defining 'truth' in a syntactical metalanguage. The reason why the definition of truth is beyond the scope of syntactical techniques is, he explained, that

> truth and falsehood are not proper syntactical properties; whether a sentence is true or false cannot be gathered in general from its design, that is to say, by the kinds and serial order of its symbols. (This fact has usually been overlooked by logicians, because, for the most part, they have been dealing not with descriptive but only with logical languages, and in relations to these, certainly, 'true' and 'false' coincide with 'analytic' and 'contradictory' respectively, and are thus syntactical terms.) (p. 216)

With the "wisdom" of hindsight many of Carnap's readers will readily grant the point Carnap has just made: truth is not a syntactical predicate and that is why semantics was such a major discovery, and its acknowledgement a revolution in Carnap's philosophy. This is wrong on two counts: Carnap's argument is indefensible and his conclusion is false. Truth *can* be defined in what he called the syntax of a language.

The latter point is obvious as soon as we realize that Carnap allowed his syntax languages to include translations of their object-languages;[23] he simply had no idea that this portion of the language could serve any interesting philosophical purpose. He thought of it, understandably, as excess baggage, a repetition of what one already had at a different level. Thus, Carnap's syntax-languages were indistinguishable from Tarski's semantic metalanguages and truth was consequently definable in them.

It follows that Carnap's argument against the possibility of defining truth in them must be wrong, and its flaw is revealing. Carnap said, to repeat, that "whether a sentence is true or false cannot, in general, be gathered from its design". In general, whether a sentence is true or false cannot "be gathered" through the techniques available in syntax. Notice that whether a sentence is analytic or not (in Carnap's sense) cannot be gathered from the techniques available in Carnap's syntax; and, more to the point, whether a sentence is true or false cannot be gathered by means of the techniques available in semantics either. What could have led Carnap to think that whether a concept C is definable on the basis of certain techniques depends on whether those techniques allow us to identify the instances of C? The obvious answer is: a verificationist

175

prejudice.

Carnap had radically departed from verificationist dogmas both in his endorsement of tolerance and in his rejection of the idea that understanding a sentence is essentially linked to the possibility of determining its truth-value. And yet, conversion is a complicated process. It is one thing to decide to abandon a church; to take your heart away is yet another. The decisive role played by indefinite concepts in LSL is testimony to the strength of Carnap's decision; but his failure to see the weakness in his argument against the syntactic characterization of truth shows that the process of conversion was far from completed.

* * * * * * * *

There is a story that Carnap used to tell his students about the first time Tarski explained to him his ideas on truth. They were at a coffee-house and Carnap challenged Tarski to explain how truth was defined for an empirical sentence such as "this table is black". Tarski answered that 'this table is black' is true iff this table is black; and then—Carnap explained—"the scales fell from my eyes".

A superficial observer will no doubt regard this as an extraordinarily silly response to an extraordinarily trivial observation. Placed in the context of our preceding analysis one may well see what Carnap meant, however. He had been *so* close to Tarski's idea: his metalanguages were of exactly the right kind, he had introduced the crucial idea of valuation and he had implicitly (and, upon occasion, explicitly) appealed to the idea of valuation to talk of nonformal truth. He had been the first to define truth for a particularly difficult case that Tarski regarded as intractable. But the *main* issue, the very *problem* of defining truth in general, had remained totally beyond his grasp. Carnap's *mathematical truth* does not appear in his work as a species of the genus *truth* which applies to languages independently of subject matter, and his *consequence* is similarly vitiated by a shortsighted dependence on the idiosyncracies of the languages for which it is defined. The price Carnap had to pay for his philosophical prejudices was that in order to reach his goal he had to apply extraordinarily convoluted and artificial methods that made it impossible to understand exactly what was going on. One can well imagine the embarrassment with which Carnap came to see that it could all have been done far better, and far more naturally and generally, if verificationism and his horror of reference had not burdened so heavily his train of thought. Little wonder that, unlike the other truth-fearing positivists (Neurath, Reichenbach, etc.), Carnap immediately and enthusiastically embraced Tarski's ideas.

Following a well established Russellian tradition, Carnap proceeded to recant from earlier sins, including very many that he had never committed. The misleading, simplistic picture which has become part of the folk-history of analytic philosophy stems largely from Carnap's unfriendly description of his syntax days. But the rest of us must also bear some blame for not having looked more carefully at the record.

NOTES

A version of this paper was read at the Conference on Philosophy of Mathematics held in October 1984 at Indiana University.

[1]It is possible to reconstruct part of the story of Carnap's writing of LSL from his correspondence with Schlick. On 7.12.31 Carnap writes that the text of his *Metalogic* (as he called it) is almost ready, consisting of 160 typed pages; on 15.3.32 he refers to a second part, not yet written; on 30.6.32 he reorts that part II is now ready except for a final, philosophical chapter (Gödel's comments in his letter of 11.9.32—see footnote 2 below—refer to part II also, which must therefore have included the treatment of Language II); on 28.11.32 Carnap sends to Schlick the concluding philosophical chapter of *Semantik* (as he now calls it, following a suggestion of Gödel and Behmann); on 19.6.33 he reports having devoted the preceding months to a revision of the book and to the writing of a new chapter (presumably, Part IV) on syntax for arbitrary languages. The manuscript of LSL was sent to Springer on 14.12.33 (letter of 21.12.33). (This Carnap-Schlick correspondence is in [RCC] 029-32.) It may be worth noting that on 11.10.34 Ernst Nagel wrote from Poland reporting that Adjukiewicz had received a copy of Carnap's book, and added: "He also asked me to express his regret at being unable to send you the German translation of Dr. Tarski's book. . ." ([RCC] 029-05-21), referring presumably to Tarski's [1936].

[2]We refer, of course, to published statements. A letter from Gödel to Carnap of 11.9.32 ([RCC] 102-43-05) indicates that—as one might expect—Gödel had also seen and perhaps even solved the problem. In that letter Gödel announces that in the sequel to [1931] he will include a characterization of truth.

[3]Tarski had claimed that since the rule requires infinitely many premises, "it cannot easily be brought into harmony with the current view of the deductive method, and finally that the possibility of its practical application in the construction of deductive systems seems to be problematic in the highest degree." ([1933] p. 249) Carnap, on the other hand, thought that there is "nothing to prevent the practical application of this rule." ([1934] p. 173) It would be impossible to prove infinitely many theorems in the object-language one by one; but a metatheorem may prove them all at once.

[4]The key reason for Tarski's negative conclusion was his commitment to the theory of semantic categories. The link will be explained below.

.   In his 1935 contribution to the Paris conference Tarski withdrew these negative conclusions and offered a definition of consequence (see chapter XVI of Tarski [1956]). After stating that "the first attempt to formulate a precise definition of the proper concept of consequence was that of R. Carnap," (p. 413) he added in a footnote (concerning Tarski's [1933]): "My position at that time is explained by the fact that when I was writing the article mentioned, I wished to avoid any means of construction which went beyond the theory of logical types in any of its classical forms. . ." He then went on to indicate that "in his extremely interesting book" Carnap has properly emphasized the need to distinguish derivability from consequence." (p. 413)

[5]It is here that Gödel added: "On the basis of this thought I will give a definition of "true" in the second part of my work. . ." (He was probably referring to part II of his [1931].)

[6]In LSL Carnap gave a different reason for abandoning the substitutional interpretation: "it may happen that although all these sentences [the 'M(P$_i$)'s] are true, 'M(F) is nevertheless false—insofar as M does not hold for a certain property for which no predicate

can be defined in II. . . . We will follow Gödel's suggestion and define 'analytic' in such a way that 'M(F)' is only called analytic if M holds for every numerical property, irrespective of the limited domain of definitions which are possible in II." (pp. 106-107)

In [1931] Tarski had discussed the problem of definability, noticing that, in view of the enumerability of the propositional functions in any given language, there had to be non-denumerably many classes of reals which are not definable in any given system.

[7]There is no indication in the correspondence with Gödel or in the notes Carnap took of conversations with him that the idea of valuation was suggested by Gödel—or by anyone else. This, together with the fact that Carnap was extremely careful, indeed, generous in attributing ideas and suggestions to others, makes it likely that the idea is entirely his own.

[8]I.e., in the fashion that has become obvious since Tarski.

[9]Let it be emphasized that the reason why we are justified in making this statement is not *primarily* that it is true but that we can *prove* it in the language on which we are standing as we make our syntactical considerations. Carnap had explained that "the method of derivation [as opposed to the method of consequence] always remains the fundamental method; every demonstration of the applicability of any term is ultimately based upon a derivation. Even the demonstration of the existence of a consequence-relation. . . can only be achieved by means of a derivation (a proof) in the syntax-language." ([1934] p. 39)

[10]See footnote 1 of [1936], p. 386.

[11]The exclusion of semantic primitives establishes that if the OL is a system of physical theory then all one needs to define truth for it is the primitives of physics and those of syntax. This was intended to make the notion of truth acceptable to physicalists.

[12]It is worth emphasizing that Tarski's convention (T) demands the provability (in ML) of $O(X) = \cdot p$, not its truth. Tarski would likely have agreed with Carnap's remark in LSL concerning deducibility and consequence (see above, p. 70). Hence Tarski's decision to endow ML with an axiomatization M.

[13]Anil Gupta has recently shown that the scope of Tarski's claim is narrower than previously thought. See his "Truth and Paradox," *Journal of Philosohical Logic 11* (1982) 1-60.

[14]*We* may see Tarski's languages as the conjunction of a formal language plus an interpretation and then abstract from what he is doing the notion of "truth in an interpretation." But, in the Fregean spirit, Tarski considers only two types of linguistic structures: (i) languages endowed with a fixed, unique interpretation and (ii) "sciences to the signs and expressions of which no contentual sense is attached" for which "the problem here discussed has no relevance." (p. 166)

[15]That Tarski's focus was on axiomatized OL's and, indeed, on mathematical theories is emphasized by the fact that when Tarski discusses the possibility of offering a "structural definition" of truth, i.e., one in terms of a meta-linguistic axiomatic system, he fails to include among objections the otherwise obvious and devastating one that the procedure couldn't possibly work for contingent sentences (see pp. 163-4). Even so, Tarski was far more keenly aware than Carnap of the fact that the definition of truth "alone gives no general criterion for the truth of a sentence" and that in this respect it "does not differ at all from the greater part of the definitions which occur in the deductive sciences". (p. 197) As we shall see below, Carnap had explained essentially the same point to Schlick; but he had lost his grasp of it once again by the time he raised the question whether one can hope to define truth in syntax.

[16]Kleene may have been the first to see that Carnap's definition could be simplified by reformulating it in a way that parallels the definition of well formed formula; see his review of LSL in [1939a] (especially p. 83), and also in his [1939b] (available from the author; I am grateful to Frank Pecchioni for bringing this item to my attention). In his review of this latter work in [1939b] Carnap wrote: "When Tarski constructed his method for defining the semantical concept of truth, it became clear that an analogous method could be used in syntax for the definition of 'analytic' (or 'provable') in systems with indefinite rules. It seems that the simplified definition proposed by Kleene is this analogue." (p. 158)

[17]The notion of semantic category is similar to Wittgenstein's idea of *form*; it is even closer to Carnap's concept of Genus (*Gattung*; see [1934] sec. 46). Carnap did not endorse the theory of semantic categories and was thereby in a position to offer in LSL a definition of truth for the infinitary language II. Carnap apparently learned the use of transfinite levels from Hilbert and Gödel. ([1934] p. 189)

[18]Having abandoned the idea of defining truth explicitly for infinitary languages, Tarski

explored the possibility of introducing the concept of truth in the appropriate metalanguages by appealing to the axiomatic method, introducing 'truth' as an undefined metalinguistic primitive and adding a number of axioms (including, perhaps, all instances of convention (T)) to specify its meaning. Even though this axiomatic account is hardly philosophically illuminating, Tarski notes that, in a certain sense, the resulting theory would be categoric. If the axiom of infinite induction were added, making the metalinguistic theory ω-complete, then any two interpretations of Tr would be coextenstional.

[19] We know that Tarski had read carefully and corrected a number of errors in Carnap's LSL before its publication in English in 1937. (Carnap thanked Tarski in 1935 for some of the corrections—see, e.g., [1934] p. 88.) It would be interesting to know if Carnap's more "tolerant" attitude on this semantic matter had any influence on Tarski's change of mind.

[20] Carnap's "homophonism" is discussed in my "Idealism and the *Aufbau*," in N. Rescher, ed., *The Legacy of Logical Positivism* (University Press of America, Lanham, MD, 1985) 133-155.

[21] The failure to allow for a concept of truth prevented Carnap from offering a coherent version of Lesniewski's argument. Presumably, what he wanted to state in his condition (1), for example, is that the sentence 'Tr(N) iff S' is true in the OL; but the closest he would have wanted to get to that statement would be the claim that the sentence in question is determined by given rules. As stated, Carnap's reasoning does not observe the use-mention distinction.

[22] Carnap occasionally suggested otherwise (e.g., Carnap [1963], p. 60, third paragraph), but this is only a manifestation of his extreme reluctance (so uncommon among his Viennese contemporaries) to present his earlier work in a favorable light.

[23] See, for example, [1934] p. 228.

REFERENCES

Unpublished Sources

[RCC] Rudolf Carnap Collection, Special Collections Department, University of Pittsburgh Libraries. The numbers included in a reference to an item in the collection indicates the [RCC] folder numbers. Quotations are by permission of the University of Pittsburgh. All rights reserved.

Published Sources

Carnap, R. [1934] *The Logical Syntax of Language* (Kegan Paul, London, 1937), A. Smeaton, trans. Translation of *Logische Syntax der Sprache* (Springer, Wien, 1934).

———. [1940] Review of S. C. Kleene, "On the term 'analytic' in logical syntax," *Journal of Symbolic Logic 5* (1940) 157-158.

———. [1963] "Intellectual Autobiography," *The Philosophy of Rudolf Carnap*, P. A. Schilpp, ed. (Open Court, LaSalle Illinois, 1963) 3-84.

Gödel, K. [1931] "Über formal unentscheidbare Satze der Principia Mathematica und verwandter Systeme I," *Monatshefte für Mathematik und Physik 38* (1931) 173-198, translated by B. Meltzer in K. Gödel, *On Formally Undecidable Propositions of Principia Mathematica and Related Systems* (Oliver and Boyde, Edinburgh and London, 1962) 35-72.

Kleene, [1939a] Review of R. Carnap, *The Logical Syntax of Language*, *Journal of Symbolic Logic 4* (1939) 82-87.

———. [1939b] "On the term 'analytic' in logical syntax," preprint for members of Fifth International Congress for the Unity of Science, Cambridge, Mass., 1939.

Tarski, A. [1931] "On Definable Sets of Real Numbers," Chapter IV of Tarski [1956], supplemented translation of "Sur les ensembles definissables de nombres reels. I. "*Fundamenta Mathematicae 17* (1931) 210-239.

———. [1932] "Der Wahrheitsbegriff in den Sprachen der deduktiven Disziplinen," *Akad. der Wissenschaften in Wien, Mathematisch-naturwissenschaftliche Klasse LXIX* (1932) 33-35.

———. [1933] "Some Observations on the Concepts of ω-consistency and ω-completeness," Chapter IX of Tarski [1956]; translation of "Einige Betreachtungen über die Begriffe ω-Widerspruchsfreiheit und der ω-Vollstandigkeit," *Monatshefte fur Mathematik und Physik 40* (1933) 97-112.

———. [1935] "Some Methodological Investigations on the Definability of Concepts," Chapter

X of Tarski [1956], supplemented translation of "Einige methodologische Untersuchungen über die Definierbarkeit der Begriffe," *Erkenntnis 5* (1935) 80-100.

_____. [1936] "The Concept of Truth in Formalized Languages," Chapter VII of Tarski
[1956]; translation of "Der Wahrheitsbegriff in den formalisierten Sprachen," *Studia
Philosophica 1* (1936) 261-405.

_____. [1956] *Logic, Semantics Metamathematics*, J. H. Woodger, translator (Oxford Univers-
ity Press, London, 1956).

Wittgenstein, L. [1921] *Tractatus logico-philosophicus* (Routledge & Kegan Paul, London, 1922),
C. K. Ogden, trans.; first published in German in 1921.

THE JOURNAL OF PHILOSOPHY

VOLUME LXXXVIII, NO. 10, OCTOBER 1991

THE RE-EVALUATION OF LOGICAL POSITIVISM*

I T is now well over half a century since the heyday of the philosophical movement known as logical positivism or logical empiricism. Depending on how one counts, it is now approaching
half a century since the official demise of this movement.[1] Since that
demise it has naturally been customary to view logical positivism as a
kind of philosophical bogeyman whose faults and failings need to be
enumerated (or, less commonly, investigated) before one's favored
"new" approach to philosophy can properly begin. And such an
attitude toward logical positivism and its demise has been widely
prevalent, not only in the narrower community of philosophers of
science (who have characteristically proceeded against the background of Thomas Kuhn's well-known critique), but also in the
broader philosophical community as well. With our increasing historical distance from logical positivism, however, a more dispassionate attitude has also inevitably begun to emerge. No longer threatened or challenged by logical positivism as a live philosophical option, it is becoming increasingly possible to consider this movement
as simply a part of the history of philosophy which, as such, can be
investigated impartially from a historical point of view. Indeed, we
have seen in recent years a veritable flowering of historically oriented reconsiderations of logical positivism.[2]

---

* To be presented in an APA symposium on Logical Positivism, December 28.
David Weissman and Richard Creath will comment. See this JOURNAL, this issue,
520-1 and 522-3, respectively, for their contributions.
[1] I assume that this event took place sometime between the publication of W. V.
Quine's "Two Dogmas of Empiricism" (1951), and that of Thomas Kuhn's *The
Structure of Scientific Revolutions* (1962).
[2] I have in mind, for example, work in this country by A. Coffa, R. Creath, B.
Dreben, W. Goldfarb, D. Howard, J. Lewis, T. Oberdan, A. Richardson, T. Rick-
etts, T. Ryckman, H. Stein, T. Uebel, Z. Zhai; work in Great Britain by S. Haack,
A. Hamilton, J. Skorupski; work on the Continent by H.-J. Dahms, R. Haller, R.

0022-362X/91/8810/505-19
© 1991 The Journal of Philosophy, Inc.

In the course of these reconsiderations it has become clear—not at all surprisingly, of course—that the above-mentioned post-positivist reaction gave birth to a large number of seriously misleading ideas about the origins, motivations, and true philosophical aims of the positivist movement. (One can hardly expect philosophical critics, concerned largely with their own agendas rather than with historical fidelity, to generate anything other than stereotypes and misconceptions.) I shall discuss what I take to be some of the most important of such misleading ideas in what follows. But I also hope to show—or at least to suggest—that achieving a better understanding of the background, development, and actual philosophical context of logical positivism is not merely of historical interest. For the fact remains that our present situation evolves directly—for better or for worse—from the rise and fall of positivism, and what I want to suggest is that we will never successfully move beyond our present philosophical situation until we attain a properly self-conscious appreciation of our own immediate historical background.

I

Perhaps the most misleading of the stereotypical characterizations views logical positivism as a version of philosophical "foundationalism." The positivists—so this story goes—were concerned above all to provide a philosophical justification of scientific knowledge from some privileged, Archimedean vantage point situated somehow outside of, above, or beyond the actual (historical) sciences themselves. More specifically, they followed the lead of the logicist reduction of mathematics to logic—where the latter is also understood as fundamentally foundationalist in motivation and import. Just as the logicists attempted to justify mathematical knowledge and place it on a secure foundation by means of a derivation from (supposedly more certain) logical knowledge, so the positivists attempted to justify empirical science and place it on a secure foundation by logically constructing the concepts of empirical science on the basis of the (supposedly more certain) immediate data of sense. Thus, formal logic furnished the foundational enterprise with the required Archimedean standpoint located outside of the actual (historical) sciences themselves, and phenomenalist reductionism carried out rigorously

---

Hegselmann, M. Heidelberger, K. Hentschel, A. Kamlah, D. Koppelberg, U. Majer, V. Mayer, C. Moulines, P. Parrini, J. Proust, H. Rutte, W. Sauer, W. Vossenkuhl, G. Wolters. In addition, the *Vienna Circle Collection*, under the general editorship of R. Cohen, has recently made available in English many previously untranslated works of Schlick, Neurath, Reichenbach, Menger, and Waismann. Carnap's earlier works are unfortunately not yet available in English translation.

using the tools of formal logic (as epitomized in Carnap's *Der lo-gische Aufbau der Welt*[3] of 1928) then provided the desired episte-mological justification of the sciences.

This conception of the aims and posture of philosophy vis-à-vis the special sciences represents an almost total perversion of the actual attitude of the logical positivists, who rather considered their intellectual starting point to be a *rejection* of all such philosophical pretensions. An eloquent example is found in the first paragraph of a paper by Schlick[4] from 1915 on the need for philosophy to adapt itself to the new findings of relativity theory:

> We have known since the days of Kant that the only fruitful method of all theoretical philosophy consists in critical inquiry into the ultimate principles of the special sciences. Every change in these ultimate ax-ioms, every emergence of a new fundamental principle, must therefore set philosophical activity in motion, and has naturally done so even before Kant. The most brilliant example is doubtless the birth of mod-ern philosophy from the scientific discoveries of the Renaissance. And the Kantian Critical Philosophy may itself be regarded as a product of the Newtonian doctrine of nature. It is primarily, or even exclusively, the principles of the exact sciences that are of major philosophical importance, for the simple reason that in these disciplines alone do we find foundations so firm and sharply defined, that a change in them produces a notable upheaval, which can then also acquire an influence on our world-view (*ibid.*, p. 153).

Schlick goes on to argue that neither of the two prevailing philo-sophical systems—neither the "neo-Kantianism" of Cassirer and Natorp nor the "positivism" of Petzoldt and Mach—can do justice to Einstein's new theory and *therefore both systems must be aban-*

---

[3] *The Logical Structure of the World*, R. George, trans. (Berkeley: California UP, 1967) [hereafter *Aufbau*].

The most explicit example of this kind of characterization of logical positivism of which I am aware is found in R. Giere, *Explaining Science: A Cognitive Approach* (Chicago: University Press, 1988), pp. 22–8. This kind of conception is certainly aided and abetted (although with important reservations in each case) by Quine's "Epistemology Naturalized" (1968) and A. J. Ayer's *Language, Truth and Logic* (1936). Richard Rorty has recently portrayed the entire modern philo-sophical tradition as based on "foundationalist epistemology," under which ru-bric the logical positivists are also clearly thought to fall: see *Philosophy and the Mirror of Nature* (Princeton: University Press, 1979), pp. 59, 332–3 (and it is thus clear for Rorty that Kuhn's critique is predicated on a rejection of such foundationalist epistemology).

[4] "Die philosophische Bedeutung des Relativitätsprinzips," *Zeitschrift für Phi-losophie und philosophische Kritik*, CLIX (1915): 129–75—P. Heath, trans. "The Philosophical Significance of the Principle of Relativity," in H. Mulder and B. van de Velde-Schlick, eds., *Moritz Schlick: Philosophical Papers* (Dordrecht: Reidel, 1979), Vol. I (part of the above mentioned *Vienna Circle Collection*).

doned. Entirely new philosophical principles—based on the work of Einstein himself and of Poincaré—are necessarily required.

For Schlick, then, philosophy as a discipline is in no way foundational with respect to the special sciences. On the contrary, it is the special sciences that are foundational for philosophy. The special sciences—more specifically, the "exact sciences"—are simply taken for granted as paradigmatic of knowledge and certainty. Far from being in a position somehow to justify these sciences from some higher vantage point, it is rather philosophy itself that is inevitably in question. Philosophy, that is, must *follow* the evolution of the special sciences so as to test itself and, if need be, to reorient itself with respect to the far more certain and secure results of these sciences. In particular, then, the central problem of philosophy is not to provide an epistemological foundation for the special sciences (they already have all the foundation they need), but rather to redefine its own task in the light of the recent revolutionary scientific advances that have made all previous philosophies untenable.[5]

Moreover, this conception of the proper stance of philosophy vis-à-vis the special sciences is not peculiar to Schlick, it is in fact characteristic of the logical positivists generally. This is so even—and indeed especially—of Carnap's *Aufbau* (which work is, of course, standardly supposed to be most representative of foundationalist epistemology). It is true that the *Aufbau* presents a phenomenalist reduction of all concepts of science to the immediately given data of experience. Yet the point of this construction has little if anything to do with traditional foundationalism. First of all, Carnap shows no interest whatever in the philosophical skepticism that motivates (for example) Russell in *Our Knowledge of the External World* (1914), nor does the text of the *Aufbau* at any point engage the traditional vocabulary of "certainty," "doubt," "justification," and so on.[6] Secondly, and more importantly, Carnap is perfectly

---

[5] As Schlick indicates, this is also a fruitful way of viewing Kant's philosophical stance with respect to the special sciences. For Kant, too, it is philosophy itself rather than the special sciences which is in question and which therefore requires justification. Kant's aim is not to ground the sciences in something firmer and more secure (for it is they that are paradigmatic of certainty) but rather to reform *metaphysics* in accordance with the already achieved successes of the exact sciences. See the "Preface" to the second edition of the *Critique of Pure Reason* (especially Bxxii–xxiii), §VI of the "Introduction," and §40 of the *Prolegomena.* This undercuts decisively, it seems to me, Rorty's portrayal (*op. cit.*) of Kant as the *echt*-foundationalist.

[6] Carnap does employ this language in retrospectively describing the motivation of the *Aufbau* at several points in his "Intellectual Autobiography," in P. Schilpp, ed., *The Philosophy of Rudolf Carnap* (La Salle: Open Court, 1963), pp. 51, 57. Nevertheless, the contrast between this retrospective account and the language of the *Aufbau* itself is striking indeed.

explicit that the particular constructions he employs depend entirely on the actual "results of the empirical sciences" (§122) and, accordingly, that the particular constructional system he presents is best viewed as a "rational reconstruction" of the actual (empirical) process of cognition (§100).[7] Thus, for example, Carnap's choice of holistic "elementary experiences" rather than atomistic sensations as the basis of his system is grounded in the empirical findings of Gestalt psychology (§67), his definition of the visual sense modality depends on the (supposed) empirical fact that it is the unique sense modality having exactly five dimensions (§86), and so on. In particular, then, the findings of the special sciences (such as empirical psychology) are in no way in question in the *Aufbau*: once again, it is philosophy that must adapt itself to them rather than the other way around.

The aim of the *Aufbau*, therefore, is not to use logic together with sense data to provide empirical knowledge with an otherwise missing epistemological foundation or justification. Its aim, rather, is to use recent advances in the science of logic (in this case, the Russellian type theory of *Principia Mathematica*) together with advances in the empirical sciences (Gestalt psychology, in particular) to fashion a scientifically respectable *replacement* for traditional epistemology. Carnap's depiction of the construction of scientific knowledge from elementary experiences via the logical techniques of *Principia Mathematica* enables us to avoid the metaphysical excesses of the traditional epistemological schools—"realism," "idealism," "phenomenalism," "transcendental idealism" (§177)—while, at the same time, capturing what is correct in all of these schools: allowing us to represent, in Carnap's words, the "neutral basis" (*neutrale Fundament*) common to all (§178).[8]

II

According to the standard picture of logical positivism briefly sketched above, that movement is to be understood not only as a species of foundationalist epistemology but also as a version of *empiricist* epistemology in the tradition of Locke, Berkeley, Hume, Mach, and Russell's external-world program.[9] The immediately

---

[7] Cf. Schilpp, *op. cit.*, p. 18.

[8] For more on this kind of approach to the *Aufbau*, see my "Carnap's *Aufbau* Reconsidered," Noûs, XXI (1987): 521–45; A. Richardson, "How Not to Russell Carnap's *Aufbau*," PSA, I (1990): 3–14, and A. Richardson, *Epistemology Purified: Objectivity, Logic, and Experience in Carnap's* Aufbau (Doctoral Dissertation: University of Illinois at Chicago, 1990).

[9] This is, of course, the conception promulgated especially in Ayer's *Language, Truth and Logic:* compare the first sentence of the "Preface" to the first edition, where the views put forward are characterized as "the logical outcome of the empiricism of Berkeley and David Hume."

given data of sense are viewed as the primary examplars of knowledge and certainty, and all other putative claims to knowledge are judged to be warranted—or, as the case may be, unwarranted—in light of their relations to the immediate data of sense. (When empiricist reductionism is acknowledged to have failed, the positivists' naive empiricism is then thought to express itself in the doctrine of an epistemically privileged and theory-neutral observation language against which all scientific claims are to be tested.) I have dealt with the conception of the positivists as foundationalists above, and this question is, I think, rather easily disposed of. The question of empiricism—and, in particular, of the relationship between the positivists and the traditional empiricism of Locke, Berkeley, Hume, and Mach —is, however, considerably more delicate.

The first point to notice is that the positivists' main philosophical concerns did not arise within the context of the empiricist philosophical tradition at all. The initial impetus for their philosophizing came rather from late nineteenth-century work on the foundations of geometry by Riemann, Helmholtz, Lie, Klein, and Hilbert—work which, for the early positivists, achieved its culmination in Einstein's theory of relativity.[10] The principal philosophical moral the early positivists drew from these geometrical developments was that the Kantian conception of pure intuition and the synthetic a priori could no longer be consistently maintained—consistently, that is, with the situation now presented to us by the exact sciences. In particular, Hilbert's logically rigorous axiomatization of Euclidean geometry (1899) shows conclusively that spatial intuition has no role to play in the reasoning and inferences of pure geometry, and the development of non-Euclidean geometries, together with their actual application to nature by Einstein, shows conclusively that our knowledge of geometry cannot be synthetic a priori in Kant's sense. All the early positivists were thus in agreement that the strictly Kan-

---

[10] Most of the early writings of the positivists focused on these revolutionary mathematical-physical developments. In addition to the 1915 paper of Schlick cited above, see his *Raum und Zeit in der gegenwärtigen Physik* (1917)—*Space and Time in Contemporary Physics*, H. L. Brose, trans. (1920) reprinted in H. Mulder, *op. cit.*; H. Reichenbach, *Relativitätstheorie und Erkenntnis apriori* (1920)—*The Theory of Relativity and A Priori Knowledge*, M. Reichenbach, trans. (Berkeley: California UP, 1965); R. Carnap, *Der Raum: Ein Beitrag zur Wissenschaftslehre.* Ergänzungsheft Nr. 56 of *Kantstudien* (1922)—*Space: A Contribution to the Theory of Science*, M. Friedman and P. Heath, trans., forthcoming. Interest in these themes was also stimulated by Weyl's 1919 edition of Riemann's 1854 dissertation and the 1921 edition of Helmholtz's geometrical writings by Schlick and the physicist P. Hertz—the latter is translated by M. Lowe as *Hermann von Helmholtz: Epistemological Writings*, R. Cohen and Y. Elkana, eds. (Dordrecht: Reidel, 1977), as part of the *Vienna Circle Collection.*

tian conception of the a priori must be rejected, and this rejection of the synthetic a priori constituted a centrally important element in what they came to call their "empiricism."

Yet it is equally important to notice, in the second place, that the positivists did not react to the demise of the Kantian synthetic a priori by adopting a straightforwardly empiricist conception of physical geometry of the kind traditionally imputed to Gauss (who is reported to have attempted to determine the curvature of physical space by measuring the angle sum of a terrestrial triangle determined by three mountain tops). On the contrary, all the early positivists also strongly rejected this kind of empiricist conception (which they attributed to Gauss, Riemann, and, at times, Helmholtz) and rather followed the example of Poincaré in maintaining that there is no direct route from sense experience to physical geometry: essentially nonempirical factors, variously termed "conventions" or "coordinating definitions," must necessarily intervene between sensible experience and geometrical theory. The upshot is that it is in no way a straightforward empirical matter of fact whether space is Euclidean or non-Euclidean.[11]

This radically new conception of physical geometry—neither strictly Kantian nor strictly empiricist—was formulated by Reichenbach in an especially striking fashion in his first book, *The Theory of Relativity and A Priori Knowledge* (1920). Reichenbach maintains a sharp distinction, within the context of any given scientific theory, between two intrinsically different types of principles: "axioms of connection" are empirical laws in the traditional sense recording inductive regularities involving terms and concepts that are already sufficiently well-defined; "axioms of coordination," on the other hand, are nonempirical statements that must be antecedently laid down before the relevant terms and concepts have a well-defined subject matter in the first place. (Thus, for example, Gauss's attempt empirically to determine the curvature of space via a terrestrial triangle inevitably fails, because it tacitly presupposes that light rays travel in straight lines and, therefore, that the notion of "straight line" is already well-defined—but how, independently of the geometrical and optical principles supposedly being tested, can this pos-

---

[11] See esp. *La Science et L'Hypothèse* (1902)—*Science and Hypothesis*, J. Larmor, trans. (New York: Dover, 1952); compare the famous remark (at the conclusion of ch. V.5): "Whichever way we look at it, it is impossible to discover in geometrical empiricism a rational meaning." Einstein himself frequently expressed admiration for Poincaré's doctrine: see, e.g., "Geometrie und Erfahrung" (1921)—"Geometry and Experience," G. Jeffrey and W. Perret, trans. in *Sidelights on Relativity* (New York: Dutton, 1923).

sibly be done?) These nonempirical axioms of coordination—which include, paradigmatically, the principles of physical geometry—are thus "constitutive of the object of knowledge," and, in this way, we can therefore vindicate *part* of Kant's conception of the a priori.

For, according to Reichenbach, the notion of aprioricity had two independent aspects in Kant: the first involves necessary and unrevisable validity, but the second involves only the just-mentioned feature of "constitutivity." The lesson of modern geometry and relativity theory, then, is not that the Kantian a priori must be abandoned completely, but rather that the constitutive aspect must be separated from the aspect of necessary validity. Physical geometry is indeed nonempirical and constitutive—it is not itself subject to straightforward observational confirmation and disconfirmation but rather first makes possible the confirmation and disconfirmation of properly empirical laws (viz., the axioms of connection). Nevertheless, physical geometry can still evolve and change in the transition from one theoretical framework to another: Euclidean geometry, for example, is a priori in this constitutive sense in the context of Newtonian physics, but only topology (sufficient to *admit* a Riemannian structure) is a priori in the context of general relativity.

Reichenbach concludes that traditional empiricism is in error in not recognizing the a priori constitutive role of axioms of coordination, and it is clear, moreover, that, despite some terminological wrangling on this point, the other logical positivists are in substantial agreement.[12] One of the central themes of Schlick's *General Theory of Knowledge* (1918), for example, is a sharp dichotomy between raw sensible acquaintance and genuine objective knowledge. Immediate contact with the given is both fleeting and irredeemably subjective. Objective knowledge therefore requires *concepts* and *judgments*, which are to be carefully distinguished from intuitive sensory presentations. Concepts and judgments are in fact only possible in the context of a rigorous formal system, of which Hilbert's axiomatization of geometry is paradigmatic. More

---

[12] The terminological wrangling in question arises in correspondence between Schlick and Reichenbach in November of 1920 concerning Reichenbach's book. Schlick criticizes Reichenbach for conceding too much to the Kantian side and argues that Reichenbach's a priori constitutive principles should rather be understood as conventions à la Poincaré. This correspondence is discussed in A. Coffa, *The Semantic Tradition from Kant to Carnap* (New York: Cambridge, 1991), ch. 10. Schlick himself alludes to this correspondence in "Kritizistische oder empiristische Deutung der neuen Physik?" *Kantstudien*, XXVI (1921)—"Critical or Empiricist Interpretation of Modern Physics?" P. Heath, trans., in H. Mulder, *op. cit.*

specifically, the Hilbertian notion of "implicit definition" of scientific concepts via their logical places in a formal system (which notion is explicitly associated by Schlick with Poincaré's conventionalist philosophy of geometry) can alone explain how rigorous, exact, and truly objective representation is possible. We are here obviously very far from traditional empiricism and very close indeed to the supposedly antipositivist doctrine of the theory ladenness of observation.[13]

Once again, the same is true even—and indeed especially—of Carnap's *Aufbau*. Carnap outlines an elaborate construction of all scientific concepts from the immediately given data of sense, using the logical machinery of *Principia Mathematica*. Yet he also holds that the objective meaning of scientific concepts can in no way depend on merely ostensive contact with the given. On the contrary, intersubjective communication is possible only in virtue of the *logical structure* of the concepts in question arising from their logical places within the total system of scientific knowledge. More specifically, intersubjective meaning must derive entirely from what Carnap calls "purely structural definite descriptions" rather than from sensory ostension (§§12-5). (Thus, for example, the visual sense modality is defined in terms of the purely formal properties of its dimensionality rather than in terms of its phenomenal content.) In this sense, the elaborate logical structure erected above the basic elements of the system (the elementary experiences) is actually more important than the basic elements themselves. Objective meaning flows from the top down, as it were, rather than from the bottom up.[14]

Carnap illuminatingly articulates the precise meaning of the resulting "empiricism" in the "Preface" to the second edition (1961!) of the *Aufbau*:

> For a long time, philosophers of various persuasions have held the view that all concepts and judgments result from the cooperation of experience and reason. Basically, empiricists and rationalists agree in

---

[13] For further discussion of Schlick, see my "Moritz Schlick's *Philosophical Papers*," *Philosophy of Science*, L (1983): 498–514. Theory ladenness follows from the circumstance that *all* scientific concepts—including those used to report the results of experiments—acquire objective meaning from their places in a theoretical system. For further discussion of this point, see my "Philosophy and the Exact Sciences: Logical Positivism as a Case Study," in J. Earman, ed., *Empiricism, Explanation and the Philosophy of Science* (Berkeley: California UP, forthcoming).

[14] For further discussion see again the references cited in fn. 8 above.

this view, even though both sides give a different estimation of the two factors, and obscure the essential agreement by carrying their viewpoints to extremes. The thesis which they have in common is frequently stated in the following simplified version: The senses provide the material of cognition, reason works up [*verarbeitet*] the material so as to produce an organized system of knowledge. There arises then the problem of finding a synthesis of traditional empiricism and traditional rationalism. Traditional empiricism rightly emphasized the contribution of the senses, but did not realize the importance of logical and mathematical forms. . . . I had realized, on the one hand, the fundamental importance of mathematics for the formation of a system of knowledge and, on the other hand, its purely logical, purely formal character to which it owes its independence from the contingencies of the real world. These insights formed the basis of my book. . . . This orientation is sometimes called "logical empiricism" (or "logical positivism"), in order to indicate the two components (*op. cit.*, pp. v–vi).

In thus emphasizing the central importance of a priori formal elements in first providing objective meaning for the otherwise undigested immediate data of sense, logical positivism has, it seems to me, broken decisively with the traditional empiricism of Locke, Berkeley, Hume, and Mach—which tradition is understood by the positivists themselves as mistakenly giving epistemic centrality to precisely such undigested immediate sensory data. Perhaps the best way to put the point is that the logical positivists have staked out an entirely novel position that is, as it were, intermediate between traditional Kantianism and traditional empiricism: it gives explicit recognition to the constitutive role of a priori principles, yet, at the same time, it also rejects the Kantian characterization of these principles as *synthetic* a priori.

III

The novel philosophical position briefly sketched above is, however, faced with several fundamental problems. The heart of the positivists' new conception is the idea of constitutive but nonsynthetic a priori principles underlying the possibility of genuinely objective scientific knowledge. There is a sharp distinction, in particular, between conventions, coordinating definitions, or axioms of coordination, on the one hand, and empirical principles properly so-called, on the other. On the one side lie the principles of pure mathematics and (at least in the context of some physical theories) the principles of physical geometry, on the other side lie standard empirical laws such as Maxwell's equations and the law of gravitation. But what is the basis for this distinction, and how, more generally, are we sharply to differentiate the two classes of principles? On a strictly Kantian view, such a distinction is, of course, grounded in the fixed constitution of our cognitive faculties, and the question of differentiation is correspondingly straightforward: the a priori principles are precisely those possessing necessary and unrevisable validity.[15] Now, however, we have explicitly acknowledged that a priori constitutive principles possess no such necessary validity and, in fact, that these principles may evolve and change—in response to empirical findings—with the progress of empirical science. So what exactly distinguishes our a priori principles from ordinary empirical laws properly so-called?

A second problem is perhaps even more fundamental. The logical positivists, I have argued, strongly rejected a foundationalist conception of philosophy vis-à-vis the special sciences. There is no privileged vantage point from which philosophy can pass epistemic judgment on the special sciences: philosophy is conceived as rather following the special sciences so as to reorient itself in response to their established results. But what then *is* the peculiar task of philosophy, and how, in general, does it relate to the special sciences? Is philosophy itself simply one special science among others, and, if not, from what perspective does it then respond to and rationally reconstruct the results of the special sciences? The positivists are nearly unanimous in explicitly rejecting a naturalistic conception of philosophy as simply one empirical science among others—a branch of psychology, perhaps, or of the sociology of knowledge.[16] On the whole, they instead prefer to think of philosophy as in some sense a branch of *logic* and to conceive the peculiarly philosophical task as that of "logical analysis" of the special sciences. Yet the perspective or point of view from which such logical analysis is to proceed remains radically unclear.

---

[15] Cf. Kant's well-known "criterion by which to distinguish with certainty between pure and empirical knowledge" articulated in §II of the "Introduction" to the *Critique of Pure Reason.*

[16] The conspicuous exception here is Neurath, who articulates a version of naturalism bearing some similarities to well-known views of Quine's. Neurath, unlike the other members of the Vienna Circle, approaches philosophical questions from a background in the social sciences rather than in the mathematical exact sciences, and, accordingly, he shows very little interest in either the problem of a priori principles or the problem of elucidating the peculiar position and role of philosophy vis-à-vis the special sciences. See the papers by R. Haller in T. Uebel, ed., *Rediscovering the Forgotten Vienna Circle: Austrian Studies on Otto Neurath and the Vienna Circle* (Dordrecht: Kluwer, 1991), and also T. Uebel, "Neurath on Naturalistic Epistemology," forthcoming.

No real answer to these questions was forthcoming until Carnap's *Logische Syntax der Sprache*[17] of 1934. Here Carnap is once again responding to recent developments in the exact sciences: to Heyting's formalization of intuitionistic arithmetic, for example, and, above all, to Hilbert's program of "metamathematics." Moreover, he is once again attempting to neutralize the philosophical disputes arising in connection with these developments by showing how all parties involved are in possession of *part* of the truth; the remaining part that appears to be in dispute is then argued not to be subject to rational debate at all. More specifically, the dispute is declared to be a matter of convention in precisely Poincaré's sense: there is simply no fact of the matter concerning which party is "correct," and thus the choice between them is merely pragmatic.[18]

The dispute in question arises from increasing appreciation of how fundamentally the program of *Principia Mathematica* is threatened by the paradoxes and involves the three traditional schools in the foundations of mathematics: logicism, formalism, and intuitionism. Carnap responds to this dispute by declaring that each side is simply putting forth a proposal to construct a formal system or calculus of a certain kind: logicism proposes to construct an axiomatization of mathematics using the rules of classical logic, intuitionism proposes to construct an axiomatization using the more restrictive rules of intuitionistic logic, and so on. The essential point is that no such formal system or calculus is more "correct" than any other—indeed, the notion of "correctness" is entirely inappropriate here. Considered simply as proposals to construct formal systems of various kinds all the apparently opposing philosophies are then equally "correct," and the choice between them can only be a purely pragmatic question of convenience. In the "Foreword" to *Logical Syntax*, Carnap calls this standpoint the "principle of tolerance" and then remarks:

> The first attempts to cast the ship of logic off from the *terra firma* of the classical forms were certainly bold ones, considered from the historical point of view. But they were hampered by the striving after "correctness." Now, however, that impediment has been overcome,

[17] *The Logical Syntax of Language*, A. Smeaton, trans. (New York: Routledge, 1937).
[18] Cf. again a well-known passage from *Science and Hypothesis* (end of ch. III): "What, then, are we to think of the question: Is Euclidean geometry true? It has no meaning. . . . One geometry cannot be more true than another, it can only be more convenient."

192

and before us lies the boundless ocean of unlimited possibilities (*op. cit.*, p. xv).

It is thus with an exuberant sense of liberation that Carnap extends Poincaré's conventionalism to logic itself.

But how is it possible for Carnap to maintain a stance of neutrality with respect to logic itself? Here is where the fundamental insights of Hilbert's metamathematics come into play. For we are to describe the logical rules governing the formal systems or calculi under consideration within the *metadiscipline* of logical syntax: each system is viewed simply as a set of strings of symbols together with rules for manipulating such strings (entirely independently of any question concerning their "meanings"), and any and all formal systems can then be specified from the neutral standpoint of a purely syntactic metalanguage. More precisely, the syntactic metalanguage need employ only the limited resources of primitive recursive arithmetic, and we can thus describe the rules of classical systems, intuitionist systems, and so on from a standpoint that is neutral between them. Our aim is not to justify one system over others as inherently more "correct," but simply to describe the *consequences* of choosing any such system.

Carnap is now in a position precisely to articulate the method and stance of logical analysis. This paradigmatically philosophical enterprise is simply a branch of logical syntax: specifically, the logical syntax of the language of science.[19] We are thus concerned with what Carnap calls "the physical language" (§82). The physical language, as opposed to purely mathematical languages, is characterized by two essentially distinct types of rules: *logical* rules represent the purely formal, nonempirical part of our scientific theory, whereas *physical* rules represent its material or empirical content. Moreover, this purely syntactical distinction between logical and physical rules is Carnap's precise explication for the traditional distinction between analytic and synthetic judgments (§§51–2). Finally, and what most concerns us here, it is also clear that these logical

---

[19] This way of conceiving logical analysis is entirely impossible within the conception of logic of *Principia Mathematica*, where there is no distinction between object language and metalanguage and the language of logic is essentially interpreted. And it is for this reason that earlier writers such as Schlick and Wittgenstein characterized logical analysis as an activity rather than a doctrine—Carnap here explicitly rejects such "mysticism" (§73). Compare W. Goldfarb, "Logic in the Twenties: The Nature of the Quantifier," *The Journal of Symbolic Logic*, XLIV (1979): 351–68; T. Ricketts, "Frege, the *Tractatus*, and the Logocentric Predicament," *Noûs*, XIX (1985): 3–15.

rules or analytic sentences of the language of science represent Carnap's precise explication for the constitutive—but nonsynthetic—notion of aprioricity discussed above: these logical rules, in other words, syntactically represent Poincaré's (and Schlick's) conventions and Reichenbach's axioms of coordination.[20] Accordingly, while logical rules, just as much as physical rules, can indeed be revised in the progress of empirical inquiry, there is still a sharp and fundamental distinction, within the context of any given stage of inquiry, between the two types of rules.

Carnap's solutions to the two problems depicted above as lying at the basis of the logical positivists' radically new philosophical position are therefore as follows: the distinction between conventions or coordinative definitions and empirical laws properly so-called is just the distinction between logical and physical rules, analytic and synthetic sentences; the standpoint and method of philosophy—now conceived as logical analysis—is just the logical syntax of the language of science. Unfortunately, however, it proves to be impossible to implement both of these solutions simultaneously. More precisely, it proves impossible to implement both simultaneously together with what I take to be another linchpin of Carnap's distinctive philosophical stance: the claim to thoroughgoing philosophical neutrality. For it is a consequence of Gödel's incompleteness theorem that, for any language containing classical arithmetic among its logical rules or analytic sentences, the distinction between logical and physical rules can itself be drawn only within a metalanguage essentially richer than classical arithmetic.[21] Implementing Carnap's analytic/synthetic distinction for such a classical language therefore results in a metalanguage that, in particular, is in no way neutral between classical mathematics and intuitionism. It follows that there is no philosophically neutral metaperspective within which Carnap's distinctive version of conventionalism can be coher-

[20] Thus, the first example of §50 parallels Reichenbach's account of the metric of physical space in *The Theory of Relativity and A Priori Knowledge:* it counts as "logical" in the flat space of classical physics but is "descriptive" (empirical) in the variably curved Riemannian space (-time) of general relativity.
[21] For details, see my "Logical Truth and Analyticity in Carnap's *Logical Syntax of Language*," in W. Aspray and P. Kitcher, eds., *History and Philosophy of Modern Mathematics* (Minneapolis: Minnesota UP, 1988). Carnap himself is perfectly clear about the technical situation here, but he fails to recognize its consequences for his claim to philosophical neutrality. Under Tarski's direct influence he later abandoned the definition of analyticity—and indeed the program—of *Logical Syntax.*

ently articulated, and it is precisely here, it seems to me, that the ultimate failure of logical positivism is to be found.

Yet the implications of this failure for our contemporary, post-positivist philosophical situation have not, I think, been sufficiently appreciated. What I want to call attention to here are the very substantial parallels between central aspects of our post-positivist situation and basic elements of the positivists' own philosophical position. Thus, for example, it is now clear, I hope, that, far from being naive empiricists, the positivists in fact incorporated what we now call the theory ladenness of observation as central to their novel conception of science—a conception neither strictly empiricist nor strictly Kantian. Accordingly, they also explicitly recognized—and indeed emphasized—types of theoretical change having no straightforwardly rational or factual basis. In Carnap's hands, these conventionalist and pragmatic tendencies even gave rise to a very general version of philosophical "relativism" expressed in the "principle of tolerance." If I am not mistaken, then, Cassirer's well-known characterization of the Romantic reaction against the Enlightenment—that the battle proceeded largely on the basis of weapons forged by the earlier movement itself[22]—is perhaps even more true of the contemporary reaction against logical positivism. Since it proved ultimately impossible to combine all the elements of positivist thought into a single coherent position, it would serve us very well indeed, I suggest, to become as clear as possible about the true character and origins of our own philosophical weapons.

MICHAEL FRIEDMAN

University of Illinois/Chicago

---

[22] E. Cassirer, *The Philosophy of the Enlightenment,* F. Koelln and J. Pettegrove, trans. (Princeton: University Press, 1951), ch. V, p. 197.

ALAN W. RICHARDSON

LOGICAL IDEALISM AND CARNAP'S
CONSTRUCTION OF THE WORLD

If there is a *locus classicus* of logical positivism, it is by all odds Rudolf Carnap's *Der logische Aufbau der Welt* (1928a).[1] It is here and in the contemporaneous work *Scheinprobleme in der Philosophie* (1928b) that Carnap enunciated and developed what are taken to be the defining concerns of logical positivism – the translational reduction of all significant discourse to discourse about private sense-data via the techniques of modern logic and the correlative rejection of metaphysical discourse as cognitively meaningless or irrational due to its nonreducibility to sensory discourse.

On this received view of its significance, logical positivism as exemplified in the project of the *Aufbau* is a direct outgrowth of Bertrand Russell's External World Program of the 1910s as put forward in such works as *Our Knowledge of the External World* (1914) and *Mysticism and Logic* (1981). This is a view of the *Aufbau* familiar from Quine (1961; 1969), Goodman (1951; 1963), Putnam (1983), and virtually any other account of the work published before 1977 and a good many thereafter. On this account of logical positivism, it and its successor, logical empiricism, formed one of the most thoroughly mistaken programs in modern epistemology: marrying an uncritical and radical empiricism to a fervent belief in an epistemically important analytic/synthetic distinction, logical positivism was completely refuted by the twin towers of epistemology in the early 1950s, Wilfred Sellars' 'Empiricism and the Philosophy of Mind' (1953) and Quine's 'Two Dogmas of Empiricism' (1961). And, of course, it has been refuted innumerable times since.

Implicit in this account of logical positivism and logical empiricism is the idea that their leading philosophical concerns can, as it were, be read off from their names: the logical positivists understood their task to consist in the application of the powerful new techniques of modern logic as developed by Frege and Russell in pursuit of the solution to the traditional epistemic problems of radical empiricists, such as Hume, Mach, and Mill.[2] The new logic provided empiricism both with the tools it needed to give the translational reduction of all discourse into

Synthese **93**: 59–92, 1992.
© 1992 *Kluwer Academic Publishers. Printed in the Netherlands.*

discourse about those elements with which we have immediate acquaintance and an account of the nature of pure mathematics consistent with the necessity of the propositions of mathematics, i.e., logicism.[3]

In recent years, however, several commentators have noted the incompleteness of the received view of the *Aufbau* and also have claimed that there are significant Kantian overtones to be found in the work.[4] Susan Haack (1977) put forward the view that there are some rather striking similarities between the *Aufbau* and Kant's *Critique of Pure Reason* itself. While some of the particular similarities that Haack finds are interesting and suggestive (such as similarities between the "physico-qualitative relation" in Carnap (§136) and at least one of the roles she sees for the transcendental object in Kant), they are also sketchy and of necessity incomplete. Any attempt to find a significant degree of agreement between Kant and early Carnap runs afoul of the one hundred and fifty or so years of truly revolutionary scientific and philosophical developments that stretches between them.

Other authors have also found general thematic similarities between Kantian and Carnapian epistemology. Coffa (1991, p. 208) claims that "the first major project that Carnap chose to undertake was to place Kant's own philosophy on [the] secure path" of science in the work that led to the *Aufbau*. Moreover, Coffa (see, e.g., 1991, p. 231) stresses the important point that the general goal of constructing (constituting) the world of science from experience does not in and of itself make the *Aufbau* an empiricist work – such a goal is equally consistent with and crucial to Kantian epistemology. Indeed the constitution of reality from experience by the formal conditions of the mind is the heart and soul of Kantianism.[5] Joelle Proust (1986) has recently also placed Carnap in the Kantian tradition due to his adherence to the philosophical importance of the analytic/synthetic distinction – a distinction which Proust argues, as did Ernst Cassirer before her, is not a dogma of empiricism, but only achieves its epistemological significance in the context of the transcendental philosophy.

Michael Friedman also has urged a place for the *Aufbau* within a broadly Kantian orientation in epistemology. He has claimed that the constructional order and the importance of the unity of science in the *Aufbau* indicate that

Carnap's project has less affinity with traditional empiricism and more with Kantian and neo-Kantian conceptions of knowledge. The primary problem is to account for the objectivity of scientific knowledge and the method of solution is based on a form/content distinction. (Friedman 1987, p. 529)

Friedman is quick to point out, however, that Carnap's rejection of *a priori* intuition as a separate source of knowledge disqualifies the *Aufbau* from being strictly Kantian.[6]

Friedman finds general Kantian themes throughout the development of logical positivism – not only in the early works of Carnap but also very explicitly in early Reichenbach and more subtly and critically in the work of Schlick. Perhaps the key idea in Friedman's assessment of the problems faced by these writers as they struggled toward logical empiricism is that they did not simply use the new logic of Frege and Russell as a new tool for the solution of old empiricist problems, but rather the development of the new logic made possible a precise, holistic, and hence Kantian account of meaningfulness and also yielded a radically altered philosophical landscape with important new philosophical problems. Indeed, Friedman (1983a, p. 513) claims that the formal, Kantian notion of objective meaning based on the new logic and an empiricist commitment to experience as the foundation of knowledge "stand in a kind of dialectical opposition" in the early works of the logical empiricists.

In this paper I would like to follow an approach to the *Aufbau* that is substantially influenced by Friedman's work but is, perhaps, more closely related in theme to recent works of German-speaking authors who have attempted to view the *Aufbau* in its relation not to Kant himself but to the scientific neo-Kantian epistemology that was developed by Hermann Cohen, Otto Liebmann, and their followers, in the last quarter of the nineteenth and the first quarter of the twentieth centuries.[7] Carlos Moulines (1985), for example, has attempted to place the *Aufbau* within the wider context of German scientific epistemology of the nineteenth and early twentieth centuries. Moulines claims that the received view of the *Aufbau* is not so much wrong as radically incomplete due to its lack of regard for the German epistemic tradition in which Carnap was writing – a tradition that included various varieties of neo-Kantians from transcendental phenomenologists (Husserl, Hans Driesch), to historically oriented neo-Kantians such as Heinrich Rickert, to scientifically oriented neo-Kantians such as Cassirer and Bruno Bauch – in addition to the Machian positivists usually mentioned as intellectual forerunners of Carnap. Even closer to the project at hand

is Werner Sauer's (1985; 1989) attempt to exhibit the similarities between the *Aufbau* and Cassirer's *Substance and Function* (1910).

Before undertaking my attempt to exhibit certain neo-Kantian features of the *Aufbau*, a word of caution is in order. For it is clear
that any simple assimilation of the *Aufbau* to Kantian transcendental
idealism runs afoul of many of the same problems attending the attempt
to assimilate the *Aufbau* to traditional empiricism. For example, one
might be tempted to quote Carnap's apparent endorsement of the
epistemic theses of transcendental idealism in §177 as evidence that
Carnap was explicitly elaborating a Kantian line in the *Aufbau*. But to
do this would be to make the same mistake made by the empiricist
interpretation, for the main point of §§177–78 is that the outlined
autopsychologically based constructional system is the entirety of epistemology and that *all* traditional epistemic schools agree with it and
hence with each another when they concern themselves strictly with
epistemology and do not introduce any metaphysical elements. Thus,
Carnap takes himself to be giving the scientifically acceptable core of
all the traditional epistemic programs – laying out what they all had
right – while subtracting the metaphysics introduced into epistemology
by those schools. And clearly Carnap takes transcendental idealism
to be as infected with metaphysical elements as the other epistemic
schools.[8]

Carnap saw his program as an attempt to improve upon *all* the
traditional epistemological schools by eliminating the "irrational" metaphysical elements that they contained. Moreover, he tells us in the
preface to the 1961 second edition of the *Aufbau* (1969, p. vi) that he
was seeking a synthesis of the insights of both traditional empiricism
and traditional rationalism. This synthesis was to be effected by an
emphasis on the importance of mathematics in the formation of knowledge combined with the idea that mathematics is purely logical. These
facts should make one hesitant to assimilate the *Aufbau* to any of the
traditional epistemic schools.

However, it is beyond question that Carnap began his career as an
unabashed neo-Kantian, embracing both intuitive space and the synthetic *a priori* in his dissertation *Der Raum* (1922), and that all his
papers on the methodology of empirical science (especially, theoretical
physics) before the *Aufbau* show a great concern with Kantian and
conventionalist themes. Also, the very antimetaphysical synthesis of
empiricism and rationalism that Carnap sought to bring forward in the

Aufbau reminds one of the motives stimulating Kant and the more scientifically minded neo-Kantians. Thus, I would venture to put forward the interpretive claim that approaching the *Aufbau* as a work that is both informed by and continuous with the main currents of German neo-Kantian epistemology illuminates those aspects of the work that are most philosophically central to it.[9]

I would like to give some plausibility to this claim in this paper by considering the *Aufbau* in relation to some of the central tenets of the scientific neo-Kantianism of the Marburg school and also in relation to the motivations present in Carnap's pre-*Aufbau* works.[10] The neo-Kantian work that I will direct most attention to is Cassirer's *Substance and Function* (1910) because Cassirer was the most logically and methodologically sophisticated of the neo-Kantians and this work in particular is the high point of Cassirer's scientifically oriented "logical idealism". The most striking convergence between Carnap and Cassirer is hinted at in the fundamental insights that Carnap claimed in 1961 to have guided the *Aufbau*: the purely logical nature of mathematics and the role of mathematics – and, hence, logic – in the formation of a system of objective knowledge even in the empirical realm. The first insight indicates how far the neo-Kantians had moved from Kantian orthodoxy, whereas the second defines their dispute with both logicism and empiricism and points to the importance they see in Kant's basic epistemological orientation. A better understanding of these issues can also lead toward a re-evaluation of the influence of Russell's External World Program on Carnap's thought.

1. CASSIRER'S LOGICAL IDEALISM

I would like to gather some of the guiding threads of Cassirer's thought on logic, mathematics, and empirical science from the multiply connected manifold that is *Substance and Function*.[11] I will by no means give an exhaustive account of the philosophical richness of this work, but seek only to isolate some themes that bear scrutiny when considering the relations of the philosophical concerns of the neo-Kantians to those of the founders of logical empiricism.

Perhaps the most important idea for Cassirer is that the development of pure mathematics and mathematical physics in the nineteenth century exhibits a new form of concept formation that makes evident the functional nature of objective concepts and stands opposed to the traditional notion of concept formation via the process of abstraction. In the work in the foundations of analysis, the theory of the manifold, and the generalization of geometry in projective geometry and Grassmann's *Ausdehnungslehre*, Cassirer sees a consistent element in the understanding of the mathematical concept. This common element is the view that the form of the mathematical concept is the serial relation which gives the law of connection of the entire series and which determines the transition from one element to the next. Thus, for example, Cassirer (1953, p. 36) writes of Dedekind's deduction of the concept of number that the

"things," which are spoken of in the further deduction, are not assumed as independent existences present anterior to any relation, but gain their whole being, so far as it comes within the scope of the arithmetician, first in and with the relations which are predicated of them. Such "things" are terms of relations, and as such are never "given" in isolation but only in ideal community with each other.

Thus, arithmetical objects are constructed from the forms of the relations in which they stand. These relations induce a systematic connection in the manifolds over which they are determined and this systematic connection allows us to claim objective understanding of the manifold and the elements of the manifold generated by those relations.

Cassirer emphasizes that this method of concept formation stands opposed to the traditional abstractive method which yields generic concepts. Such a traditional view of concept formation, he claims (ibid., p. 19), cannot begin to make sense of the process of mathematics:

When a mathematician makes his formula more general, this means not only that he is *to retain* all the more special cases, but also be able *to deduce* them from the universal formula. The possibility of deduction is not found in the case of the scholastic concepts, since these, according to the traditional formula, are formed by neglecting the particular, and hence the reproduction of the particular moments of the concept seems excluded.

This new, functional understanding of the logic of concept formation as exemplified in the concepts of mathematics and geometry indicates why the Marburg neo-Kantians were happy to depart from Kant's own reliance on pure, *a priori* intuition in accounting for the necessity of mathematical knowledge. For the functional concepts of mathematics go beyond the syllogistic logic of Aristotle in precisely the important way that Kant recognized that mathematical and geometrical knowledge must go. Thus, Cassirer (1907, p. 32) writes in his review of Russell and Couturat that the division of understanding and sensibility is, in the way it is introduced in the transcendental aesthetic, in the first instance thoroughly convincing: because here it is a matter only of distinguishing mathematical concepts from the general *species-concepts*, which are defined by genus and difference, of traditional logic.[12]

Thus, neo-Kantians like Cassirer, Natorp (1910), and Cohen (1902) were happy to see a new logic at work in mathematics that allowed them to dispense with pure intuition as a separate origin of mathematical knowledge, so long as the important ways in which Kant showed mathematical knowledge to go beyond Aristotelean logic were acknowledged.

This acknowledgment of one of the fundamental philosophical points of logicism is, however, in stark contrast to the assessment Cassirer gives to logicism as an epistemological program. Cassirer himself seeks to show in *Substance and Function* that the self-same logical function played by the concepts of pure mathematics is played by the concepts of any truly objective empirical science. Thus, Cassirer sees in the nature of energism in mathematical physics the correlate of the type of mathematical thinking exhibited by Riemann, Dedekind, Cantor, and others. Of the fundamental importance of energism in physics Cassirer (1953, p. 190) writes:

The structure of mathematical physics is in principle complete when we have arranged the members of the individual series [e.g., quantity of heat, motion, electric charge, etc.] according to an exact numerical scale, and when we discover a constant numerical relation governing the transition from one series to the others. . . . Only then it becomes clear how all the threads of the mathematical system of phenomena are connected on all sides, so that no element remains without connection.

In this way Cassirer is concerned to show that the type of concept formation that is operative in pure mathematics can be extended to the empirical realm in a way that guarantees the objectivity of both the empirical laws and the applied mathematical principles. This role for the logical principles underlying mathematics is missing in the logicism of (Frege and) Russell according to Cassirer. As he writes (1907, p. 44 ff.) in his review of Russell:

Thus there begins a new task at the point where logistic ends. What the critical philosophy seeks and what it must demand [*fordern*] is a *logic of objective knowledge* . . . . Only when we have understood that the same foundational syntheses [*Grundsynthesen*] on which logic and mathematics rest also govern [*beherrschen*] the scientific construction of experiential knowledge, that they first make it possible for us to speak of a strict, lawful ordering among appearances and therewith of their objective meaning: only then is the true justification of the principles attained.

It is this requirement that the principles of logic and mathematics receive their justification through the way in which they ground the objectivity of empirical science that captures what is of lasting importance in Kant's transcendental turn and guarantees these principles their synthetic *a priori* status. For the most important role for the propositions of pure mathematics are as guarantors of objectivity within empirical science by providing the logical syntheses needed for the creation of actual empirical objects. In opposition to this critical view, Cassirer (ibid., p. 43) sees a radical separation of epistemic tasks in Russell's logicism:

> According to the fundamental view of logistics the task of thought has ended when it has succeeded in making a strict deductive connection among its structures [*Gebilden*] and creations [*Erzeugungen*]. The problem of the lawfulness of the world of objects, on the other hand, is left completely to direct observation, which alone within its own, very narrow limits is able to teach us whether there are also here certain regularities or whether pure chaos reigns.

Cassirer (ibid., p. 48) thus leaves the logicist with the challenge to give an account of "the role that [mathematical principles] play in the construction of our concept of an "objective" reality". Failing that, the logicist has not even addressed the central epistemic question raised by transcendental philosophy and the critique of knowledge.[13]

In *Substance and Function* and also in his monograph *On Einstein's Theory of Relativity* (1921), the principal problem Cassirer sees in extending logical and mathematical principles to the empirical realm – thereby grounding the objectivity of the empirical knowledge – is to give content to the Planckian dictum that in physics everything that can be measured exists. Cassirer's problem becomes one of finding the "logical conditions of the operation of measurement itself" (Cassirer 1953, p. 361) and the relations that those conditions stand in with respect to scientific knowledge and its empirical basis. So in his discussion of the theory of relativity he puts the puzzle this way (ibid., p. 363):

> If not only place but the velocity of a material system is to signify a magnitude that entirely depends on the choice of a reference body and is thus infinitely variable and infinitely ambiguous, there seems no possibility of an exact determination of magnitude and thus no possibility of an exact objective determination of the state of physical reality.

This problem leads to the search for a *non plus ultra* that grounds the objective reality of empirical knowledge.

It is here that Cassirer finds the distinction between empiricism and his logical idealism. As he sees it, empiricism grounds the objectivity of scientific knowledge through a "simple registration of facts" of experience, which are then conceptually 'worked up into' scientific theories. Logical idealism, on the other hand, sees a conceptual element in experience itself, i.e., that only by virtue of having certain logical forms does experience first yield matters of fact. Thus, Cassirer's answer to his own question of the objective determination of physical reality is this (ibid., p. 365):

[I]n the multiplicity and mutability of natural phenomena, thought *possesses* a relatively fixed standpoint only by *taking* it. In the choice of this standpoint, however, it is not absolutely determined by the phenomena, but the choice remains its own deed for which it alone is responsible. The decision is made with reference to experience, i.e., to the connection of observations according to law, but it is not prescribed in a definite way by the mere sum of observations. For these in themselves can always be expressed by a number of intellectual approaches between which a choice is possible only with reference to logical "simplicity", more exactly, to systematic unity and completeness of scientific exposition.

The idea here is that by the methodologically governed fixing of the conceptual reference point and system of measurement we block the infinite ambiguity of knowledge by fixing a set of objective meaning relations. This in turn yields the content of the facts of experience. With the principles of measurement in hand we can express the facts of experience in a form that will allow the formulation of mathematically expressed laws that will connect the manifold of experience into a law-governed and predictable manifold.

Moreover, in an important passage in *Substance and Function*, Cassirer calls the theory of experience as given in logical idealism "the universal invariant theory of experience" (ibid., p. 268). Likening the procedure to that of a geometer investigating the invariants of figures under transformations, Cassirer (ibid., pp. 268ff.) claims that in critical philosophy "the attempt is made to discover those universal elements of form, that persist through all changes in the particular material content of experience". He gives a preliminary list that includes space, time, magnitude, and the functional dependency of magnitudes. These play the role of Kantian categories – they are *a priori* conditions for objective knowledge in the empirical realm in general:

The goal of critical analysis would be reached, if we succeeded in isolating in this way the ultimate common element of all possible forms of scientific experience [i.e., facts of experience as conceptually comprehended from within some scientific theory]; i.e., if we succeeded in conceptually defining those elements which persist in the advance from theory to theory because they are conditions of any theory.... Only those ultimate logical invariants can be called *a priori*, which lie at the basis of any determination of a connection according to natural law. (Ibid., p. 269)

Although it is not obvious that Cassirer was completely clear about this, these two notions of the conceptual element of objective experience amount to closely related but rather different views. On Cassirer's first account, where the important conceptual element in experience is the imposition of particular metric stipulations to yield both mathematically expressible laws of physics and matters of fact of experience expressed within that mathematical language, the synthetic *a priori* element is in the conventional conditions of the construction of some particular physical theory. The conventions are not determined by experience, rather experience is objectified when the mathematical physics constructed on the basis of those conventions is in hand. On the second view – the universal invariant theory of experience – the important formal moment in experience is the form that experience must have across both the actual streams of experience of individuals and the various possible conventions of measurement which guarantee the possibility and nonambiguity of those conventions themselves. Moreover, this formal element in experience must yield the necessary result that there is something about experience that allows the various systems of physics formulated by these differing conventions to be about the same experiential basis. On the first view, the synthetic *a priori* conventions are principles of objectification through mathematization from within a particular theoretical framework and, hence, change with change of framework; they form a *relativized* notion of the synthetic *a priori*.[14] The second view entails a notion of a universal *a priori* formal element in experience which grounds the possibility that such conventions can be imposed.

Here then we have, due to the conventionalist and relativist lessons of the day, a disentangling of two theses that were intertwined for Kant. Since Kant thought that the synthetic *a priori* conditions imposed by the forms of sensibility and understanding were of sufficient strength to impose one particular physics and physical geometry upon the manifold of experience, he explicitly conjoined two very different claims under the heading of the synthetic *a priori*: the synthetic *a priori* was considered both as the conditions of the possibility of experience and

as the conditions of the possibility of theoretical understanding. In
this sense, experience was already a type of theoretically informed
knowledge for Kant; objective experience was already of physical objects governed by the laws of Euclidean geometry and Newtonian physics and must be strictly separated from the mere play of sensations.

However, a neo-Kantian who has absorbed the lessons of Poincaré
and Einstein wants to maintain the following. First, with Kant, he will
maintain that real objective knowledge is found within mathematical
physics, not simple, theoretically uninformed experience. Second, with
Poincaré, mathematical physics relies on conventions in a way that
requires that alternative conventions are equally experientially possible
(and, hence, a multiplicity of experientially equivalent systems of physics is possible). Third, these conventions go beyond mere experience
to provide the mathematical structure that is necessary for objectivity.
This leads to two distinct resolutions to the problem of the formal
components of knowledge, each of which requires, however, that we
make a distinction between conditions of experience and conditions of
objective (theoretical) knowledge.

First, one could – as in the first understanding of the synthetic *a priori*
we gleaned from Cassirer – decide that the conventions underlying any
one particular system of physics form the synthetic *a priori* component
of knowledge. This is done because those conventions allow the mathematization of the manifold of experience and, hence, the formulation
of the mathematical natural laws that are the hallmark of objective
understanding. Since systems of alternative conventions are possible,
this notion of the synthetic *a priori* is then relativized to particular
systems of physics constructable from those conventions. This was
Reichenbach's famous reorientation of Kantian doctrine in his first
book, *Relativitätstheorie und Erkenntnis A Priori* (1921). Reichenbach
was led to understand synthetic *a priori* principles as conditions of the
possibility of objects of knowledge but to deny Kant's claims that they
are apodeictically necessary principles.[15] Moreover, they cannot be
conditions of the possibility of experience if we want to consider the
possibility that alternative conventions are possible on the same experiential basis. Nevertheless, one could consider them conditions of the
objectification of experience inasmuch as without such conventions the
mathematical principles that permit the individuation of objects via
their mutual relations are inapplicable to experience.

Alternatively, one could decide that the synthetic *a priori* lies in the conditions of the possibility of the conventions themselves and construe such judgments then as conditions of the possibility of (objectifiable) experience. Since these conditions underlie any possible conventions and hence any possible objectifying mathematical physics but don't limit us to any one physics, these principles are universal and necessary; but they are not sufficient to be conditions of the possibility of objective knowledge. This idea is what lies behind Cassirer's universal invariant theory of experience and is, I believe, Cassirer's considered view of the formal components of experience itself.

Of course, on both the relativized, conventional and the universal invariant understanding of the *a priori* element of experiential knowledge, the mathematico-logical principles that underlie the possibility of the mathematical expression of the laws of natural science are pure *a priori* principles. As we have seen, however, Cassirer considers these also to be synthetic *a priori* because mathematics isn't justified as a pure discipline about abstract or ideal objects but via the role it plays in theoretical, empirical knowledge and the individuation of objects in the empirical realm.

2. CARNAP'S NEO-KANTIAN ORIGINS

We now have a broad understanding of some of the central epistemological themes of mature Marburg neo-Kantianism: the importance of mathematical physics as the paradigm of objective knowledge; the correlative role of the "logic of objective knowledge" in going beyond the logicism of Russell in showing how logical and mathematical principles are applied to and make possible empirical knowledge; and the corresponding notion of a conceptual formal element in experience itself which is necessary for the establishment of experiential matters of fact. The latter was outlined in two different ways: either as the objectifying conventions underlying mathematical physics or as the formal elements of experience which make such conventions possible.

In what follows I hope to show that the early Carnap was pursuing a program that is very similar to that of Cassirer. The crucial distinction is that Carnap was very much a logicist and, therefore, was pursuing the program with a Russellian view of logic. In essence, then, I see Carnap as trying to answer the neo-Kantian demand that logicism only becomes an important epistemic program if it can show how its conception of logic can illuminate the formal prerequisites of objective, empirical knowledge as given in mathematical physics.

It is clear that at the beginning of his career Carnap was more interested in and influenced by Kantian epistemology than by any form of radical empiricism. This is most evident in his dissertation *Der Raum* (1922), where he endorses Russell's logicism with respect to geometry but still seeks to find a place for intuition and intuitive space as well as a synthetic *a priori* element in knowledge. This endorsement of broadly Kantian themes in the work of Carnap depends more explicitly than in Cassirer on the conventionalism put forward by Poincaré and Dingler, as well as the notion of essential insight that Carnap takes from Husserl's early writings.[16] But his distance from radical empiricism in this period is perhaps best expressed by the lines with which Carnap opens his 1923 essay 'Über die Aufgabe der Physik' ('On the Task of Physics': Carnap 1923, p. 90):

After a long time during which the question of the sources of physical knowledge has been strenuously debated, perhaps it may be said already today that pure empiricism has lost its dominance [*Herrschaft*]. That the construction of physics cannot be founded on experimental results alone, but rather must employ nonempirical axioms [*nichterfahrungs-mässige Axiome*], has been proclaimed already for a long time by philosophy. However, only after representatives of the exact sciences had begun to investigate the particular nature of physical methodology, and had in so doing been led to a nonempiricist conception, were solutions produced that could satisfy even the physicists.

However, let's begin with *Der Raum*. In this work Carnap was in some ways more Kantian than Cassirer inasmuch as Carnap felt the need to introduce an intuitive space which provided *a priori* intuitional constraints on the possible structure of physical space. Our discussion of Cassirer above sheds some light on Carnap's need to introduce intuitive space and the relations of it to mathematical and physical space.

Very broadly speaking, Carnap's main philosophical point in *Der Raum* is to argue that various philosophers, mathematicians, and physicists who had been arguing about the nature of space and our knowledge of it were only apparently in disagreement. This was because they each had a different notion of space in mind in their discussions. So, in this work Carnap (1922, p. 5) undertook "to clarify the situation [by] present[ing] here a survey of the various meanings of space and the types of space that emerge in connection with each meaning".[17] The three meanings of space are for Carnap: formal space, intuitive space, and physical space. Each of these meanings has a variety of different spatial structures associated with it: topological, projective, and metrical spaces of various dimensionality. Carnap maintains that in their disputations the various parties were maintaining theses that were in the main true of the particular type of space they had in mind, but all misunderstood their opponents to be talking about the same type of space that they were. In fact, however, the mathematicians were principally speaking about formal space, the philosophers about intuitive space, and the physicists about physical space.

In *Der Raum*, Carnap seems to concede the principal epistemic point leveled by Cassirer against Russell's logicism. For, while Carnap does endorse logicism with respect to pure geometry, he also feels the need to introduce a version of intuitive space that mediates between the formal space of the mathematicians and the physical space of empirical knowledge. This is because the absolute logical generality of formal space allows of multiple applicability in the empirical realm and cannot serve to ground the objectivity of determinate spatial relations in the empirical world. He introduces the notion of intuitive space with these words (ibid., p. 22):

Intuitive space is an order structure whose formal type we can certainly delimit conceptually but, like everything intuitible, not its particular nature [*Sosein*]. Here we can only point to contents of experience, namely to intuitively spatial forms and relations. . . .

The problem Carnap is faced with is clear enough. The very generality of formal space requires that it is applicable to not merely those elements that are experienced as spatial – lines, planes, etc. – but also those elements that although not experienced as spatial are orderable into order structures subsumable under those of formal space, e.g., colors, tones, etc. The problem is then how to single out this one order structure which is the space of experience from all the other structures; for this purpose Carnap brings in intuition as a separate source of knowledge and, with it, intuitive space. But this solution requires that Carnap also distinguish intuitive space and the knowledge we have of it from the space of actual empirical experience and the empirical knowledge we thereby have of it. Carnap does this by claiming of the axioms that govern intuitive space that (ibid., p. 22):

[T]hey are independent of the "quantum of experience": i.e., their cognition is not, as with experiential propositions, made ever more secure by often repeated experience. For here, as Husserl has shown, we are certainly not dealing with facts in the sense of experiential reality, but rather with the essence [*Wesen*] (*Eidos*) of certain data which can already be grasped in its particular nature [*Sosein*] by being given in a single instance.

What then is the structure of intuitive space, and how does this structure provide objective grounding of the empirical knowledge of physical spatial relations? Essentially, intuitive space is a connected three-dimensional manifold that has the structure common to all Riemannian manifolds of that dimensionality. The important point about this construction is that such a manifold allows a metric. That is, although intuitive metrical space does not have a specific global metrical structure, it is so constructed that it can be given such a structure via a metrical stipulation. The metrical stipulation is, however, the province of physical space in Carnap's scheme. The epistemic point is, however, that the metrizable manifold is intuitive for Carnap and, therefore, is a synthetic *a priori* ground for the possibility of objective experience. This is synthetic *a priori* knowledge in the sense of Cassirer's universal invariant theory: formal conditions of experience that guarantee the possibility of metric conventions and, hence, of mathematical physics.

Whereas intuitive space is constructed from the essential insight of spatial forms, "[t]he theory of physical space . . . has the task of establishing which [spatial] relations hold for the particular things that confront us in experience" (ibid., p. 32). Carnap's main purpose in his presentation of physical space is to argue that the establishment of any spatial relations for physical space over and above topological relations on the basis of experience alone is impossible. Carnap is led in this way to a version of conventionalism.

Of particular interest in Carnap's account of physical space is a distinction he makes which he says is related to but not identical with Kant's distinction between the form and matter of experience. Carnap seeks not to indicate two forces through the combined efforts of which experience first becomes possible, as did Kant, but rather to analyze completed experience into two parts. For this he makes "a division within the realm of form between necessary and optional form" (ibid., p. 39). This distinction allows Carnap to draw a connection between topological space and matter subject solely to necessary form. Similarly, he associates metric space with optional form. Carnap introduces the idea as follows (ibid., p. 39):

Let matter which is certainly not unformed, but appears only in necessary form, be called "*matter of fact*" of experience. This can be subjected to a still further formation in terms of optional form. In order to test an experiential statement for whether it is a statement of matter of fact or not, and, in the latter case, what in it pertains to matter of fact and what depends on the form determined by choice, we have to investigate whether the experiential statement remains valid for all possible formations, which means, for our investigation, for all possible types of spatial transformations.

As Carnap then points out, all and only topological statements are thus invariant for arbitrary continuous one-one spatial transformations. Thus, physical topological space is the only physical spatial structure determined by the facts of experience. Moreover, the global metrical structure of intuitive space guarantees *a priori* that a metrical structure can be added to this physical topological structure.[18]

This function for intuitive space is crucial because physical topological space is not the space that Carnap is most interested in. Carnap's main interest is the space used by the physicist in developing physical theories and mathematically expressible laws of kinematics and dynamics which require measurement of spatial intervals and, hence, metric geometry. As Carnap puts it (ibid., p. 40):

Although S_{3t}^u [three-dimensional physical topological space] has the advantage over [the various versions of metrical physical space] of emerging uniquely from the matters of fact, the structures S_{3m}^u [three-dimensional physical metrical space] are still incomparably more important for both natural science and everyday life, because it is here a question of measurement.

Thus, I think we can see that Carnap is concerned with some of the very same issues we saw above in the discussion of Cassirer. Mathematical physics forms the paradigm case of knowledge for Carnap as well. Moreover, on Carnap's account, although themselves deprived of any intuitive content, the spaces of mathematics by virtue of their generality apply rigorously to spatial structures in physical experience, but experience is itself given its necessary and most general form via its connection to intuitive space, S_{nt}' (intuitive topological space of arbitrarily many dimensions). Thus, the matters of fact of experience themselves require a minimal intuitive form. This is given by the synthetic *a priori* offerings of topological intuitive space as conceptually comprehended by formal topological space.[19] This, when combined with the Riemannian metrical intuitive space discussed above, allows the metrical conventions to induce a fully objective structure on the facts of experience by first making possible the mathematically expressible laws of nature. Carnap stresses repeatedly (e.g., ibid., pp. 44ff.) that once the metric stipulation is made the actual results of measurement are always facts of experience, i.e., purely topological facts, such as the coincidence of two points. Nevertheless, when imbued with the optional form derived from the metrical stipulation those facts take on a new significance. For now metrical relations are applicable to them and they now make possible the achievement of mathematically expressible laws. Thus, intuition yields a metrizable manifold with a topological structure which is the necessary form of experience, and objective science is guaranteed by the further imposition of optional (metrical) form which allows for precise metrical relations to be ascertained and for laws to be expressed.

Carnap's thinking about the synthetic *a priori* in *Der Raum* is more akin to Cassirer's universal invariant notion than to Reichenbach's relativized conventional notion. Both the structure of intuitive space given by essential insight and the argument from the necessary form of experience point toward a universal, necessary understanding of the synthetic *a priori*.' Moreover, Carnap never calls the metrical conventions he discusses in the chapter on physical space synthetic *a priori*. Indeed he leaves them out of his table of the cognitive sources of spatial judgment (ibid., p. 47) entirely, thereby suggesting that the conventions aren't judgments at all. This is also suggested by the rather terse refusal to place the conventional stipulations within the *a priori/a posteriori* and synthetic/analytic distinctions (ibid., p. 47):

Therefore, apart from the determinations added by means of freely chosen stipulations, the propositions governing formal, physical, and intuitive space are analytic *a priori*, synthetic *a posteriori*, and synthetic *a priori*, respectively.

After *Der Raum*, Carnap no longer brings in intuitive space or intuition as a separate source of knowledge to mediate between the conceptual forms of logic and physical concepts. Thus, he moves closer to Cassirer's problematic. Moreover, he becomes increasingly interested in giving an account of the conventional component of theoretical empirical knowledge. This comes out especially in 'Über die Aufgabe der Physik' (Carnap 1923) and his monograph *Physikalische Begriffsbildung* (Carnap 1926).

In 'Über die Aufgabe der Physik', Carnap considers the various places in which conventional decisions are needed in constructing physical theories: the construction of the spatial system, the temporal system, and the force law (*Wirkungsgesetz*). That the systems of space and time require conventional choices and cannot be read off directly from experience Carnap takes to be relatively well known and uncontroversial. This is, however, not so for the idea that the force law requires a further convention choice. Carnap writes (1923, pp. 92ff.):

That, however, when [the conventions for space and time] already are fixed, the laws of nature cannot be deduced from mere experiential findings requires a new deliberation, which Dingler undertook. In the laws of nature there appear magnitudes whose measurement is not immediately possible, but is reduced to space-time measurements (e.g., mass, force of gravity, electric charge, electrostatic field, etc.). This reduction, however, presupposes a general force law.

Carnap has in mind the idea that, to take his example (ibid., p. 93), if mass is defined as the quotient of force and acceleration, we cannot measure mass without presupposing a general force law because without such a law forces cannot be determined without knowing masses. Carnap refers us to the fact that the determination of the masses of the heavenly bodies are determined in classical mechanics via the use of the Newtonian force law and states that clearly another non-Newtonian force law could instead be used for this purpose without contradicting experience.

It is important to note that Carnap is not here worried about empiricist problems that we generally associate with Hume regarding the inductive justification of general laws. Rather, he is concerned with the fact that in mathematically expressible laws of nature there occur concepts of physical magnitude whose numerical values and mathematical relations cannot in principle be established without some form of conventional choice of force law. The problem that Carnap sees in grounding laws of nature solely in experience is not that they are not then immune to possible conflict with future experience, but that they cannot even be formulated; for they employ concepts whose values are structurally underdetermined by experience alone.

The sense in which the conventions go beyond the formal structure of experience itself is, of course, not new to 'Über die Aufgabe der Physik'; Carnap had argued that already in Der Raum. What is different in 'Über die Aufgabe der Physik' is the absence of an appeal to intuition. Carnap makes no effort to discuss the formal structures of experience that serve as the synthetic a priori conditions of empirical knowledge and the underpinnings of the possibility of the conventions themselves. In fact, Carnap half-heartedly applies the term "synthetic a priori" to the theoretical system of physics itself, i.e., the system of the spatial and temporal metric conventions and the laws of physics.

He (ibid., p. 97) considers this as the axiom system given in the first volume of a completed physics:

These [axioms] consist of the basic principles of space determination, time determination, and the dependence of processes on one another, in short, spatial, temporal, and force law. The deduction of the laws of nature from these axioms occurs, even though in many ways occasioned by experience, still without any foundation in experience. The first volume therefore contains synthetic judgements *a priori*, however not exactly in the Kantian transcendental-critical sense.

The reason why it isn't exactly the Kantian sense is that for Kant it would mean that the first volume could only have one form, whereas the conventionalist insists that there are a multiplicity of forms that the volume might take depending on the choice of conventions. So we have moved to the relativized notion of the synthetic *a priori* in a very radical way because Carnap is not restricting this appellation to the conventions alone, but to the entire system of theoretical physics built from them. Carnap therefore suggests that Peano's term "hypothetical-deductive systems" serves better than the traditional Kantian terminology.

After this brief and radical flirtation with the relativized synthetic *a priori*, Carnap takes his own advice and ceases to use the Kantian terminology but returns to the task of considering the logico-mathematical and experiential conditions of the possibility of metric conventions and mathematical natural science. His most detailed meditation on this topic is his 1926 monograph *Physikalische Begriffsbildung*. In it Carnap sets himself the task of giving the general conditions for the formation of objective concepts in physics. His approach to this task is indicated in the following example: How do we proceed from our own vague and private feelings of warmth or cold to a precise, useful, and intersubjectively available temperature scale, such as the Celsius scale? A radically simplified account of Carnap's deliberations of this case (cf. 1926, pp. 16ff.) goes as follows. The occasion that presents itself for the formation of a concept of temperature is given in our experiences of objects that present different thermal sensations in us either simultaneously or over time. Now certain formal aspects of this thermal experience give us the opportunity to construct a temperature scale.

First, we need to know when to assign the same number to two different objects. This can be done by assigning the same number to any two objects that undergo no thermal change on contact with one another. This is a possible assignment because of the "experiential" fact that this is a transitive relation. Moreover, it is clearly a symmetric relation. Next, we need to know when to assign a larger (smaller) number to one of two bodies. We can do this, in considering the two bodies a and b, by assigning a smaller number to a (b) if when brought into contact with b (a) it undergoes a warming up. This assignment will work because two bodies not in thermal equilibrium will not both warm up when brought into contact (provided no chemical reactions take place). Thus, this relation is transitive and asymmetric. Thus, the topological conditions that must be met in experience for the construction of metrical concepts are that we have two relations – one transitive and symmetric and one transitive and asymmetric – which allow certain experiential findings to be placed into distinct equivalence classes (by the first relation), which themselves can be ordered (by the second).

But, as Carnap says (ibid., p. 19):

Beyond this we want to now establish that [any] two [physicists] assign the same number as the temperatures of every body; we say: their temperature assertions should not agree only topologically, but also "*metrically*".

To do this we must apply certain metrical conventions. First, we must decide on a "scale form", that is, we must decide when two differences in temperature are to be considered equal. Second, we must decide on a null point for the temperature scale. Third, we must decide on a unit for the scale. For temperature, one can fulfill these by, for example and in turn, stipulating first that (ibid., p. 35) "two temperature differences [be] set as equal if mercury experiences the same increase in volume in both the corresponding heatings". Second, one can choose a natural point, such as the melting point of ice, as the zero point and a certain division of the difference between two natural processes, such as the melting of ice and boiling of water, as the unit. In this way, a temperature scale, indeed the Celsius scale, can be developed and temperature can be measured in the same way by various different agents. Thus, we achieve both precision and intersubjectivity of judgment as the result of this one process of concept formation.

Concept formation then depends on five conditions: two topological determinations [*Bestimmungen*] and three metrical conventions [*Festset-zungen*]. The topological determinations amount to certain formal conditions of experiential relations. Carnap puts the point this way (ibid., p. 22):

The prerequisite and occasion for the introduction of a type of magnitude is an empirical finding of the type that among the objects (bodies, processes) of a domain two relations obtain: one transitive and symmetric and one transitive and asymmetric. The first relation gives the occasion for the formation of a particular concept of identity, the second for the formation of the concept of a particular type of magnitude, and indeed chiefly (that is if that relation has a certain sequentiality) a one dimensional ("scalar") magnitude.

That is, objects standing in the first relation, which is transitive and symmetric, are assigned the same number for the magnitude being formed, and the second relation gives the order of the assignment of numbers to objects for this magnitude. These topological determinations, then, are what relate the quantitative, formal objects of theoretical physics to experience and allow the prediction of future experience from the laws of physics. Thus, the topological determinations play the role of the synthetic *a priori* conditions of experience that ground the possibility of various metrical conventions.

However, there is an important difference between the view put forward in *Der Raum* and that in *Physikalische Begriffsbildung*. In *Der Raum*, intuition and essential insight were brought in as a separate source of knowledge that provided a philosophical account of the *a priori* guarantee that objective theoretical knowledge is possible by providing the form that guaranteed that conventions could be found. The logically comprehended topological determinations of experience still stand as conditions of the possibility of objective concepts in the sciences in *Physikalische Begriffsbildung*, but they themselves no longer have an *a priori* guarantee via the invocation of intuition. Nothing guarantees any longer that experiences are suitably topologically structured to allow of conventions for all concepts one might want in science. However, when and only when such topological structure is extant in experience are fully objective, metrical concepts and mathematically expressed scientific laws possible.

The laws of physics still require, however, the complete formation of quantitative concepts and, hence, the metrical conventions. These conventions go beyond experience and require stipulations whereby individual objects are coordinated with individual numbers. Only on the basis of such stipulations can laws of nature be formulated. Hence, the conventionalist aspect of Carnap's thinking persists.

After *Physikalische Begriffsbildung* the remnants of Carnap's Kantianism can be summarized as follows. First, there remains the fundamental idea that mathematically expressible laws of nature are the paradigm of objective knowledge (now without *a priori* intuitive guarantee that such laws will be found in a given area of science). Second, these laws are not merely not derivable from experience but require concepts that go beyond experience in certain crucial formal ways. These first two points relate to the first notion of the synthetic *a priori* canvassed in our consideration of Cassirer above, where the conventions themselves, by virtue of being the formal components necessary for objective knowledge, are considered synthetic and *a priori*. Third, the relation of such mathematizable concepts to experience is a relation that involves only the logical structure of experience – its topological form, where this formal component is expressible in Russell's theory of relations – and the metric conventions. This topological structure is the formal structure of experience itself and, hence, synthetic *a priori* in the second sense. From these first three theses it follows, fourth, that logic plays some form of a transcendental role in knowledge by providing forms that both make precise laws of nature possible and that provide the relation of these laws to experience.

3. STRUCTURE AND OBJECTIVITY IN THE *AUFBAU*

Where do we stand when we return to the *Aufbau* with these considerations in hand? Of course, famously, the *Aufbau* does seek to give a definitional construction of all the concepts of science from a phenomenalist basis. Thus, the *Aufbau* seems fundamentally opposed to Carnap's previous conventionalism – in the *Aufbau* all scientific concepts are explicitly definable in terms of elementary sensory experience. It seems then that there is no room for a conventional element in concept formation that goes beyond experience and which grounds the objectivity of scientific knowledge. Moreover, no procedure for grounding the objective meanings of scientific terms seems more suitable to the radical empiricist than is the constructional system outlined in the *Aufbau*. What then do we learn from the Kantian themes canvassed above?

A closer look, however, reveals not so much a repudiation of the Kantian themes of his prior work but a reorientation of Carnap's understanding of the operative notion of form that grounds objective knowledge. For, first, there are aspects of the construction of the physical world in which conventions still play a role. Moreover, the construction of the intersubjective world mirrors certain methodological lessons Carnap stressed in the formation of metrical conventions and their role in objective knowledge in *Physikalische Begriffsbildung* and elsewhere.

Finally, of course, the second notion of the synthetic *a priori* (the universal invariant theory of experience), while explicitly stripped of that Kantian title, continues in the *Aufbau* and plays the fundamental epistemic role that allows the program of objective knowledge in the *Aufbau* to proceed.

On the first point, we should note that the conventional element does not disappear entirely from the *Aufbau*. Instead, and rather surprisingly, we get a very brief indication in §136 of how we construct the world of physics from a private perceptual world, and it is clear that Carnap still believes that this procedure is guided by physical laws from systems that are empirically equivalent and which are chosen ultimately by appeal to methodological principles. Thus, Carnap writes in §136:

> This construction [of the world of physics] has the purpose of formulating a domain which is determined through *mathematically expressible laws*. . . . The indicated purpose of this construction does not unambiguously determine which physical-state magnitudes must be chosen for the construction of the world of physics. . . . It is probable that eventually a clear decision will be made (which will be based on empirical evidence but which will be guided by methodological principles, for example, the principle of the greatest possible simplicity).

The construction of the world of physics proceeds on the basis of physical theories and physical concepts chosen for methodological reasons from a variety of alternative theories and uses perceptual qualities and methodological principles (such as least action) to project values for those physical quantities onto the points of the space-time world. The methodological principles are necessary because, as Carnap writes in §136:

> [T]he assignment of a quality to a world point in the perceptual world does not determine which structure of state magnitudes is to be assigned to the neighborhood of the corresponding physical world point of the world of physics; the assignment of this quality merely determines a class to which this structure must belong.

So Carnap's explicitly maintained view about the construction of the physical world is that neither the physical magnitudes themselves nor the values of those magnitudes at the points of physical space-time are fully determined by perceptual qualities and their distribution over space-time; rather, we require the laws of physics themselves and methodological principles to give the magnitudes and find their values. These facts are clearly related to the problems Quine and Goodman raised about the status of the *at* relation in the construction of the perceptual world, which raise serious questions about Carnap's understanding of explicit definition in the *Aufbau*.[20] However, my worries are somewhat different from Quine's: Quine (1961, p. 40) is concerned about the *at* relation itself and whether it can be defined in terms of logic and sensation. My point is that, even if Quine's worries can be finessed, the physical quantities we construct and their values at the points of space-time require the use of the laws of physics and, thus, the domain of the *at* relation for physical space-time isn't constructable from logic and sensation alone – and this is a consistent feature of Carnap's early philosophy.[21]

Moreover, this construction of the world of physics is crucially important for the ultimate lesson of the constructional system. For, as Carnap tells us (§136), "only this world, but not the perceptual world, can be made intersubjective in an unequivocal, consistent manner". Since the *Aufbau* is meant to solve the problem of the possibility of intersubjective knowledge, this point is absolutely fundamental (cf. §16). This intersubjectivizing function (§§146–49) depends crucially on the construction of the worlds of physics of each epistemic agent and holds the methodological key to the solution of the problem of intersubjective knowledge.

The idea is that each epistemic agent constructs for herself both a private perceptual world of sense qualities and then a private physical world of physical state magnitude values applied to each point of space-time. The construction of the private physical worlds is necessary here because these assignments are such that each epistemic agent can transform via continuous one-one coordinate transforms her private physical world into that of any other epistemic agent's world in such a way that she can find a certain physical object in her own world that corresponds to a physical object in the other agent's world. This fact forms the occasion for the construction of the intersubjective object which is the class of all such intersubjectively corresponding objects (cf. §§146–49). The important point is that the perceptual worlds have such wildly differing assignments of qualities to space-time points that such an intersubjectivizing function performed through topological transforms of the perceptual space-time worlds would come up empty (§133), whereas the rule-governed nature of the assignment of physical state magnitudes leads to great agreement among agents in their assignments in their respective worlds of physics.[22]

The relative lack of discussion of the conventional element in the construction of metric concepts is correlative with the other crucial change in the *Aufbau*; Carnap's understanding of the form necessary for objective knowledge in the *Aufbau* is no longer mathematics but logical form in general. Thus, Carnap puts forward the view that all scientific concepts are definable from purely structural descriptions of relations among elements of experience. And here the notion of structure is logical structure as given in the logic of relations regardless of whether that structure is rich enough to yield metrical relations.

Moreover, as was implicit in the point just made, Carnap explicitly demands in the *Aufbau* that it is the form – not the content – of private experience that grounds the objectivity of knowledge. He puts the point this way in §16:

From the point of view of construction theory, this state of affairs is to be described in the following way. The series of experiences is different for each subject. If we want to achieve, in spite of this, agreement in the names of entities which are constructed on the basis of these experiences, then this cannot be done by reference to the completely divergent content, but only through the formal description of the structure of these entities.

Thus, the logical form of experience is identical for all agents and this guarantees that the logical constructions yield well-formed concepts for all agents also. Therefore, the synthetic *a priori*, considered in the sense of the formal conditions of experience common to all agents and across possible choices of physical state magnitudes, still plays the fundamental epistemic role as the basis of the constructional system.

The basic idea behind this structuralist notion of objectivity is that we can define each scientific concept uniquely by looking at its purely structural relations to all other scientific concepts – only by embedding the concepts into this holistic system can we hope that logical structure will single out relations uniquely. If successful this program would solve the problem of the possibility of objective knowledge and, moreover, solve the problem Carnap found with the application of formal logic to physical geometry in *Der Raum*. There Carnap interposed intuition between logic and experience to overcome the problem that mathematical geometry for the logicist captured too many structures and failed to pick out physical space uniquely. Carnap's holist semantics now attempts to solve this problem by reconceiving the relation of logic and the concepts of empirical science. By using the power of type theory to capture all the logical relations among all the concepts of science Carnap hopes to uniquely define each empirical concept by reference to its unique place within the whole system of logical interconnections among such concepts. This logical holism recalls at the level of type theory Cassirer's account of the role of energism in the construction of mathematical physics noted in Section 1 above and shows how the logicist account of logic can play the objectifying role Cassirer demanded.

Unfortunately, the type theoretical framework was in a sense too strong and too weak to carry out Carnap's guiding insight. It was too strong in the sense that Carnap could not actually guarantee the uniqueness claims necessary for his constructional definitions; type theory allows too many relations to be constructed. It was too weak in the sense that Carnap's ultimate idea was to extend the notion of analytic truth to sentences crucially containing empirical terms, and the logicist, type theoretic understanding of logic does not allow for this extension. The structuralist program and its ultimate failure are well detailed by Michael Friedman (1987), so I will just mention certain features which are of special interest when considered against the background of neo-Kantianism.

First, as mentioned above, although Carnap explicitly denies the possibility of synthetic *a priori* knowledge in the *Aufbau* (§106), this structuralist notion of objectivity is clearly related to neo-Kantian concerns about the objective structure of concepts. Indeed Carnap attributes the fundamental insight that relational concepts among the elementary experiences must form the basis of the constructions to the neo-Kantians, making special reference to *Substance and Function* (§75). Furthermore, in a section entitled "The Basic Relations as Categories" (§83), Carnap explicitly raises the question of what might count as the categories within construction theory and puts forward the following suggestion:

[P]erhaps we are in better agreement with established usage (which is not very clear) if we call the basic relations categories. The following fact would seem to support this: in a certain sense, every statement about any object is, *materialiter*, a statement about the basic elements. But, *formaliter*, it is a statement about the basic relations.

Beyond the striking use of Kantian language, this passage serves to connect Carnap's use of the basic relation as the single category and the formal structure of experience to the universal invariant theory of experience in Cassirer. The basic relation has certain formal properties common to all streams of experience that drive all the subsequent constructions and make possible objective concepts via logical definitions that only make reference to these structural features.

Second, the movement away from precise, mathematized concepts, such as those in physics to concepts definable in type theory is already foreshadowed in *Physikalische Begriffsbildung*. There Carnap noted the topological determinations of sensory qualities which guaranteed that metric conventions were possible for them. These topological properties were structural properties in the sense of type theory: transitivity, symmetry, etc. Carnap's fundamental idea in the *Aufbau* is that if we loosen the idea of objective structure from mathematical structure – the notion he took over from Helmholtz, Planck, etc., and put forward in *Physikalische Begriffsbildung* – to logical structure as given in the theory of relations in type theory, then one can use this notion of structure to drive objectivity down to the sensory level. That is, the prerequisite topological properties of sensory experience do not allow mathematization without additional metrical conventions; they are therefore on the Helmholtz account subjective. On Carnap's new notion of structure, however, precisely those topological properties indicate that even "subjective", sensory experience has certain structural features and can therefore be made objective.

This loosened notion of objective structure threatens to collapse our crucial distinction between formal conditions of the possibility of objective knowledge and formal conditions of the possibility of experience. The formal structure of experience is now adequate by itself and without any additional structural conventions to guarantee objective concepts in science. This leads to the reductionist and positivist tone of the work. Having made logical structure the locus of objectivity and claimed that all concepts are derivable by logical means from the structure of experience, Carnap can now say that the concepts of science are abbreviational terms for talk about experience alone (§160) and that all concept construction is done by convention (§179).

However, this should not be read as a conversion to radical empiricism. We have already noted the role that the laws of physics play in the construction of the physical world; this belies any foundationalist motives for construction. Second, it is clear that from a scientific perspective the question to be answered is how can science make intersubjective claims given the subjective nature of experience. The answer is given in the construction of the intersubjective world, the world of science (§149). Of course, this answer is now tied to an epistemic reduction that allows all substantive questions in science to be answered by appeal to experience, but this is, as Carnap notes (§80), as congenial to transcendental idealism as it is to empiricism.[23]

Rather than a movement to radical empiricism, I think that Carnap is driven to reconceive the operative notion of objective structure by a heightened concern about the status of epistemology and because he feels he can make the important methodological distinctions, now as distinctions within the constructional system. On the first point, loosening objective structure to logical structure makes objective talk of experience itself possible, and this, on Carnap's view, is not merely a precondition for scientific psychology but of epistemology itself. This is because, in order to understand the important difference between experience and knowledge, we must as epistemologists be able to talk about experience and this requires an objective sense of experience. Indeed it was a primary focus of the *Aufbau* as well as *Scheinprobleme* to find a place for epistemology within the scientific world-view and thus to provide an account of the meaningfulness of epistemological statements themselves.

But, to take up the second point, the crucial distinction between the objectifying knowledge of mathematical physics, its conventional components, and its privileged role in the construction of the intersubjective world on the one hand, and the private autopsychological realm of experience on the other, can still be captured within the constructional system. This requires the construction of the private autopsychological realm for each agent, a methodological distinction between it and the physical realm, which we noted in talking about the construction of the intersubjective world, and is aided by the syntactic restrictions of type theory which prevent the sensible affirmation of any attribute of the intersubjective world of science to the autopsychological realm. That is, the order and method of the construction indicates the privileged role that physical concepts play in the objectification of experience. Moreover, despite the definition of all terms from the sensory basis, the restrictions on meaningfulness of type theory mean that we cannot ascribe any properties of the autopsychological realm to the objects of physics. Thus, Carnap hopes to both preserve what was right in his methodological views before the *Aufbau* and carve out a perspective from which to say them.

Here, I would argue, is where the decisive influence of Russell's logic and his External World Program can be felt. The use Russell gives to the logic of relations in epistemology shows that it is a fertile tool even where it is not used to derive traditional mathematics. The notion of structure Carnap uses is taken over fully from Russell's notion of relation number (§107). Moreover, the power and precision of Russell's logic provides Carnap with the means to actually advance his structuralist notion of objectivity in a way that outstripped anything that Cassirer or the other neo-Kantians could provide as results of their epistemic perspective.

Cassirer's complaints against logicism clearly have no force against the program of the *Aufbau*. For Carnap clearly does not base his epistemology of empirical knowledge in acquaintance as Russell does. Rather, Carnap joins a long tradition in German philosophy – itself going back to Kant and including Cassirer as well as Schlick and Reichenbach – in distinguishing mere acquaintance [*kennen*] from true knowledge [*erkennen*]. Moreover, as we saw in the quote from §16, Carnap founds his system on certain structural coincidences in streams of experience rather than the material content of experience. The most important point made about objective knowledge in the *Aufbau* is not found in the ability of scientific statements to be translated into pure sensory terms, but in our ability with the constructional system in hand to translate the sentences of science into completely intersubjective terms after the construction of the intersubjective world: "[S]cience aims to produce a supply of exclusively intersubjective statements" (§149) and the logic of Russell shows how it can do so.

Rather than an attempt to fulfill the empiricist leanings of the External World Program, the *Aufbau* is the first detailed systematic attempt to show how a logicist understanding of logic and mathematics can provide an account of objectivity within the empirical realm. This view is more consistent with Carnap's early avowedly Kantian views and also, I would claim,[24] with his philosophy's subsequent development than is the standard account of logical empiricism. The forms of logical definition applied to the fundamental form of experience provide science with the capacity to achieve objective knowledge despite its subjective origin in personal experience. The place of the *Aufbau* in Cassirer's demand for a logic of empirical knowledge is borne out in the titles Carnap[25] suggested for the *Aufbau* in his correspondence with Schlick in 1926: *Der logische Aufbau der Erkenntnis* or, simply, *Erkenntnislogik*.[26]

NOTES

[1] All references to this work will be by section number only; all quotes will be from Rolf George's translation (Carnap 1969).

[2] Hylton (1990) argues persuasively that even Russell's early epistemology ought not be uncritically assimilated to traditional empiricism. This means that a more informed understanding of the External World Program may illuminate our understanding of (especially Carnap's) logical empiricism. My concern is to combat the received view of logical empiricism and its relation to standard empiricism. I'd be happy if the received view turned out to be doubly false due to insufficient attention to what is new and interesting in the External World Program.

[3] The dominance of this view of logical empiricism may be traceable to its early expression in the English-speaking world in Ayer (1936) and in the German-speaking world in Neurath (1929). But the logicism of Russell and Frege can scarcely be considered congenial with empiricism. Neither Russell, at least in the early stages of his logicist phase, nor Frege was an empiricist. See again Hylton (1990).

[4] Sauer (1989) traces this line of thinking to Beck (1965).

[5] Friedman (1992, Section 2) takes up the neo-Kantian notion of constitution and argues persuasively for its consistency with one reading of Russell's supreme maxim of scientific philosophizing.

[6] Friedman (1992) further elaborates the continuities between Carnap and the neo-Kantians of the prior generations and also puts forward a view that takes more seriously Carnap's claim to be giving an epistemology neutral between Kantianism and all other traditional epistemic programs. See my cautionary note in the text below.

[7] Friedman's latest work (1992), which I was only able to read in the final revision stages of my paper but which has aided me in sharpening several of the main points, has moved in this direction.

[8] Sauer (1989, p. 117) in fact does cite precisely this passage from §177. His citation is part of a more subtle argument that Carnap's notion of experience itself is closer to the neo-Kantians' than it is to traditional Machian empiricists'. Thus, I don't mean to be covertly criticizing Sauer at this point. Sauer's position substantially informs my account of Carnap's work in Sections 2 and 3 below.

[9] This is not meant to denigrate other aspects of the epistemological background to logical positivism. Moulines (1985) is quite right in emphasizing the complexities of the entire spectrum of German epistemology early in this century and I would welcome the illumination that would certainly accompany a detailed study of works such as Theodor Ziehen's 1913 and Hans Driesch's 1912 monographs, both of which Carnap mentions as programmatically akin to the Aufbau in §3.

[10] For other aspects of the relation between the Aufbau and neo-Kantianism, see especially Friedman (1987, 1991, 1992), Sauer (1985, 1989), and Richardson (1990, forthcoming).

[11] All quotations from this work are taken from the translation (Cassirer 1953). Other aspects of Cassirer's work are discussed in Sauer (1985, 1989), and, with particular reference to his relations with Schlick's early philosophy, in Coffa (1991, pp. 197–204) and Ryckman (1991).

[12] All translations from previously untranslated works are mine.

[13] Compare here the remarks about Cassirer and logicism in Sauer (1989, Section II).

[14] This notion of the relativized *a priori* comes from Friedman.

[15] Reichenbach's relativized *a priori* is discussed illuminatingly in Friedman (1983a, chapter one; 1983b, 1991). See also Coffa's discussion (1991, chapter ten) of Schlick, Cassirer, and Reichenbach.

[16] The most famous conventionalist arguments can be found in Poincaré's *Science and Hypothesis* (1902). Dingler's arguments can be found in, e.g., his *Die Grundlagen der Physik* (1919). Husserl's notion of essential insight is propounded in Husserl (1913a, 1913b).

[17] All translations from *Der Raum* are from the forthcoming Friedman and Heath translation *Space*. I am very grateful to Friedman and Heath for the opportunity to use their manuscript.

[18] This idea puts Carnap in the same camp with other philosophers interested in geometry and physics and who soon became founders of logical empiricism, notably Schlick and Reichenbach. See again the literature referred to in note 15 above. There is some vagueness in Carnap's account of the synthetic *a priori* status of intuitive space in *Der Raum*. His first desire was to make the structure of intuition and intuitive space sufficient to provide a nonempirical, *a priori* guarantee of the possibility of a determinate global metrical structure for physical space. This will give the synthetic *a priori* the methodological role it should have; it is an *a priori* condition of the possibility of objective physical knowledge. His second idea is to tie the synthetic *a priori* to the necessary form of matters of fact of experience. Unfortunately his test for necessary form yields a synthetic *a priori* structure for intuitive space that is (barely) too weak to provide a guarantee that we can find metrical conventions and hence can have a mathematical theory of physical space. In the chapter on intuitive space, Carnap calls the structure of intuitive space "three dimensional metrical [because metrizable] intuitive space"; and in the chapter on physical space and in the conclusion only n-dimensional topological intuitive space is considered the synthetic *a priori* intuitive component of geometrical knowledge. But, as Michael Friedman has suggested to me, Carnap can be read as presupposing the full structure of intuitive space argued for in chapter II when he gives the argument from necessary form of matters of fact of experience.

[19] This is argued explicitly in Part Vb of *Der Raum*.

[20] For Goodman's original objections, see Goodman (1951) and Quine (1961). For more on peculiarities in Carnap's understanding of definition in the *Aufbau*, see Coffa (1991, pp. 218–22). For more on the details of the construction of the physical world and its relations to the External World Program, see Friedman (1992).

[21] I am indebted to Michael Friedman for pressing me to clarify the claims of this paragraph.

[22] In fact, there is another, in some ways more fundamental, sense in which intersubjective knowledge requires the construction of the physical world for Carnap. Within the construction theoretic framework we only recognize one another as epistemic agents after we have constructed one another as physical objects in our respective worlds of physics. After we do this, certain linguistic and other expressive behavior leads us to impute psychological states to one another and, hence, to view one another as epistemic agents. Thus, the problem of intersubjective knowledge can't even be raised within the constructional system until a multiplicity of epistemic agents is available and this requires the construction of the physical world.

227

[23] Carnap (§179) does note that his reduction conflicts with Natorp's view of the infinite task of determining the object of experience. One way to begin to understand Natorp's point of view is to consider Cassirer's universal invariant theory of the structure of the object of experience. If we have no other way to understand this structure than to wait and see what forms are in fact continuously used in all change of physical theory, then clearly this task won't end and we shall never know the object of experience in its simplest form. Carnap's use of Russell's logic as the locus of structure also gives an independent understanding of form which solves this problem. But, I think that Carnap is mistaken in thinking that Natorp's view was in fact the general Marburg view.

[24] See Richardson (1992, forthcoming) for further elaboration on this point.

[25] For some remarks on the history of the title of the *Aufbau*, see Coffa (1991, p. 403, n. 11).

[26] Ideas leading to this paper were presented at Northwestern University, the University of Minnesota, the University of Tulsa, and the 9th International Congress of Logic, Methodology, and Philosophy of Science, in Uppsala; I'd like to thank my audience in each case for helpful discussion. My thinking about Cassirer and Carnap has benefited from helpful conversations with Richard Creath, Geoffrey Hellman, Mark Notturno, Charles Nussbaum, Thomas Ricketts, Thomas A. Ryckman, and David Sullivan. My deepest intellectual debt is, however, to Michael Friedman, who first introduced me to the riches of Carnap's early philosophy and whose work in history of philosophy of science serves as a constant inspiration.

REFERENCES

Ayer, A. J.: 1946, *Language, Truth, and Logic*, 2d ed., Gollancz, London.

Beck, L. W.: 1965, *Early German Philosophy. Kant and his Predecessors*, Harvard University Press, Cambridge.

Carnap, R.: 1922, *Der Raum* (*Kant-Studien* **56** supp.), von Reuther und Reichard, Berlin (forthcoming, *Space*, trans. by M. Friedman and P. Heath).

Carnap, R.: 1923, 'Über die Aufgabe der Physik', *Kant-Studien* **28**, 90–107.

Carnap, R.: 1926, *Physikalische Begriffsbildung*, Braun, Karlsruhe.

Carnap, R.: 1928a, *Der logische Aufbau der Welt*, Weltkreis, Berlin (trans. in Carnap (1969)).

Carnap, R.: 1928b, *Scheinprobleme in der Philosophie*, Weltkreis, Berlin (trans. in Carnap (1969)).

Carnap, R.: 1969, *The Logical Structure of the World and Pseudoproblems in Philosophy*, trans. by R. George, University of California Press, Berkeley.

Cassirer, E.: 1907, 'Kant und die moderne Mathematik', *Kant-Studien* **12**, 1–49.

Cassirer, E.: 1910, *Substanzbegriff und Funktionbegriff*, Springer, Berlin (trans. in Cassirer (1953)).

Cassirer, E.: 1921, *Zur Einstein'schen Relativitätstheorie*, Springer, Berlin (trans. in Cassirer (1953)).

Cassirer, E.: 1953, *Substance and Function*, trans. by W. C. and M. C. Swabey, Dover, New York.

Coffa, J. A.: 1991, *The Semantic Tradition from Kant to Carnap: To the Vienna Station*, Cambridge University Press, Cambridge.

Cohen, H.: 1902, *Logik der reinen Erkenntnis*, B. Cassirer, Berlin.

Dingler, H.: 1919, *Die Grundlagen der Physik*, Kraut, Leipzig.

Driesch, H.: 1912, *Ordnungslehre*, Diederichs, Jena.

Friedman, M.: 1983a, *Foundations of Space-Time Theories*, Princeton University Press, Princeton.

Friedman, M.: 1983b, 'Critical Notice: Moritz Schlick, *Philosophical Papers*', *Philosophy of Science* 50, 498–514.

Friedman, M.: 1987, 'Carnap's *Aufbau* Reconsidered', *Noûs* 21, 521–45.

Friedman, M.: 1991, 'The Re-evaluation of Logical Positivism', *Journal of Philosophy* 88, 505–19

Friedman, M.: 1992, 'Epistemology in the *Aufbau*', *Synthese* 93(1).

Goodman, N.: 1951, *The Structure of Appearance*, Harvard University Press, Cambridge.

Goodman, N.: 1963, 'The Significance of *Der logische Aufbau der Welt*', in P. A. Schilpp (ed.), *The Philosophy of Rudolf Carnap*, Open Court, La Salle, IL, pp. 545–58.

Haack, S.: 1977, 'Carnap's *Aufbau*: Some Kantian Reflexions', *Ratio* 19, 170–75.

Husserl, E.: 1913a, *Ideen zu einer reinen Phänemonologie und phänemonologischen Philosophie*, Max Niemeyer Verlag, Halle (1982, *Ideas Toward a Pure Phenomenology and Phenomenological Philosophy*, trans. by F. Kersten, Nijhoff, The Hague).

Husserl, E.: 1913b, *Logische Untersuchungen*, 2d. ed., Vol. 1, Max Niemeyer Verlag, Halle (1970, *Logical Investigations*, trans. by J. N. Findlay, Vol. 1, Routledge and Kegan Paul, London).

Hylton, P.: 1990, *Russell, Idealism, and the Emergence of Analytic Philosophy*, Oxford University Press, Oxford.

Kant, I.: 1965, *Critique of Pure Reason*, trans. by N. Kemp-Smith, St. Martin's Press, New York.

Moulines, C.: 1985, 'Hintergründe der Erkenntnistheorie des frühen Carnap', *Grazer Philosophische Studien* 23, 1–18.

Natorp, P.: 1910, *Die logischen Grundlagen der exakten Wissenschaften*, Teubner, Leipzig.

Neurath, O.: 1929, *Wissenschaftlicher Weltauffassung. Der Wiener Kreis*, Artur Wolf, Vienna.

Poincaré, H.: 1902, *Le Science et Hypothèse*, Flammarion, Paris (1952, *Science and Hypothesis*, trans. by W. J. Greenstreet, Dover, New York).

Proust, J.: 1986, *Questions de Forme*, Librairie Artheme Fayard. Paris (1989, *Questions of Form*, trans. by A. A. Brenner, University of Minnesota Press, Minneapolis).

Putnam, H.: 1983, 'Why Reason can't be Naturalized', in his *Realism and Reason*, Cambridge University Press, Cambridge, pp. 229–47.

Quine, W. V.: 1961, 'Two Dogmas of Empiricism', in his *From a Logical Point of View*, 2d ed., Harvard University Press, Cambridge, pp. 20–46.

Quine, W. V.: 1969, 'Epistemology Naturalized', in his *Ontological Relativity and Other Essays*, Columbia University Press, New York, pp. 69–90.

Reichenbach, H.: 1921, *Relativitätstheorie und Erkenntnis A Priori*, Springer, Berlin (1965, *The Theory of Relativity and A Priori Knowledge*, trans. by M. Reichenbach, University of California Press, Berkeley).

Richardson, A.: 1990, 'How not to Russell Carnap's *Aufbau*', in L. Wessels et al. (eds.), *PSA 90*, Vol. 1, Philosophy of Science Association, East Lansing, MI, pp. 3–14.

Richardson, A.: 1992, 'Metaphysics and Idealism in the *Aufbau*', *Grazer Philosophische Studien*.

Richardson, A.: forthcoming, *Carnap's Construction of the World*, Cambridge University Press, Cambridge.

Russell, B.: 1914, *Our Knowledge of the External World*, Open Court, London.

Russell, B.: 1981, *Mysticism and Logic*, Barnes and Nobles Imports, Totowa, NJ.

Ryckman, T. A.: 1991, '*Conditio Sine Qua Non*?: *Zuordnung* in the Early Epistemologies of Cassirer and Schlick', *Synthese* **88**, 57–95.

Sauer, W.: 1985, 'Carnaps *Aufbau* in kantianischer Sicht', *Grazer Philosophische Studien* **23**, 19–35.

Sauer, W.: 1989, 'On the Kantian Background to NeoPositivism', *Topoi* **8**, 111–19.

Sellars, W.: 1953, 'Empiricism and the Philosophy of Mind', in H. Feigl and M. Scriven (eds.), *Minnesota Studies in the Philosophy of Science*, Vol. 1, University of Minnesota Press, Minneapolis, pp. 253–329.

Ziehen, T.: 1913, *Erkenntnistheorie auf psychophysiologischer und physikalischer Grundlagen*, G. Fischer, Jena.

Department of Philosophy
University of Pennsylvania
3440 Market St., Ste. 460
Philadelphia, PA 19104–3325
U.S.A.

HILARY PUTNAM

REICHENBACH'S METAPHYSICAL PICTURE

I would like to begin by saying a word about Reichenbach's impact on my own philosophical formation. As an undergraduate at the University of Pennsylvania, I studied philosophy of science mainly with C. West Churchman, who taught a version of pragmatism informed by a substantial knowledge of and interest in the logic of statistical testing, and with Sidney Morgenbesser, who was himself still a graduate student, and logic and American philosophy with Morton White. My orientation on graduation was positivistic (although my knowledge of positivism came mainly from A. J. Ayer's Language, *Truth and Logic*). I did a year of graduate work at Harvard in 1948–49 (studying mainly mathematics and mathematical logic), where I came under the influence of Quine's views on ontology and his scepticism concerning the analytic/synthetic distinction. At that point, I was in a mood that is well known to philosophy teachers today: it seemed to me that the great problems of philosophy had turned out to be pseudoproblems, and it was not clear to me that the technical problems that remained to be cleared up possessed anything like the intrinsic interest of the problems in pure logic and mathematics that also interested me. It was a serious question in my mind at that point whether I should really go on in philosophy or shift to mathematics.

Within a few months of my arrival in Los Angeles in the fall of 1949 these philosophical "blahs" had totally vanished (although I did not lose my desire to pursue mathematics in some way, and eventually I succeeded in pursuing both fields simultaneously). What overcame my "philosophy is over" mood, what made the field come alive for me, made it more exciting and more challenging than I had been able to imagine, was Reichenbach's seminar, and his lecture course on the philosophy of space and time. And it was not the technical details of Reichenbach's philosophy that did this, although I pondered those details with great excitement, but the sense of a powerful philosophical vision behind those details: the vision that I am going to sketch in this essay.

A number of times some of my former students (and some who were

Erkenntnis 35: 61–75, 1991.
© 1991 *Kluwer Academic Publishers. Printed in the Netherlands.*

not my former students) have complimented me on the fact that I have
been willing to think and to write about the "big problems", and not
just about technical problems. But this is something that Reichenbach
instilled in me. Reichenbach taught me – by example, not by preaching – that being an "analytic philosopher" does not mean simply rejecting the big problems. Although Reichenbach was just as much of an
empiricist as Ayer, empiricism, for Reichenbach, was a challenge and
not a terminus. The challenge was to show that the great questions –
the nature of space and time, the nature of causality, the justification
of induction (and also, in other writings which I shall not discuss here,
the question of free will and determinism, and the nature of ethical
utterances) could be adequately clarified within an empiricist framework, and not merely dismissed. In clarifying them, Reichenbach was
guided by what many philosophers today would be comfortable calling
a metaphysical picture (even if "metaphysics" was a pejorative term
for Reichenbach himself). That Reichenbach had a "metaphysical picture" did not keep him from being an analytic philosopher at the
same time: rather, it gave direction and motivation to his philosophical
endeavors, and not just motivation but excitement. I felt that excitement, and it generated my own excitement for the subject. I think that
every one of Reichenbach's Ph.D. students would say the same thing.

REICHENBACH AND CARNAP

Although Reichenbach continues until this day to be lumped together
with the Vienna Circle, his views, as he himself often insisted, differed
in significant respects from those of the other leading representatives
of Logical Empiricism. Throughout its existence, Logical Empiricism
was characterized by a certain tension between its metaphysical and its
antimetaphysical aspirations. On the one side, the antimetaphysical
side, the self-image of the movement was that it represented the entire
renunciation of metaphysics, that in the future metaphysical questions
would be replaced by questions in the logic of science. Metaphysical
questions which could not be so replaced would thereby be seen to be
unintelligible. In this way, it was believed, philosophical questions
would for the first time become susceptible to objective resolution. On
the other side, the metaphysical side, the great Logical Empiricists –
the most famous example is Rudolf Carnap – produced "rational reconstructions of the language of science" which looked for all the world

like elaborate systems of metaphysics – or at least that is how they looked to philosophers who did not join the movement. Certain features of the *Logische Aufbau der Welt* have received a great deal of attention in this regard – Carnap's phenomenalism, his logicism,[1] his doctrine of the "incommunicability of content" and the "methodological solipsism" associated with that doctrine, have all been extensively discussed. Indeed, the ontology and epistemology associated with the *Logische Aufbau der Welt* are now so widely known that there is a tendency to assume that that metaphysical picture was *the* metaphysics of Logical Positivism.

In the present essay, I want to describe a totally different metaphysical picture: the picture that Hans Reichenbach defended in his major works, and particularly in *Experience and Prediction*, *The Philosophy of Space and Time*, *Axiomatization of The Theory of Relativity*, and *The Direction of Time*. I believe that that picture is of interest for at least two different reasons. One is that, like Carnap's picture, it represents the work of a brilliant and philosophically ambitious mind. Even if today one cannot accept either Carnap's or Reichenbach's solutions to the problems of philosophy, their attempts represent contributions of enormous value, and there is much to be learned from their study. The second reason is that one does not properly understand Logical Empiricism[2] itself if one simply identifies it with any of Carnap's successive programs of rational reconstruction. The fact that the movement had room for metaphysical visions as different as those of Carnap and Reichenbach is itself a fact that has to be appreciated if simplistic pictures of "positivism" are to be overcome.

The first point at which the metaphysical picture in Reichenbach's mature work stands in opposition to the metaphysical picture associated with Carnap's *Logische Aufbau* has to do with Reichenbach's realism. To be sure, Reichenbach did not see this realism as presupposing traditional realist metaphysics in any way. He insisted that the choice of "realistic language" is just that: a choice of a language. He emphasized, however, that it was not an *arbitrary choice*.[3] He once told me that he welcomed Carnap's shift from an *Eigenpsychische Basis* (a phenomenalistic basis) to a physicalistic basis in *Testability and Meaning* as an indication of convergence in their views, which is also how Carnap saw the matter. But the discussion in *Experience and Prediction* (henceforth cited as "EP") reveals deep disagreements with Carnap in spite of this "convergence". For one thing, it turns out that what Reichenbach is thinking of is not the choice of *sense-datum language* versus "thing language"; it is, instead, the choice of a language in which I can speak of things as *existing when unobserved* ("usual language") as opposed to the choice of a language in which I can speak of things as existing only when they are observed by me (an "egocentric language"). It is true that if "things" only exist when observed by me, and are ascribed only the properties that they appear (to me) to possess while observed, then these "things" will be, in a way, just as "phenomenal" as the empiricists' "impressions", and this is why Reichenbach, in the first part of EP, *identifies* the traditional empiricist problem of justifying the inference from "impressions" to "the external world" with the problem of justifying the inference from "observed things" to "unobserved things". He offers an account of this inference in Section 14 of that work; but that account, as he points out, assumes that one has already selected "usual language".[4] Reichenbach's inductive inference to the existence of unobserved things is not the *justification* for the choice of usual language – that is, for the decision to adopt a language in which I can ascribe properties to things that they do not appear to have, and in which I can ascribe properties to things at times when I do not observe them, and, indeed, when nobody observes them. The justification offered for that choice (EP, p. 150) is interesting: instead of making the usual argument that realist language enables us to formulate more successful scientific theories, Reichenbach argues that it makes nonsense of the *justification* of a great many ordinary human actions: "the strictly positivistic language . . . contradicts normal language so obviously that it has scarcely been seriously maintained; moreover, its insufficiency is revealed as soon as we try to use it for the rational reconstruction of the thought-processes underlying actions concerning events after our death, such as expressed in the example of the life-insurance policies. We have said that the choice of a language depends on our free decision, but that we are bound to the decisions entailed by our choice: we find here that the decision for the strictly positivistic langugage would entail the renunciation of any reasonable justification of a great many human actions."

What is striking, however, is that although Reichenbach does allow as possible the choice of an egocentric language in which I speak of, the table I observe as existing only when I observe it, he rejects the idea that what I *really* observe are impressions. Instead (EP, p. 163), he warmly endorses Dewey's view in *Experience* and *Nature* that impressions are inferred entities. (A view that goes back to Peirce's writings in the 1860's.) In Reichenbach's view, what we observe are *things*; this is the correct description of what we might call the "phenomenology" of observation.[5] That the appearances of things may sometimes be deceptive, that I may mistake a shadow for a human figure, or whatever, does not show that we do not directly observe things; it shows simply that observation is corrigible. But, Reichenbach insists, it would be corrigible even if we did have such a thing as an "impression language". The idea that there can be any such thing as "incorrigible observation" is rejected. Even in the extreme case in which someone decides that he was simply hallucinating (or dreaming), Reichenbach would say that he observed a physical "thing" (counting events, and even states of one's own body as "things"), namely a state of his own central nervous system, although he did not know that that was what he was observing.

For Reichenbach, to describe an observation using impression language – to say, e.g., "I had the impression of a table", or "I had the impression of a flash of light" – is always to make an inference (EP, pp. 163–169). Thus the difference between Reichenbach's position and Carnap's is not just that Reichenbach adopted "thing language" as basic rather than "impression language" – after a certain point Carnap also shifted to taking "thing language" as basic – the decisive difference is that Carnap continued to regard impression predicates as a possible, although less convenient, observation basis for the reconstruction of the language of science, whereas Reichenbach insisted that it was a mistake to think of impressions as observed entities at all. Impressions are inferred entities. For Carnap, there remains, even after the turn to physicalism, a lingering tendency to think of sensations as "epistemologically primary"; for Reichenbach it is things that are epistemologically primary.

"Things" in a wide sense, however. For, as already remarked, events are also "physical things", and when Reichenbach develops his reconstruction of modern physics in his writings on Relativity theory, the things which become fundamental are not the things we observe in everyday life, but the things we describe in our most basic physical theories.

REICHENBACH AND KANT

Although Reichenbach is justly famous for his decisive criticism of Kant's doctrine of the synthetic a priori as applied to the geometry of space and time, Reichenbach was by no means incapable of incorporating Kantian insights into his own philosophy.[6] One of the most important insights that Reichenbach takes from Kant[7] (even if he in one place credits it to Leibniz![8]), but transforms, is the inadequacy of the traditional empiricist way of accounting for the existence of an objective time order. That account presupposed the truth of some form of subjective idealism. If everything that exists is a complex of sense impressions, then, many empiricists thought, it suffices to define an objective order for sense impressions. Such an order is implicitly defined by the following principles: (1) if an impression A is simultaneous with a memory image of an impression B, then A occurred after B. (2) If two impressions A and B are experienced simultaneously, then, of course, they occur simultaneously. But, as Kant saw, the case is different once one takes the "experiences" of the external senses to be experiences of things rather than of impressions; neither of these principles holds if "impression" is replaced by "physical thing or event". If A and B are things or events rather than impressions, then I cannot infer from the fact that an experience of A occurred simultaneously with a memory of an experience of B that B occurred or existed before A occurred or existed; and if an experience of A occurred simultaneously with an experience of B, I cannot infer that A and B occurred or existed simultaneously. This is clearest in the case of astrophysics (and Kant and Reichenbach shared, of course, a profound interest in astrophysics). If I observe an explosion of a supernova A before I observe the explosion of a supernova B it does not at all follow that the A explosion occurred before the B explosion; it may be that the A explosion occurred later but that it was so much nearer to me that the light from the A explosion reached my telescope before the light from the B explosion. Similarly, if I observe (or "experience") two astronomical events simultaneously, it does not follow that they happened simultaneously. There are, however, homelier examples that make the same point. One, which was used by Kant,[9] is the following: if I look at a building, I may observe the top of the building before I observe the bottom of the building, but I cannot infer that the top of the building existed before the bottom of the building did. On the other hand, if I see a ship sailing down the river, and I first observe the ship passing under a bridge and then I observe the ship passing by a dock further down the river, I do infer that the ship's being under the bridge temporally preceded the ship's passing the dock. The Kantian explanation is that my very conception of the ship as an object is bound up with a whole system of laws ("rules"), and given my total system of knowledge I am able to interpret the two experiences in the case of the ship as corresponding to the time order in which the events occurred. When I look at a building I apply a different part of my knowledge – different laws as well as different initial conditions – and I conclude that I am simply seeing in a different order things that have continued existence. Here Kant stresses a further causal consideration: I know that I can see the different parts of the building in any order I choose; but I cannot see the events in the case of the ship in any order I choose. Kant's conclusion is that it is only by interpreting my experiences with the aid of causal laws that I am able to understand them as giving me an objective world with objective time relations.

As an empiricist, Reichenbach does *not* follow Kant in holding that we do not have a notion of an object at all without a concept of causality. He does, nevertheless, agree that it is causal relations that determine what the correct ("objective") time order is. However, Reichenbach has a problem: as a good empiricist, he does not want to take the notion of causality as primitive. On the other hand, he is convinced that one cannot simply take objective time order as primitive, that one must explain objective time order in terms of objective causal order. This leads Reichenbach to one of the most ambitious projects ever undertaken by an analytic philosopher, namely *to define causality without using temporal notions* so that he will be able to use the thus defined relation of causal precedence to define time order. In short, time order is to be reduced to causal order, and causal order is not to be taken as primitive, but to be explained. But explained using what notions?

Reichenbach's answer is that it is to be explained using the notion of *probability*.[10] In Reichenbach's metaphysical picture, the world consists of events, where an event might be either a macroscopic event or a microscopic one, say, a force-field's having a certain intensity at a certain spacetime point. Imagine, now, that I am given a sequence of events, say, that I observe a sequence of astronomical events $a_1, a_2, a_3, \ldots$. And let us suppose that I observe a second sequence of events, $b_1, b_2, b_3, \ldots$, and that I notice a statistically significant relation between events in the first sequence and events in the second sequence. For example, suppose that I observe that the probability that an event in the b-sequence will have a certain property Q is much higher if the corresponding event in the a-sequence has a certain property P. Inductively projectible statistical relations are not always a sign that one event directly influences the other. It could be that, for example, the a-events and the b-events are effects of a common cause.[11] Moreover, even if there is direct causal influence between an a-event and the corresponding b-event, we do not yet know in which direction the causality goes, that is, whether a_i's being a P causally affects b_i's chances of being a Q, or whether it is the other way around. However, Reichenbach believed that there is a method of experimental intervention, which he calls "the method of the mark", by which one can determine when the statistical relation is directly causal (i.e., by which one can rule out the hypothesis that the correlation is entirely explained by the presence of a common cause) and by which one can determine in which direction it goes. The idea is very simple: if the a-events are the causes (or partial causes) of the b-events, then small variations in the character of the a-events produced by an experimental intervention – Reichenbach calls such a variation a "mark" (*Kennzeichen*) – will bring about small variations in the character of the corresponding b-events.

Few contemporary philosophers of physics believe that Reichenbach succeeded. One difficulty is that the interpretation of an action as an "intervention" involves *already* knowing the direction of time/direction of causal processes, which is precisely what we are trying to determine. In *The Direction of Time*, Reichenbach tried to meet this objection by using thermodynamic criteria for irreversibility.[12] A further difficulty is the following: how can one know that the "small variation" in the a-event and the corresponding small variation in the b-event are not themselves effects of a common cause? If I myself "bring about" the variation in the a-events, this may be unlikely (but am I not appealing to causal knowledge when I say this?); but what if the "intervention" is Nature's, and not a human experimenter's? Presumably, Reichenbach would reply that further experimentation could rule out the common cause hypothesis; discussing this might also involve the question of the adequacy of Reichenbach's theory of the counterfactual conditional.[13]

My concern here, however, is not with the success or failure of Reichenbach's program, but with the picture which inspired him. The picture is this: the world consists of events. Those events form various sequences,[14] and there are statistical relations among the events in those sequences. Those statistical relations indicate the presence of causal relations, and the detailed nature of the statistical relations fixes the nature of the causal relations.

From this point on, Reichenbach's procedure is well known to philosophers of science. If an event A causally influences an event B, then the event A (absolutely) temporally precedes the event B: the interval separating the events A and B is time-like.[15] If A does not absolutely precede B and B does not absolutely precede A, then the separation between A and B is said to be space-like. (Such events are simultaneous in at least one coordinate system, while events whose separation is time-like are simultaneous in no admissible coordinate system.) In terms of causal (absolute temporal) precedence, one can define the notion of *temporal betweenness* and this is the first step in the construction of a topology[16] for spacetime.

With respect to the metric, as opposed to the topology, of spacetime, Reichenbach's position was a sophisticated form of conventionalism.[17] The world as it is in itself does not have a unique metric; it does, however, have a unique topology, and this is defined, as just sketched, in terms of the causal relations (and ultimately in terms of the statistical relations) between the events of which the world consists.

EPISTEMOLOGY AGAIN

At this point we may seem to have gotten very far indeed from the epistemological issues with which I started. The events of which we have been speaking are, for the most part, microevents, and thus very remote from observation. But here too there is a fundamental difference between Reichenbach's epistemological views and Carnap's. And this is true not only if one thinks of Carnap's approach in *Der Logische Aufbau der Welt*, but even if one considers Carnap's approach after the turn to "physicalism" (in "Testability and Meaning", and more completely in "Foundations of Logic and Mathematics"[18]). Carnap's successive positions might be described as follows: In the *Logische Aufbau*, Carnap hoped to reduce all talk of physical things and events to talk of the subject's own impressions by the method of logical construction.[19] In "Testability and Meaning" the aim of reduction was still present, but with two changes: (1) instead of taking impression predicates as the observation predicates to which all other descriptive predicates were to be reduced, a collection of "observable thing predicates" was taken as basic. Sentences without quantifiers in the observation vocabulary were eligible to represent what is given in experience. (2) Non-observation predicates were still to be reduced to observation predicates, but the allowed methods of reduction now included "reduction sentences", that is, operational definitions in a certain conditional form. Finally, in "The Interpretation of Physical Calculi", Carnap proposed to simply take theoretical terms as primitive and regard the adoption of a set of physical laws ("P-Postulates") as something to be justified on the following sorts of grounds: (1) an acceptable physical law was supposed to have observational consequences which are (as far as we know) true; and (2) in addition the set of laws is supposed to be simple, elegant, fruitful, etc. Physical laws and entities are, so to speak, *chosen rather than constructed or inferred*, according to late Carnap, but chosen subject to empirical constraints. Even in Carnap's last writings,[20] the whole content of the physical theory could be expressed by a single sentence in second order logic which contains only observation predicates, the so-called "Ramsey sentence" of the theory.

In Reichenbach's view, in contrast, physical laws and entities are *inferred*.[21] An example of what Reichenbach was thinking of might be the following. If I see a paddle wheel spinning, I normally observe that something is hitting the paddle wheel: air, or water, or perhaps someone is throwing stones at it, etc. If I perform a demonstration which was very popular in the secondary schools of Europe at the end of the nineteenth century, and place a cathode inside one end of a glass tube and an anode inside the other end, replace the air in the glass tube completely with an inert gas, and then turn on the electric current, I will (if all goes well) observe a glowing ray (this is, of course, the basis for our present "neon signs"). An experiment which was already performed in the nineteenth century was to place a tiny paper paddle wheel inside the tube. It was discovered that if the wheel was light enough, the presence of the ray would cause the wheel to spin. This spinning could not be due to to electromagnetism, because the wheel is made of paper. By an induction from the macroscopic phenomena I mentioned, we infer that therefore the ray consists of matter of some kind. Again, if I shoot a charged iron pellet through a magnetic field, I can observe that the pellet will be deflected from its usual trajectory by an amount which depends on the ratio of the charge on the pellet to its mass. If I put the glowing tube (this time without the paddle wheel) in a magnetic field, I can observe that the ray is also deflected, and, on the assumption that it consists of *charged* matter, I can infer the mass-charge ratio of the particles of which the ray consists. (Historically, this was one of the first determinations of the mass-charge ratio of the electron.) This is an example of an inference to the existence of what Reichenbach calls *illata*, inferred entities.

Whereas in the Carnap view (of the *Logische Aufbau* period, and even in the view of the "Testability and Meaning" period) the particles of which such a ray consists are logical constructions (*abstracta*, in Reichenbach's terminology), in Reichenbach's view they are inferred entities, and the inductive inferences by which we infer their existence are not different in kind from the inductive inferences that we find in, for example, Sherlock Holmes stories.[22] In fact, Reichenbach was famous for using Sherlock Holmes stories as the sources of his examples in his celebrated course on inductive logic.

REICHENBACH'S VINDICATION OF INDUCTION

If we picture Reichenbach's system as an arch, with the causal theory of time, the probability theory of causality, and the theory of the spacetime metric as one side, and the theory of the epistemological primacy of physical things and the inductive inferences to unobserved things and to *illata* as the other side, the keystone of the arch is his celebrated pragmatic vindication of induction (EP, pp. 339 ff). To vindicate induction, Reichenbach first reconstrued all inductive inferences as inferences whose purpose is to estimate the limit of a relative frequency in an infinite sequence.[23] Induction, for Reichenbach, always means wagering ("positing") that the limit of the relative frequency in some sequence or other is approximately equal to the relative frequency observed in the so-far examined initial section of that same sequence. With "induction" so defined, it is a tautology that *if* a sequence possesses a limit at all, then continued use of the method of induction will enable us to eventually estimate it correctly, to within any preassigned accuracy.

To see why this tautology should be regarded as a vindication of induction, we make the following thought experiment (EP, pp. 358–359): imagine that some method very different from induction leads to successful prediction, say, relying on a certain "clairvoyant". To say that the clairvoyant makes very successful predictions, say, that her predictions are true 85% of the time, is to say that the relative frequency in a certain sequence exceeds 0.85. But this is the sort of truth we can

discover using the method of induction. Thus, Reichenbach concludes
that induction will work in the long run if any method works.[24]

CONCLUSION

To recapitulate, then, for Reichenbach *probability* is the foundation
of both metaphysics and epistemology. Metaphysically, probability is
fundamental because it is the probability relations among the sequences
of events in the world that gives rise to causality, time, and space.
Epistemologically, probability is fundamental because empirical knowledge is simply knowledge of probabilities. Even knowledge of observation sentences is considered to be probabilistic knowledge by Reichenbach (EP, pp. 183–188), because Reichenbach's fallibilism leads him
to hold that no observation sentence is absolutely incorrigible, and
with the advance of scientific knowledge we need to inquire into the
probability that our singular observation judgments may be in error.
   My aim here has not been to argue that Reichenbach succeeded in
his magnificent attempt any more than Carnap succeeded in his. But I
hope to have convinced you that it *was* one of the most magnificent
attempts by any empiricist philosopher of this or of any other century,
and I believe that the effort to understand it and to master its details
will as richly repay us as the much greater effort which has been devoted
to the study of Carnap's work has already repaid us.

NOTES

[1] By "logicism" I mean the desire, expressed by Carnap late in the *Logische Aufbau*,
to eliminate empirical primitives from the system *entirely*. For a discussion of this feature,
see Michael Friedman, "Carnap's Aufbau Reconsidered", *Nous*, vol. 21, no. 4 (Dec.
1987), pp. 526–545.
[2] Reichenbach refused to call himself a positivist, because he associated that term with
phenomenalism. For this reason, when I am considering Reichenbach's contribution I
use the term "Logical Empiricism", which is the term by which he himself always referred
to the movement to which he belonged.
[3] Cf. *Experience and Prediction*, 145ff. "In spite of our reference to a free decision, we
should not like to say that the decision in question is arbitrary .... If the languages are
in question are not equivalent, if the decision between them forms a case of a volitional
bifurcation, this decision is of the greatest relevance: it will lead to consequences concerning the knowledge obtainable".
[4] In particular, one must adopt a "postulate of the homogeneity of causality". This is
described as one of a number of "postulates concerning the rules of language" (EP, p.

139) and shortly thereafter the choice between "egocentric language" and "usual language" is described as "a free decision". (Compare n. 3).
[5] Although this view reminds one of Husserl, Reichenbach (EP. p. 163) finds an impeccably empiricist forerunner in Avenarius.
[6] Indeed, Reichenbach's great book *Relativity Theory and Apriori Knowledge* (University of California Press, 1965: English translation of *Relativitätstheorie und Erkenntnis apriori*, Springer 1920), although a criticism of Kant's thesis of the apriority of geometry, is in many respects a neo-Kantian work.
[7] "Kant saw very well the close relationship between time and causality. He saw that we discover time order by examining causal order". *The Direction of Time*, pp. 14–15.
[8] Cf. Reichenbach's "Die Bewegungslehre bei Newton, Leibniz und Huyghens", *Kant-Studien*, Bd. 29 (1924), p. 421. After quoting Leibniz ("Time is the order of non-contemporaneous things. (Tempus est ordo existendi eorum quae non sunt simul.) It is thus the *universal* order of change in which we ignore the specific kind of changes that have occurred". Reichenbach remarks, "This passage seems to me to represent a depth of insight into the nature of time that has not been equalled in the whole classical period from Descartes to Kant. Even Kant's theory of time (as formulated, for example, in the 'Second Analogy of Experience') does not attain Leibniz' perspicacity. Leibniz' characterization of time as the general structure of causal sequences is superior to Kant's unfortunate characterization of time as the form of intuition presupposed by causality".
[9] In the Second Analogy, A193/B234; the interpretation of the Second Analogy is my own, but I believe it accords with Henry Allison's *Kant's Transcendental Idealism* (Yale University Press, 1983), 225ff.
[10] In some of his writing on the subject, e.g., "Die Kausalbehauptung und die Möglichkeit ihrer empirischen Nachprüfung", *Erkenntnis*, vol. III, no. 1, 1932, pp. 32–64 (English translation, "The Principle of Causality and the Possibility of its Empirical Confirmation" in Reichenbach's *Modern Philosophy of Science*, ed. Maria Reichenbach, Routledge and Kegan Paul, 1959) Reichenbach, thinking of the deterministic case, identified causality simply with the existence of functional relationships (with the proviso that the relationships must be inductively projectible to unobserved cases). But the need to replace this deterministic notion of causality by a probability structure was already explained in his 1925 paper "The Causal Structure of the World and the Difference Between Past and Future" (in *Hans Reichenbach: Selected Writings, 1909–1953, vol. 2*, ed. Maria Reichenbach and Robert S. Cohen, Reidel, 1978; this originally appeared in German in the *Sitzungsberichte der mathematisch-naturwissenschaftlichen Abteilung der Bayerischen Akademie der Wissenschaften zu München* (Nov. 1925, pp. 133–175).
[11] For Reichenbach the above are the *only* possibilities; there cannot be an inductively projectible probability relation which *always* holds and which has no basis in a causal connection between the events related. (Two events' being effects of a common cause counts as a "causal connection" for this purpose.) This might seem to be trivially wrong on the following ground: suppose two sequences of events, *a*-events and *b*-events, both follow the same periodic law. Suppose, for simplicity, that every fifth *a*-event has property P and every fifth *b*-event has property Q. Then the probability that a *b*-event has property Q *given that* the corresponding *a*-event has property P is *one*, yet it may well be that the *a*-events and the *b*-events do not causally influence one another and do not have a common cause. Reichenbach might answer that, in this case, my prediction of the character of a *b*-event can also be made without using any knowledge of the character of the corresponding *a*-event: it is enough to know the character of the *four preceding*

b-events. Perhaps it is when knowledge of the character of one or more *a*-events permits me to better predict the character of *b*-events ("better" in the sense of: better than I could do on the basis of the knowledge of earlier or later *b*-events) that Reichenbach would postulate causal connection – postulate it aparently *a priori*, judging by what he says in Chapter 10 of *The Rise of Scientific Philosophy*. However, a case, in which one sequence is a "carbon copy" of another *could* also arise from a genuine causal connection; so the answer I have just suggested on Reichenbach's behalf does not satisfy me. (Moreover, a further counterexample to Reichenbach's claim might be this: suppose the *a*-events are connected by a fortuitous relation of the kind just described – one that does not rest on any causal connection – to *c*-events, and the *c*-events *are* causes of the *b*-events. Then knowledge of the character of an *a*-event *may* enable me to better predict the character of the corresponding *b*-event whether or not there is any causal connection between the *a*-events and the *b*-events.)

[12] *The Direction of Time* (University of California Press, 1956), p. 198. Note that the concept of entropy to which Reichenbach appeals itself involves *spatial* notions in its definition; but the spatial metric is explained by Reichenbach in terms which assume that time order has already been established. The difficulty that I am worried about is not determining the direction of time, given that the topology of time (= the structure of the causal net) has already been established; the problem is whether purely statistical relations *can* determine even the topology of time if no temporal notions are anywhere "smuggled in".

[13] Cf. *Nomological Statements and Admissible Operations* (North-Holland, 1954). The theory Reichenbach presents in this book presupposes the view I questioned in note 11, that there cannot be a true but non-lawlike universal statement of the form: whenever an *a*-event has property P, the corresponding *b*-event has property Q. (In an unpublished note dating back to about 1954, Carnap also expressed scepticism as to the possibility of ruling out non-lawlike truths of this form.)

[14] A problem which arises in connection with this picture is just what "sequence" means. Reichenbach does not mean *arbitrary* sequence: to see why not, let A be a random sequence of heads and tails produced by flipping a coin, and let B be a second, causally independent sequence of heads and tails, also random. Let C be a subsequence of B which consists of heads and tails in exactly the order that they occur in A. (Every random sequence contains a subsequence which is a "carbon copy" of any sequence you like.) Then the probability that c_i is a head if the corresponding a_i is a head is one: but this is not because there is any causal connection between the a_i's and the c_i's. Reichenbach was aware of this objection, and in his lectures at UCLA he tried to rule it out by requiring that the sequences be "extensionally given". When I asked him what this meant, he explained that the probability relation we find must not be an artifact of the way the sequences are *defined*. But this raises the following problems: (1) It seems, then that causal connections exist only between definable sequences of events (and with respect to definable attributes of the members of those sequences). This seems counterintuitive, and it raises a host of questions: e.g., in what *language* must the sequences be definable? (2) What exactly is meant by a relation's being an *artifact* of the way something is defined? The intuitive idea is clear, but it seems difficult to capture in a definition. Reichenbach's theory of nomological statements, presented in the book cited in note 13, may be designed to deal with this problem.

[15] Strictly speaking, I should have written "time-like or light-like". Once the space-time metric has been constructed, one can identify a class of "fastest causal signals"; in our

world, these are the signals that travel at the speed of light. In the customary metric of relativity theory, the world lines of these fastest signals are considered neither time-like nor space-like.

[16] To define such a topology, one may proceed as follows (here, let P, P', Q, R be variables over point events and S a variable over sets of point events): define P *is a limit point of* S to mean that for every Q, R such that P is temporally between Q and R, there is a P' in S such that P' ≠ P and P' is temporally between Q and R.

[17] For a discussion see my "Refutation of Conventionalism", in my *Mind, Language and Reality: Philosophical Papers, Vol 2.* (Cambridge University Press, 1975).

[18] Reprinted in *International Encyclopedia of Unified Science: vol. 1, part 1*, ed. by Otto Neurath, Rudolf Carnap, and Charles W. Morris (University of Chicago Press, 1938); see, especially pp. 198–211.

[19] Whether this was the principle *aim of the Logische Aufbau*, is, however, a matter of disagreement among interpreters. An alternative reading is that the phenomenalistic construction is offered as only one of a number of alternative reconstructions of the language, and the primary purpose of the work is to argue that it is only by being willing to formulate philosophical positions in terms of such reconstructions that we can see exactly what the issues are. Cf. the article by Friedman cited in note 1.

[20] See, for example, Carnap's "The Methodological Character of Theoretical Concepts", in *Minnesota Studies in the Philosophy of Science, vol. 1: The Foundations of Science and the Concepts of Psychology and Psychoanalysis*, ed. by Herbert Feigl and Michael Scriven (University of Minnesota Press, 1956).

[21] There is, however, an element of choice in Reichenbach's view as well as in Carnap's: by adopting different coordinating definitions one can arrive at different but empirically equivalent metrics for spacetime, for example. The inductive inferences needed to infer the existence of illata depend on these coordinative definitions, and thus rest on a certain element of convention. For a discussion of this element of Reichenbach's thought see "Equivalent Descriptions", in my *Realism and Reason: Philosophical Papers, Vol. 3* (Cambridge University Press, 1983) and my essay cited in n. 17.

[22] In a way, then, Reichenbach may have originated what was later called the "inference to the best explanation" account of scientific reasoning – an account which was often thought to constitute a *rejection* of Logical Empiricism!

[23] Reichenbach idealized very long finite sequences as infinite sequences. I discussed this idealization and its bearing on Reichenbach's pragmatic vindication in my Ph.D. thesis, written under Reichenbach's direction in 1951. This has now been published by Garland Publishing Company (*The Meaning of the Concept of Probability in Application to Finite Sequences*, Garland, 1990).

[24] For a criticism of Reichenbach's vindication argument, see my "Introduction added some years later" to the work cited in note 23.

Department of Philosophy
Harvard University
Emerson Hall
Cambridge, MA 02136
U.S.A.

ABNER SHIMONY

ON CARNAP: REFLECTIONS OF A
METAPHYSICAL STUDENT

1. INTERACTION WITH CARNAP

Rudolf Carnap was my teacher in a number of classes and seminars at the University of Chicago in the spring of 1948 and the academic year 1948–49. The classes were on semantics, philosophy of mathematics, and inductive logic; the seminars (held in his apartment) were wide-ranging, including readings of Wittgenstein's *Tractatus*, C. I. Lewis's *An Analysis of Knowledge and Valuation*, and Frege's *Die Grundlagen der Arithmetik*, but there were numerous digressions into politics, since it was an election year. There were able and enthusiastic students in these classes and seminars, including Howard Stein, Richard Jeffrey, Norman Martin, Robert Palter, Stanley Tennenbaum, John W. Lenz, and Fanchon Aungst (Fröhlich). I was never a disciple of Carnap, since I came to the University of Chicago strongly influenced by Peirce, Whitehead, and Gödel (to whom my teachers at Yale University, Paul Weiss and Frederick Fitch, had introduced me), and never abandoned the idea that metaphysical inquiry is possible and fruitful. However, Carnap did not demand discipleship as a condition for friendship and a warm student-teacher relation.

After writing a master's thesis at the University of Chicago (on Whitehead, under the direction of Charles Hartshorne), I returned to Yale University, where I eventually wrote a doctoral dissertation on inductive logic, submitted in 1953. By this time Carnap was a visitor at the Institute for Advanced Study in Princeton, where I visited him to discuss problems in probability and induction, and received much encouragement.

Thereafter I saw Carnap only a few times and had occasional correspondence with him. He sent me some papers that he had written around 1953–54 on information theory and entropy, arguing that the frequency concept of probability rather than the epistemic concept should be used in the formulation of statistical mechanical entropy. I did not feel competent to comment on these papers at the time, but after his death I edited them with an introduction, and they were

Synthese **93**: 261–274, 1992.
© 1992 *Kluwer Academic Publishers. Printed in the Netherlands.*

published as *Two Essays on Entropy*. My last encounter with Carnap occurred in 1965, when he was en route to Los Angeles from a visit to Germany: Peter Hempel organized a two-day meeting in Princeton for a few people interested in Carnap's work on inductive logic. The high point of the meeting was the debate between Carnap, defending his logical concept of probability, and Leonard Savage, who advocated personal probability. It was clear in this debate that Carnap had moved considerably from his original vision of the aim of inductive logic, for he had come to think less in terms of a global confirmation function and more in terms of specialized confirmation functions for special disciplines and investigations. I was able to talk briefly to Carnap about my idea of tempered personalism, which I was then working out. From his reply to Hilary Putnam in the Schilpp volume (to which I shall return in Section 5), I later learned that Carnap was receptive to the idea of making the assignment of prior probabilities depend upon the the set of seriously proposed hypotheses (a pragmatic consideration, in his terminology), and I regret that we did not have time to explore this idea further.

Carnap deeply influenced me even though I was not his disciple. His seriousness about the enterprise of philosophy, his strong feeling about the mutual relevance of philosophy and the natural sciences, the high standard of clarity which he achieved in his own work and tried to maintain for his students, his openness of mind and his exploratory attitude (which led him to a number of liberalizations during his career – syntax to semantics, relaxation of meaning criteria, extensions of the apparatus of inductive logic), and his personal dignity and benevolence have all left a permanent impression on me.

In the remainder of this paper I shall review four areas in which I was influenced by Carnap but went in a different direction from his, for reasons to be stated. I shall recount some actual discussions that I had with him, but some of the most important discussions were in the privacy of my mind – What would he say? How could I answer him?

2. INTERNAL AND EXTERNAL QUESTIONS OF EXISTENCE

Carnap's strong opposition to metaphysics in the early days of the Vienna Circle, summed up in the title of a famous paper, 'Überwindung der Metaphysik durch logische Analyse der Sprache' (1932), was softened with the passage of time, largely because of his struggle with the difficult problem of criteria of meaningfulness (see especially his 1963, pp. 887–81). A significant residue of his opposition, however, is apparent in the famous late paper, 'Empiricism, Semantics, and Ontology' (1950a), which contains the following passage:

To accept the thing world means nothing more than to accept a certain form of language, in other words, to accept rules for forming statements and for testing, accepting, or rejecting them. The acceptance of the thing language leads, on the basis of observations made, also to the acceptance, belief, and assertion of certain statements. But the thesis of the reality of the thing world cannot be among these statements, because it cannot be formulated in the thing language or, it seems, in any other theoretical language.

The decision of accepting the thing language, although itself not of a cognitive nature, will nevertheless usually be influenced by theoretical knowledge . . . . The purposes for which the language is intended, for instance, the purpose of communicating factual knowledge, will determine which factors are relevant for the decision. The efficiency, fruitfulness, and simplicity of the use of the thing language may be among the decisive factors. And the questions concerning these qualities are indeed of a theoretical nature. But these questions cannot be identified with the question of realism. They are not yes-no questions but questions of degree. (Carnap 1950a, pp. 23–24)

One thing that troubles me about this passage is that the facts of efficiency, fruitfulness, and simplicity, which Carnap points to with approbation, should be not merely noted but explained as deeply as possible. And the search for explanations can lead to questions about the structure of the world and the place of knowers and language users in the world – questions hardly different from those of traditional metaphysics and epistemology. I think Carnap would not consider this remark to be a criticism of his position, since he would be hospitable to the traditional investigations if the questions were clearly posed and the means of investigation were satisfactory (i.e., using empirical data and correctly making inferences from them).

I would persist, however, and ask how he knows that the thesis of the reality of the thing world "cannot be formulated . . . in any other theoretical language". Do we not have here an instance of a vague 'explicandum' which stimulates the search for an 'explicatum' – comparable to the situation which he found and tenaciously studied in inductive logic? I can imagine that Carnap might agree, but with the proviso that if the "thesis of the reality of the thing world" were formulated in some theoretical language of the future, it would ipso facto turn into an internal statement of existence. I would not be entirely content with this answer, because it would fail to take sufficient account of the constraints which the world imposes on the procedure of finding a fruitful explication, and these constraints are external in a certain sense to the frame of a language.

The position which I have just taken with regard to the question of internal and external existence is an application of a principle of philosophical methodology which I have in the past somewhat facetiously formulated as follows: "In every neighborhood of a *Scheinproblem* there is a real problem". I can imagine that Carnap, with his genuine open-mindedness, might have been sympathetic with this principle. But I can also imagine him calmly asking, "Abner, what is a neighborhood?". And of course I would have no answer.

3. PROCEDURES FOR INTRODUCING THEORETICAL TERMS

In *Der logische Aufbau der Welt* (1928) Carnap had an ambitious program of constructing the objects of ordinary life and of physics on a phenomenalistic basis, the construction consisting of explicit definitions in terms of the phenomenalistic vocabulary. He later found explicit definition to be unfeasible even for the more modest program of defining the terms of physics by means of the vocabulary of ordinary life (the physicalistic or thing language). The difficulty already arises at the level of disposition terms, like 'fragile' and 'soluble', which are in a sense intermediate between theoretical and observational terms. In 'Testability and Meaning' (1936–37), Carnap proposed that disposition terms be introduced by the following scheme of 'reduction pairs':

$$P_1x \supset (P_2x \supset Qx),$$
$$P_3x \supset (P_4x \supset \sim Qx),$$

where Q is a disposition predicate, while P_1, P_2, P_3, and P_4 are observation predicates. A further extension of this idea is the reduction chain, in which terms increasingly remote from those of direct observation are introduced by a finite sequence of reduction pairs. In *Foundations of Logic and Mathematics* (1939) Carnap proposed the device of 'partially interpreted syntactical systems', in which there are two subvocabularies: an observational subvocabulary, whose terms are taken to be understood from ordinary usage; and a theoretical subvocabulary, to which meanings are attached only via connections to the observational terms. The connections are provided by the axioms of the system, some of which are formulated only in terms of one or the other of the two subvocabularies, but some of which are formulated in terms of both and hence serve as 'rules of correspondence'. The totality of axioms, not just those of mixed vocabulary, provide an incomplete interpretation of the theoretical terms, but one which is sufficient for 'understanding' in the sense of ability to use the terms in practical applications.

I was sympathetic with both of these devices proposed by Carnap for introducing terms that do not refer directly to observable features of the world, but this sympathy is connected with my plea for realism in Section 2. In order to give a coherent account of ordinary experience, we are driven to acknowledge a world with a career that is largely independent of our experience, but to which our experience is causally linked. The set of causal links is open because of the contingencies of orientation of the observing subject relative to objects of the environment, the contingencies of the carriers of information, and the contingencies of the sensory apparatus of the subject; and even the class of types of linkage is open, because the arts of experimentation and instrumentation have introduced new causal paths between objects of interest and observing subjects and can be expected to yield further innovations. Consequently, it is impossible in principle to give an exhaustive set of criteria in observational terms for the applicability of a dispositional or theoretical term. Carnap's liberalization of the procedures for introducing non-observational terms is effectively an acknowledgment of this openness, and therefore of the place of knowing subjects in nature.

From this point of view two proposals follow naturally. The first is that this envisagement of causal linkage – vague though this concept may be, and dependent though it may be on our current scientific picture of the world, especially as regards human cognition – should serve as the heuristic of any semantical device for introducing terms in the languages of science. The second is that the semantics of theoretical and observational terms be considered primarily in the context of actual scientific theories, rather than of models. Actual scientific theories have subtleties and complications that have commonly been lost in models, for example, the role of 'phenomenological physics' as an intermediary between observational reports and high-level physical theories (never mentioned in Carnap's discussion of observational and theoretical vocabularies). And a study of actual scientific theories can relieve some of the anxiety produced by subtle and complicated formal considerations of scientific language. For example, in actual practice, theory seems not to be so grossly underdetermined by observation as many philosophers of science hold.

4. THE CHARACTER OF MATHEMATICAL TRUTH

Throughout Carnap's career he maintained that mathematical truths are analytic in the same sense as those of logic (Carnap, 1963, p. 47) and that logical truth is "truth based on meaning" (ibid., p. 916) – though he refined these theses over time. At the time that I became Carnap's student in 1948 I was acquainted with the views of Poincaré and Gödel, which were very different from his, and was impressed by both. According to Poincaré, "[m]athematical induction, that is, demonstration by recurrence . . . imposes itself necessarily because it is only the affirmation of a property of the mind itself" (Poincaré, 1913, p. 40). And thus he regards the principle of mathematical induction as a synthetic a priori truth. Gödel maintained that the truths of mathematics are analytic, in that they hold because of relations among the concepts involved in them, where 'concepts' is to be understood not linguistically but Platonically, as forms or structures (Gödel, 1944, 1947).

I recall a discussion that I had with Carnap in 1948. I cited Poincaré's position approvingly. He replied that the principle of mathematical induction holds by the definition of 'finite cardinals'. I do not recall that he defined this term on the occasion, but I knew that the set theoretical definition requires steps like the following, which I take from Rosenbloom (1950):

> '$Nc(\alpha)$' for '$x \ni (\alpha \; sm \; x)$'   (where 'sm' is the relation of
> one-one correspondence),
> 'Nc' for '$x \ni ((\exists y) \cdot x = Nc(y))$',
> '0' for '$Nc(\Lambda)$'   (where 'Λ' denotes the empty class),
> '1' for '$x \ni ((\exists y) \cdot x = \iota y)$'   (where '$\iota y$' designates the
> class whose sole element is y),
> '$u + v$' for '$x \ni ((\exists y, z): Nc(y) = u \cdot \bigwedge \cdot Nc(z) = v \cdot \bigwedge \cdot y \cap z$
> $= \Lambda \cdot \bigwedge \cdot x \; sm \; (y \cup z))$',
> 'Fin' for '$x \ni (0 \in y \cdot \bigwedge \cdot z \in y \supset_z (z + 1) \in y: \supset_y : x \in y)$'.

In this list of definitions it is clear that: '$Nc(\alpha)$' designates the cardinality of the class α; 'Nc' designates the class of cardinal numbers; '0' designates the cardinal number zero; '1' designates the cardinal number unity; '+' designates the operation of adding two cardinal numbers; and 'Fin' designates the class of finite cardinal numbers – the concept at which Carnap was aiming. I replied that for this construction of the finite cardinals the principle of mathematical induction is indeed a consequence of the definition, but that there is another way of characterizing the finite cardinals for which this claim cannot be made – that is, as the cardinals of classes that cannot be put in one-one correspondence with any of their proper subclasses. Carnap then said that it would be interesting to prove the equivalence of these two characterizations of the finite cardinal numbers, and, as I recall, the discussion ended at this point. Later I learned that the equivalence can indeed be established, if one uses the axiom of choice (Zermelo, 1908; Russell, 1906) as Carnap surely knew at the time of our discussion. I suspect that his initial contention that the principle of mathematical induction follows from the definition of 'finite cardinal' would not have been affected by an acknowledgment of the need to use the axiom of choice in order to establish the equivalence of the two concepts of finite cardinal, because he wrote: "Later I came to the conviction that the axiom of choice is analytic, if we accept that concept of class which is used in classical mathematics in contrast to a narrower constructivist concept" (Carnap, 1963, p. 47). But I would not regard this answer to be satisfactory. If the axiom of choice is to be used to settle the status of the principle of mathematical induction, it can only be because that axiom is analytic in Gödel's sense, and, if this is so, then the thesis that mathematical truths are based on meanings – i.e., consequences of the rules of a language – is seriously undermined.

5. INDUCTIVE LOGIC

Carnap regarded the concept of degree of confirmation $c(h, e)$ of an hypothesis h on evidence e to be the primary tool of inductive logic. He offers three explicanda for which the concept of degree of confirmation is intended to be the explicatum (Carnap, 1950b, pp. 164–68):

(a)   the evidential support of h by e;
(b)   a fair betting quotient for bets on h, given e as the only evidence; and
(c)   an estimate of relative frequency of a certain property associated with h in a reference class determined by e.

253

Given the rules of a metalanguage that incorporates rules for c, a statement '$c(h, e) = r$' (where 'r' designates a real number) is analytically true or false. The dependence upon rules of the metalanguage suggests a conventionalist attitude towards the concept of degree of confirmation, but Carnap held that conceptual inquiry could narrow the choice of a reasonable set of rules to a quite small range. It seemed to me that a kind of Platonism regarding relations among propositions was the tacit metaphysical assumption of his hope of finding a small range of rationally acceptable confirmation functions. One of Carnap's important predecessors in the program of logical probability theory, John M. Keynes, was quite explicit about his Platonism (which he probably learned from G. E. Moore); he wrote: "What particular propositions we select as the premise of our argument naturally depends on subjective factors peculiar to ourselves; but the relations in which other propositions stand to them, and which entitle us to probable beliefs are objective" (Keynes, 1921, p. 4). Whatever Carnap's attitude may have been towards this kind of Platonism, I was surely initially favorable towards it because of the mathematical Platonism that I had learned from Fitch and Gödel.

Carnap had a certain amount of success in narrowing the range of acceptable confirmation functions. It seemed inevitable to him, in view of the explicanda of c, that the function c should satisfy some standard rules of the calculus of probability, though he referred to these rules as "conventions on adequacy" (Carnap, 1950b, p. 285):

(i)   if e and e' are logically equivalent, then $c(h, e) = c(h, e')$;
(ii)   if h and h' are logically equivalent, then $c(h, e) = c(h', e)$;
(iii)   (General Multiplication Principle): $c(h_1 \& h_2, e)$
         $= c(h_1, e) \times c(h_2, e \& h_1)$; and
(iv)   (Special Addition Principle): if e logically implies the falsity of $h_1 \& h_2$, then $c(h_1 \lor h_2, e) = c(h_1, e) + c(h_2, e)$.

Then, by the skillful use of intuitively reasonable symmetry principles, Carnap (1952) was able to show that all the acceptable c-functions on a certain (rather simple) language could be parametrized by a single parameter λ in the closed interval $[0, \infty]$. The meaning of this parameter λ is exhibited by the value assigned by c_λ, the c-function parametrized by λ, when the evidence e tells how the members of a sample of s individuals is partitioned among the k Q-predicates of the language (these being the predicates of maximal strength within the limits of logical possibility), and h_j asserts that an individual not in the sample has the jth predicate. Then

$$c_\lambda(h_j, e) = \frac{s_j + \lambda/k}{s + \lambda},$$

where s_j is the number of members of the sample having the jth predicate. Clearly, for $\lambda = 0$, the value obtained by Carnap's formula is s_j/s, which is a ratio in the sample, while for $\lambda = \infty$, the value is $1/k$, the inverse of the number of Q-predicates. In general, as λ increases, Carnap's formula assigns less and less weight to the 'empirical factor' s_j/s, and more and more to the 'logical factor' $1/k$. If rather narrow limits are set upon λ, then the range of acceptable c-functions would be well delimited.

There is one flaw in the program as described so far which was corrected in a manner very congenial to Carnap. Kemeny (1951) constructed a case in which he claimed that the General Multiplication Principle (iii) fails to hold; and even though careful analysis removed the force of his objection, he had clearly opened the important question of how Principles (i)–(iv) could be justified by more than convention. In the summer of 1952 I worked on this problem, partly by reading through the bibliography of Carnap's *Logical Foundations of Probability* (1950b). Fortunately, 'De Finetti' was early in the bibliography, and I noticed that his method of justifying the principles of the calculus of probability (independently discovered by F. P. Ramsey) applied to Carnap's logical concept of probability as well as to their own personalist concept. De Finetti showed that a necessary condition for the beliefs to be 'coherent' (i.e., such that no set of bets compatible with those beliefs would lead to the bettor's net loss, regardless of outcomes) is that the beliefs should satisfy the Principles (i)–(iv). By explicandum (b) above, the concept that Carnap was trying to explicate should yield fair betting quotients, and hence must satisfy the condition of coherence. This was the one time that I made Carnap happy, and indeed his later writings on probability (e.g., 1971) incorporated De Finetti's and Ramsey's justification of the principles of the calculus of probability.

The serious flaw of Carnap's program, however, is that there seems to be no intuitive and a priori method for choosing a small range of acceptable values of λ out of the half-infinite interval $[0, \infty]$. In spite of my proclivities for Platonism in mathematics, I became discouraged with Platonism regarding c–functions for several reasons. The first was a penetrating remark of Ramsey:

> But let us now return to a more fundamental criticism of Mr. Keynes' views, which is the obvious one that there really do not seem to be any such things as the probability relations he describes . . . . I do not perceive them, and if I am to be persuaded that they exist it must be by argument; moreover I shrewdly suspect that others do not perceive them either, because they are able to come to so little agreement as to which of them relates to any two given propositions. (Ramsey, 1931, p. 161)

The second reason was a conversation with Howard Stein in 1948 or 1949, in which I pointed out that the values of the c-function which Carnap was seeking are indeed fairly well fixed by reasonable considerations in a number of special cases, and these might suffice to determine the function globally; after all, I argued, the values of an analytic function at a denumerable set of points including a limit point suffice to determine the function over its whole domain of analyticity. Yes, Howard replied, but think how strict are the conditions that an analytic function must satisfy. I realized immediately that nothing like the Cauchy–Riemann conditions exists for the family of c-functions. A third reason came to me somewhat later: the program of logical probability, being a priori, should provide a guide to reasonable degrees of belief in any possible world. But is not this program over-ambitious? We want an inductive logic which works in the actual world; or better, in a class of worlds including our actual world. Might we not escape from the difficulty posed by the immense multiplicity of Carnap's c-functions parametrized by λ by paying attention to the constraints which we know to hold in the actual world?

In my later work on inductive logic (1970) I deviated from Carnap's program of logical probability in two major ways. The first was to propose the 'tempered personalist' concept of probability, which is similar to the personalist concept of Ramsey, De Finetti, and Savage in being the degree of belief that a person actually has in a proposition, given a body of evidence and/or assumptions, constrained, however, not merely by the condition of coherence (accepted by Ramsey et al.) but also by a condition of open-mindedness called 'the tempering condition': this gives to each seriously proposed hypothesis in an investigation sufficient prior probability that it has a non-negligible chance of obtaining preponderant posterior probability as a result of the outcomes of investigations that are planned. This condition is similar to Harold Jeffreys's 'simplicity postulate' (1961, pp. 245–46). It has the consequence that when a new hypothesis is proposed in the community of investigators, a revision must be made in the assignment of the prior probabilities – a pragmatic departure from Carnap's procedure of making the assignment of probabilities a matter of semantical rules, and also a departure from Bayesianism, in that a modification of probabilities is made in a way that is not governed by Bayes's theorem. This departure is pragmatic not only in the sense of referring to the users of language, but also in the sense of attentiveness to the conditions of inquiry. Peirce had already pointed out the social character of inquiry (1932, p. 129) and the importance of optimism regarding human abductive powers (1958, p. 137), and these aspects of his pragmatism are embedded in the tempering condition.

The second deviation is the admission of a posteriori principles in inductive logic, for these may be needed for a logic which works well in the actual world, without pretensions to utility in all possible worlds. It seems to me likely, in fact, that the tempered personalist concept of probability could not be applied in actual reasoning without some a posteriori principle, since the crucial phrase 'seriously proposed hypothesis' cannot be explicated a priori in a non-arbitrary way. There is an obvious problem of circularity when a posteriori principles are incorporated into inductive logic. Various authors who have advocated this strategy, such as Max Black (1958, 1963) and Michael Friedman (1979), have tried to show that the circularity of inductive support of induction can be non-vicious. Exactly how viciousness is to be avoided seems to me to be a deep problem not only of inductive logic but of epistemology generally. Certainly a necessary condition for non-viciousness is that inquiry not be blocked, so that the possibility is kept open that the principle under consideration be rejected or revised as a result of investigation. Some sections of my papers (1970, pp. 158ff.; 1987, pp. 30–34) concern the dialectical procedure which aims at non-vicious circularity, but obviously more remains to be worked out.

I said earlier that these are deviations from Carnap's program of inductive logic. But there are occasional passages in his writings which show anticipations or expressions of sympathy with these departures. In answering Putnam's contribution to the Library of Living Philosophers volume on Carnap (Schilpp, 1963), Carnap is remarkably receptive to a pragmatic suggestion: "I think it is worthwhile to consider the possibility of preserving one interesting suggestion which Putnam offers, namely, to make inductive results dependent not only on the evidence but also on the class of actually proposed laws" (Carnap, 1963, p. 986). And he makes the following remarkable appeal to experience in deciding on the choice of a *c*-function:

> An inductive method is . . . an instrument for the task of constructing a picture of the world on the basis of observational data and especially of forming expectations of future events as a guidance for practical conduct. X may change this instrument just as he changes a saw or an automobile, and for similar reasons . . . after working with a particular inductive method for a time, he may not be quite satisfied and therefore look around for another method. He will take into consideration the performance of a method, that is, the values it supplies and their relation to later empirical results, e.g., the truth-frequency of predictions and the error of estimates; further, the economy in use, measured by the simplicity of the calculations required . . . . If X feels that the method he has used does not give sufficient weight to the empirical factor in comparison with the logical factor, he will choose a method with a smaller λ . . . . On the other hand, if he wishes to give more influence to the logical factor and less to the empirical factor, he will move up his mark on the λ-scale. Here, as anywhere else, life is a process of never ending adjustment: there are no absolutes, neither absolutely certain knowledge about the world, nor absolutely perfect methods of working in the world. (Carnap, 1952, p. 55)

Here Carnap is engaged in a dialectical method, in his own way. His open-mindedness and exploratory attitude are evident.

6. CONCLUDING REMARKS

Two unwavering commitments of Carnap's philosophy throughout his career were to empiricism regarding the basis of human knowledge, and to clarity in reasoning and exposition. I am not convinced, however, that he solved the difficult problems of how to be a good empiricist and how to achieve clarity. He certainly made some important contributions to the solution of both problems. He was an empiricist not merely in matters of detail, within a chosen linguistic framework, but also with regard to the choice of frameworks. After all, the criteria of fruitfulness and success which he used in evaluating frameworks are themselves empirical criteria. But he was negligent of both the history of the natural sciences and the details of the most successful scientific theories, and therefore lost the lessons that these can teach regarding the fine-tuning of empiricism. The strategy of Gödel and Hao Wang of looking at mathematical practice (empirically, as a special kind of psychology), in order to judge whether and how human beings are endowed with powers of mathematical intuition, is outside the range of Carnap's allowable strategies. Nor did he fully internalize the lesson from the history of science that premature formalization can be sterile. Despite these criticisms, I believe that Carnap's insistence upon empiricism and clarity is permanently valuable, and should not be forgotten at a time of proliferation of relativisms, anti-methodologies, and incommensurabilities. His healthy insights should be preserved, but refined in the light of further experience.

Carnap's vision of a world free of obscurantism, authoritarianism, irrational sectarianism, and obstacles to knowledge of nature and the self went hand in hand with an ethical and political vision of a world in which human beings cooperate rationally and benevolently. Carnap belongs in the company of two of my other heroes, David Hume and Denis Diderot – non-religious men for whom morality was based upon a benevolent feeling towards others, and whose allegiance was to 'the party of mankind'. Carnap's opposition to Nazism and his dedication to democratic socialism were the manifestations of his profound benevolence and sense of justice.

ACKNOWLEDGMENT

The work for this paper was supported by a fellowship from the National Endowment for the Humanities.

REFERENCES

Black, M.: 1958, 'Self-Supporting Inductive Arguments', *Journal of Philosophy* 55, 718–25.

Black, M.: 1963, 'Self-Support and Circularity. A Reply to Mr. Achinstein', *Analysis* 23, 43–44.

Carnap, R.: 1928, *Der logische Aufbau der Welt*, Weltkreis-Verlag, Berlin-Schlachtensee.

Carnap, R.: 1932, 'Überwindung der Metaphysik durch logische Analyse der Sprache', *Erkenntnis* 2, 219–41.

Carnap, R.: 1936–37, 'Testability and Meaning', *Philosophy of Science* 3, 419–71, and 4, 1–40.

Carnap, R.: 1939, *Foundations of Logic and Mathematics*, University of Chicago Press, Chicago.

Carnap, R.: 1950a, 'Empiricism, Semantics, and Ontology', *Revue Internationale de Philosophie* 11, 20–40.

Carnap, R.: 1950b, *Logical Foundations of Probability*, University of Chicago Press, Chicago.

Carnap, R.: 1952, *The Continuum of Inductive Methods*, University of Chicago Press, Chicago.

Carnap, R.: 1963, 'Intellectual Autobiography' and 'The Philosopher Replies', in P. A. Schilpp (ed.), *The Philosophy of Rudolf Carnap*, Open Court, La Salle, Illinois, pp. 3–84, 859–1013.

Carnap, R.: 1971, 'A Basic System of Inductive Logic, Part 1', in R. Carnap and R. C. Jeffrey (eds.), *Studies in Inductive Logic and Probability*, Vol. I, University of California Press, Berkeley, California, pp. 33–165.

Carnap, R.: 1977, *Two Essays on Entropy*, University of California Press, Berkeley, California.

De Finetti, B.: 1937, 'La prévision: ses lois logiques, ses sources subjectives', *Annales de l'Institut Henri Poincaré* 7, 1–68.

Friedman, M.: 1979, 'Truth and Confirmation', *Journal of Philosophy* 76, 361–82.

Gödel, K.: 1944, 'Russell's Mathematical Logic', in P. A. Schilpp (ed.), *The Philosophy of Bertrand Russell*, Open Court, La Salle, Illinois, pp. 125–53.

Gödel, K.: 1947, 'What is Cantor's Continuum Problem?', *American Mathematical Monthly* 54, 515–25.

Jeffreys, H.: 1961, *Theory of Probability*, 3rd ed., Clarendon Press, Oxford.

Kemeny, J.: 1951, 'Carnap on Probability', *Review of Metaphysics* 5, 145–56.

Keynes, J. M.: 1921, *A Treatise on Probability*, Macmillan, London.

Peirce, C. S.: 1932, *Collected Papers*, Vol. 2, Harvard University Press, Cambridge, Massachusetts.

Peirce, C. S.: 1958, *Collected Papers*, Vol. 7, Harvard University Press, Cambridge, Massachusetts.

Poincaré, H.: 1913, *Science and Hypothesis*, The Science Press, Lancaster, Pennsylvania.

Putnam, H.: 1963, ''Degree of Confirmation' and Inductive Logic', in P. A. Schilpp (ed.), *The Philosophy of Rudolf Carnap*, Open Court, La Salle, Illinois, pp. 761–83.

Ramsey, F. P.: 1931, *The Foundations of Mathematics and Other Logical Essays*, Routledge and Kegan Paul, London.

Rosenbloom, P.: 1950, *Elements of Mathematical Logic*, Dover, New York.

Russell, B.: 1906, 'On Some Difficulties in the Theory of Transfinite Numbers and Order Types', *Proceedings of the London Mathematical Society*, Series 2, 4, 29–53.

Savage, L. J.: 1954, *Foundations of Statistics*, John Wiley, New York.

Shimony, A.: 1955, 'Coherence and the Axioms of Confirmation', *Journal of Symbolic Logic* 20, 1–28.

Shimony, A.: 1970, 'Scientific Inference', in R. G. Colodny (ed.), *The Nature and Function of Scientific Theories*, University of Pittsburgh Press, Pittsburgh, Pennsylvania, pp. 79–172.

Shimony, A.: 1987, 'Integral Epistemology', in A. Shimony and D. Nails (eds.), *Naturalistic Epistemology: A Symposium of Two Decades*, D. Reidel, Dordrecht, pp. 299–318.

Wang, H.: 1986, *Beyond Analytic Philosophy*, MIT Press, Cambridge, Massachusetts.

Zermelo, E.: 1908, 'Neuer Beweis für die Möglichkeit einer Wohlordnung', *Mathematische Annalen* 65, 107–28.

Departments of Philosophy and Physics
Boston University
Boston, MA 02215
U.S.A.

RICHARD C. JEFFREY

CARNAP'S INDUCTIVE LOGIC

Carnap pretty much agreed with Keynes:

Part of our knowledge we obtain direct; and part by argument. The theory of probability is concerned with that part which we obtain by argument, and it treats of the
different degrees in which the results so obtained are conclusive or inconclusive.

(J. M. Keynes, *A Treatise on Probability*, Ch. 1)

Of course, Carnap would speak of the theory of *inductive* probability
where Keynes speaks simply of the theory of probability. But Carnap's
main deviation from Keynes at this point concerns the assertion that
probabilities give degrees of *conclusiveness* of the results of inductive
arguments. There, Carnap adopted Ramsey's view, that

the kind of measurement of belief with which probability is concerned... is a measurement of belief *qua* basis of action.

(F. P. Ramsey, 'Truth and Probability,' §3.)

Carnap thought it essential that we describe people's beliefs, not in an
all-or-none fashion, but in terms of degrees of *credence*. Jones's beliefs at
time t are to be given by a function $cr_{Jones,t}$ which obeys the laws of the
elementary probability calculus. (Either or both of the subscripts may be
dropped, where the context makes clear what is intended.) With Ramsey,
Carnap interpreted statements of form '$cr(h)=x$' as dispositional (or
perhaps theoretical) statements about Jones's behavior. In a rough and
ready way, both men took $cr(h)$ to be the largest part of a dollar that
Jones would be willing to pay, at time t, in order to receive a dollar if and
only if h is true. Carnap followed Ramsey in taking this to be a rough
indication of the full account, in which Jones's degrees of belief are
deducible from certain characteristics of his preference ranking.

So much for actual credence. Carnap took the task of *applied* inductive
logic to be that of telling people like Jones what their credence functions
ought to be, given their experience. Carnap's paradigm was the case in
which we consider a sequence of moments, $t=0, 1, 2,...$, and a corresponding sequence of sentences $e_1, e_2,...$, where for each positive t, e_t

Jaakko Hintikka (ed.), Rudolf Carnap, Logical Empiricist, 325–332. All rights reserved.
Copyright © 1975 by D. Reidel Publishing Company, Dordrecht-Holland.

261

expresses the information which Jones got directly from experience between times $t-1$ and t. Then Jones's total evidence at time t is $e(t) =_{DF} e_1 \cdot e_2 \cdot \ldots \cdot e_t$. In such a case, Carnap thought, Jones's degree of credence in a statement h at time t ought to depend only on h and $e(t)$, i.e., $cr_t(h)$ ought to be a *function* of h and $e(t)$. Carnap called that function 'c': $cr_t(h) = c(h, e(t))$ if Jones's beliefs at time t are determined in a logically satisfactory manner by his experience up to time t. The function c should have the characteristic that for any sentence h in Jones's language, the number $c(h, e(t))$ will seem (after sufficient reflection) to be the degree of credence that anyone ought to have in h at time t, provided his total directly experimential evidence at time t is expressed by the sentence $e(t)$. To carry out that task, Carnap thought it important that the function c be defined even when the second argument is not the sort of sentence that could represent someone's total direct observational experience. Thus, for any sentences h and e (with certain exceptions, e.g. where e is self-contradictory), Carnap wanted $c(h, e)$ to be a definite number. He took the task of *pure* inductive logic to be that of providing a mathematical definition of the function c.

I. IN WHAT SENSE IS c 'LOGICAL'?

Carnap had two different, Pickwickian senses in mind for the word 'logical' in this question. Briefly,

> $logical_1$ = nonfactual
> $logical_2$ = in agreement with inductive intuition

Let me explain and evaluate the situation as I see it.

'c' *ought to be a logical$_1$ functor*, just as 'sin' and 'log$_{10}$' are logical$_1$ functors. This is a true description of Carnap's program, which is to identify a satisfactory c-function in such a way that, once h and e are identified either ostensively (as in quotation-mark names) or by structural descriptions, it is a matter of calculation to determine the value which c assigns to the pair (h, e): no empirical investigation is needed. Thus, in his paper 'On Inductive Logic' (1945) Carnap defined a functor 'c^*' in such a way that it is a matter of calculation to discover that $c^*('Pa_2'$, '$Pa_1') = \frac{2}{3}$. Note that it is the functor, 'c', not the function, c, which it denotes, which is logical$_1$. It is entirely possible for one and the same function to be denoted by two functors, one of which (say, 'c'') is logical$_1$, while the other (say, '$cr_{Jones.\ V\text{-}E\ day}$', thought of as a function of two arguments, where $cr(h, e) = cr(h \cdot e)/cr(e)$) is factual. (To discover Jones's degrees of credence you must study Jones, but the values of 'c' can be determined with pencil and paper.)

The function c is logical$_2$: Its values will be in agreement with our inductive intuitions *as they will exist at the time when the program has been carried out*. At the moment, we are often at a loss to say what the values $c(h, e)$ ought to be. Carnap's method for carrying out his program was to consider cases where we are fairly clear about what numbers we want, and then ask how various unclear cases might be decided so as to get the required results in the clear cases. If the program ever succeeds, it will be because we have been able to apply that method on a broad front, thus developing and sharpening our inductive intuitions – creating them, one might every say – while at the same time narrowing the set of candidates for the role of c. Then the processes of finding c and of developing our inductive intuitions are interwoven in such a way that the chosen c function must agree with inductive intuition – if the program ever succeeds.

Then I take it to be a true description of Carnap's program, that *if it succeeds, the functor 'c' will be logical$_1$, and the function it denotes will be logical$_2$*. Note that as I have been using and mentioning it, the symbol 'c' has no referent unless Carnap's program is eventually carried to a successful conclusion. Then my claim about the two senses of logicality can be put more precisely and less technically as follows:

If Carnap's program ever succeeds, we shall then be in posession of a method for computing values $c(h,e)$ when h and e are specified either ostensively or by structural descriptions; and these values will agree with our logical intuitions (as they shall then be) about the degree of credence in h that would be appropriate for someone whose total observational experience is summarized by the sentence e.

This is what I take to be the content of Carnap's claim that c is *logical*. Interpreted in this way, the claim is clearly true – but uninteresting, unless Carnap's program is in fact destined to succeed. Carnap hoped and expected that others would continue his work on the program, and would eventually bring it off. Therefore he took the claim to be interestingly explicative.

It seems clear to me that Carnap explicitly distinguished two senses of 'logical', essentially as I have done here – except (*a propos* logical$_2$) for the latitudinarianism which I shall now discuss, and except for a tendency on Carnap's part to speak of the inductive intuitions which will emerge as the program proceeds as already existing within us (if only covertly).

II. LATITUDINARIANISM

Carnap was prepared to admit the possibility that different people might have somewhat different inductive intuitions, e.g. when, ca. 1951, he thought the right c-function might be found somewhere in the continuum of inductive methods, he thought that different people might discover that they had somewhat different values of λ and hence that their inductive intuitions were describable by somewhat different functions c_λ. He thought it possible that these differences were irreducible, so that his program, as described above, might fail in the mild sense that there might be no such thing as *the* c-function which represents *our* inductive intuitions. Rather, there might prove to be a range of inductive intuitions among us, which might be represented by different c-functions. The failure would be mild – so mild as not to deserve the name, failure – if the various inductive intuitions are sufficiently similar so that their differences are swamped out by experience. Thus, suppose that the functions c and c' represent your inductive intuitions and mine, respectively, and that at time t our experiential evidence is represented by sentences $e(t)$ and $e'(t)$, for you and me, respectively. Then the difference between c and c' is swamped out by experience if my rational credence function at time t would have been pretty much the same if I had used c instead of c', and similarly yours would have been pretty much the same if you had used c' instead of c. Thus, the numbers $c(h, e(t))$ and $c'(h, e(t))$ may be very close, for all h, and so may be the numbers $c(h, e'(t))$ and $c'(h, e'(t))$. If so, the differences between our inductive logical intuitions are all but irrelevant at time t, having been 'swamped out' by our experience.

III. SUBJECTIVISM

Carnap's latitudinarianism is suggestive: perhaps he is describable as a special sort of subjectivist. Differences among the c functions which represent our various inductive intuitions are matters of individual psychology, while the respects in which all of our c functions agree represent human nature – general psychology. Perhaps Carnap would accept this subjectivist reading, but would think it appropriate to dignify with the name 'logical' the common features of all *c* functions, since the agreement concerns *norms* for credence. Note that de Finetti's subjectivist program seeks (*Foresight*, Ch. 6),

to show that there are rather profound psychological reasons which make the exact or approximate agreement that is observed between the opinions of different individuals very natural,

and here I take his position to be compatible with Carnap's. But de Finetti then urges

that there are no reasons, rational, positive, or metaphysical, that can give this fact any meaning beyond that of a simple agreement of subjective opinions.

Presumably the disjunction of *rational, positive,* and *metaphysical* is *philosophical*: de Finetti declines to try to derive the similarities from more basic, philosophical considerations. But then, so does Carnap. Inductive logic is a *description* of the norms we shall find ourselves accepting, once we have thought the matter through. It is no part of Carnap's program, to base those norms on still deeper philosophical considerations.

But if Carnap is a subjectivist, he is a very exigent one: he thinks it possible and worthwhile to try to describe someone's inductive temperament quite explicitly and generally, *via* a *c*-function. If your experience differs from mine, that difference will appear in the second argument places of our respective *c*-functions – it will not affect the functions themselves. At times he sounds Kantian: one's inductive temperament is *evoked* but not *determined* by one's experience. But subjectivists like de Finetti despair of such neat separation of experience from temperament; anyway, they think it hopeless, and profitless, to attempt the separation along Carnapian lines.

IV. RESPONSIBILITY FOR EXPERIENCE

Let me now turn to a problem which Carnap never treated explicitly (as far as I know), but which *can* be treated readily enough within the framework of his system.

Suppose that cr is Jones's actual credence function at time t, and that $e(t)$ represents his total observational knowledge at time t. Suppose further that Jones's inductive temperament is represented by the function c. Then if $cr(h)$ fails to be the same number as $c(h, e(t))$, Jones's belief about h is not fully rational, for it is not as it ought to be, given his experience and temperament. Thus, inductive logic can be applied to impugn the rationality of someone's beliefs.

Can inductive logic also be applied to *endorse* the rationality of someone's beliefs? This may not be so clear. Thus, suppose that $cr(h)$ and $c(h, e(t))$ are one and the same number. Does it follow that Jones's belief about h is rational? Surely, that belief is as it ought to be, *given his temperament and his experience.* But are his temperament and experience as *they* ought to be?

Let us suppose that the answer to this question about his temperament is 'yes', and examine his experience. Can Jones's experience be faulted as inadequate to support the thesis that $cr(h)$ is a *rational* degree of credence in h for him, at time t? The question is not simply, whether $cr(h)$ is rational for him at time t, given what his experiences have been, up to that time. We are also asking, whether his experience at time t should not, perhaps, be a bit richer than it is.

No doubt, there are situations in which Jones's experience can be so faulted, e.g. where h is the statement that it will be raining, at times during the day at which Jones will want to be outdoors; and where at time t Jones is in his bedroom dressing for bad weather, partly on the basis of a weather forecast he heard the night before, and partly through a passive, sleepy reluctance to undertake a trip to the window, look out, and see what sort of day it looks like being. Perhaps, in fact, $e(t)$ is such that $cr_t(h) = c(h, e(t)) = .7$, but by drawing the blinds, Jones could have provided himself with a bit more evidence, which might have driven his credence in h nearly up to 1, or nearly down to 0. We (or Jones himself, later in the day), might well fault him for not having bestirred himself to augment $e(t)$.

It is of no great moment, whether we put the matter as above – as a criticism of the slightness of $e(t)$ – or as a recommendation concerning $e(t+1)$, viz., that it be the conjunction of $e(t)$ with a report of what Jones sees when he looks out the window. In either case, the conditions $cr_t(h) = c(h, e(t))$ – or, as it may be, $cr_{t+1}(h) = c(h, e(t+1))$ – is seen to be necessary but not sufficient for reasonableness of Jones's credence in h at the time in question.

MORAL: Experience is only sometimes *the Given*. It is always *the Taken*. (Taking can be rather passive, as when one *accepts* a gift; but it may also be rather active – more of a *seizing*.)

Experience can be had fortuitously, or by design. I don't mean that Jones can decide to look out the window and see that it looks like being a fine day. But he *can* decide to look out the window and see what he can find out that way. In the stylized examples studied in statistical decision theory, one thinks about the matter in advance and sees that there are a number of mutually exclusive, collectively exhaustive possible outcomes, o_1, ..., o_n, of the *act* of observation which Jones is contemplating. At the time when he is considering whether to make the observation (to perform the act), Jones has definite degrees of credence $p_1, ..., p_n$ in the possible outcomes, and associates definite 'payoffs' $u_1, ..., u_n$ with them. Making the observation has a certain cost, say u; and the observation is worth making if its expected payoff, $p_1u_1 + \cdots + p_nu_n$, exceeds its cost. (In other words Jones ought to perform the act (=make the observation) if its *expected utility*, $p_1u_1 + \cdots + p_nu_n - u$, is positive.)

The utility of a possible outcome tends to be greater, the less expected that outcome is. Thus, since Jones rather expected rain anyway, seeing that it looks like rain would be a less useful outcome than seeing that it looks like being fine – not because Jones prefers fine weather to rain, but because the unexpected outcome, unlike the expected one, would lead him to take a course of action (dress for fine weather) which he would not have undertaken without the observation. Thus, the utilities of outcomes of observations depend on what acts are contemplated. But if one wishes to express the general tendency of outcomes to be more useful, the less likely they are, one can take $-\log p_i$ to be a measure of the utility of outcome o_i (insofar as that *can* be measured without knowing what actions are in view), and then take the *entropy* of the observation, $-(p_1 \log p_1 + \cdots + p_n \log p_n)$, as an action-independent estimate of its expected utility.

In this way, the element of responsibility for experience can be accomodated within the framework of Carnapian inductive logic, which then seems adequate 'in principle' to the task of endorsing, as well as impugning, the rationality of people's credence functions.

V. IDEALIZATION OR FALSIFICATION?

I see no knock-down argument (no glory, in Humpty Dumpty's sense) either for or against the thesis that Carnap's program is feasible and worthwhile. The program aims at a certain sort of 'rational reconstruction' of our notions of how we ought to proceed, in acquiring what we like to think of as knowledge. Any such enterprise abounds with elements which its proponents call 'idealizations' and its opponents call 'falsifications'. And the opponents are generally right, *to begin with*. I mean (in the case of Carnap's program) that the business of *discovering* what our inductive intuitions are is generally not so much a matter of *uncovering* preexistent, covert intuitions, as of *creating* intuitions: forging an inductive temperament out of materials which were not inductive intuitions before they passed through the Carnapian fire, and rang between that hammer and that anvil. For my part, I find it plausible that there is a human inductive temperament in the making – less plausible, of course, that it will emerge with the definite features which Carnap designed for it over the past twenty-five years or so. But then, Carnap's design kept changing, as the work went on. It would be no service to him, or to philosophy, to try to stop the process at the point it happened to have reached when he had to leave it.

University of Pennsylvania

Positivism and the Pragmatic Theory of Observation

Thomas Oberdan

Clemson University

The most influential critique of the Logical Positivists' analysis of scientific observation was posed by Paul Feyerabend in his classic essay, "Explanation, Reduction, and Empiricism". Feyerabend countered the later Positivist conception with his so-called 'Pragmatic Theory of Observation' which was founded on two ideas. The first is that observation reports are 'theory-laden', in the sense that they are always interpreted in the light of the best current theory and are subject to reinterpretation when one theory succeeds another. Feyerabend traced the development of this idea in the early writings of the Positivists, especially Rudolf Carnap, in the classic discussion of protocol sentences. In that context, Feyerabend finds that the theory-ladenness of observation leads inexorably to the second idea of the Pragmatic Theory, that what counts as an observation report does not depend on either its empirical content or its logical form, but on its causal or pragmatic features. Thus neither the meanings of observation reports nor their distinction from other scientific statements are fixed absolutely. Rather, observation reports are distinguished on pragmatic grounds and, thus demarcated, are significant only against the background of a specific theoretical context.

Although the purpose behind Feyerabend's explanations is to discredit the later Positivist account of observation, the ultimate success of his systematic program depends on the historical thesis that Carnap endorsed the fundamental tenets of the Pragmatic Theory in his Thirties' writings on protocols. This historical claim is the first subject of the discussion that follows below. It will be granted that Carnap endorsed the idea that observation is theory-laden and denied that observation reports are distinguished absolutely on the basis of their empirical content or logical form. It will even be conceded that Carnap allotted an indispensable role to pragmatic factors in the demarcation of observation reports from other scientific statements. But these admissions will be qualified by arguing that the function Carnap allotted pragmatic factors in the discrimination of observation reports is nothing like what Feyerabend has in mind. Then it cannot be said that Carnap endorsed the second idea of the Pragmatic Theory in anything like the sense in which Feyerabend intended. These considerations suffice to show that Feyerabend's historical contentions about Carnap's early philosophy are false. These results are then applied to

---

PSA 1990, Volume 1, pp. 25-37
Copyright © 1990 by the Philosophy of Science Association

Feyerabend's critique of Carnap's later account of observation, as presented in "The Methodological Character of Theoretical Concepts". Finally, in closing, it is concluded that Carnap's conception of the matter remained unaltered in essential respects throughout his philosophical career.

The basic problem afflicting Feyerabend's account is that he seems to forget what an artful blend of logical and epistemological insights early Positivism was. Even hackneyed re-tellings of its history give a recipe calling for equal portions of the Frege-Wittgenstein conception of logic, and empiricist epistemology mined from the Hume-Mach-Russell vc.n. While there is much that is misleading about this picture of early Positivism, even its simplistic rendering shows what Feyerabend misses that is central to the thought of the Vienna Circle. For Feyerabend neglects altogether the important role of Positivism's logical doctrines in its epistemology, focussing instead on its empiricism, and the pragmatic orientation it eventually received in the discussion of protocols. In particular, Feyerabend fails to realize that Carnap's epistemology of protocols was dominated throughout by his syntactic analysis of language. Indeed, the development of Carnap's view of protocols occurred in two stages, corresponding to the emergence of the two leading precepts of his conception of language. The first stage, in which he recognized the theory-ladenness of observation reports, was founded on his 'Thesis of Metalogic'. This precept grounded Carnap's idea that protocols are to be regarded, first and foremost, as *facts*.

Carnap's idea that scientific reports of observation, or 'protocols', are to be regarded as facts was first developed in his essay on "Psychology in Physical Language". There he stressed that the utterances of others are primarily acoustic phenomena, their inscriptions are concatenations of physical objects (ink and paper), and so forth. Like all physical facts, these purportedly linguistic ones provide information to an extent determined by current scientific theory. At bottom, the protocols of others are no different from other physical processes. True, they may be particularly informative about other physical occurrences and are, for this reason, given special consideration in science. "However, between the contribution of these assertions to our scientific knowledge and the contribution of a barometer there is, basically, at most a difference of degree." (Ayer 1959, pp. 180-1) Scientific observers are detectors of physical occurrences, and their reports are on a par, epistemologically, with the readings of barometers, voltmeters, and geiger counters. (Ayer 1959, pp. 183-4) The important point, however, is that "Whether the *detector is organic or inorganic* is irrelevant to the epistemological issue involved." (Ayer 1959, p. 184) And this is the essential insight for understanding the production of protocols by scientists. "The utterance of the psychologist's protocol sentence is a reaction whose epistemological function is analogous to the tree-frog's climbing and to the barking of the diagnostic dog." (Ayer 1959, p. 185)

Regarded as facts, observation reports are to be interpreted from the standpoint of current theory or, in Carnap's terminology, they are to be translated into statements of physical language. Carnap rests this point directly on the Thesis of Metalogic, the claim that, loosely speaking, all philosophical contentions —that are not nonsense— are metalinguistic and are concerned solely with the logical analysis of language. (1934, p. 38fn.)[1] Using the Metalogic Thesis, Carnap argued that even the most extreme type of observation report, like a phenomenal protocol reporting immediate experience, must be translatable into physical language.

This argument was presented in the companion essay on "The Unity of Science", where Carnap assumed, for the sake of argument, that phenomenal protocols are not translatable into physical statements. Then the protocol "Thirst now" of an experimental subject, call him 'S', would not be equivalent to any intersubjective statement describing S's physical state, like "S is thirsty". Further, S's own protocols would not count as evidence for his physical state of thirst. And it would follow that each observer is locked into a protocol language of his own, and his observation reports would be unintelligible to all others.

In general, every statement in any person's protocol language would have sense for that person alone, would be fundamentally outside the understanding of other persons, without sense for them. Hence every person would have his own protocol language. Even when the same words and sentences occur in various protocol languages, their sense would be different, they could not even be compared. (1934, p. 80)

This conclusion leads to the counterintuitive result that perceptual reports cannot count as evidence for any physical state of affairs. But, Carnap argued,

... an inferential connection between the protocol statements and the singular physical statements must exist for if, from the physical statements, nothing can be deduced as to the truth or falsity of the protocol statements there would be no connection between scientific knowledge and experience. Physical statements would float in a void disconnected, in principle, from all experience. (1934, p. 81)

And this result "cannot therefore be reconciled with the fact that the physical descriptions can be verified empirically". (ibid.) Supposing protocol languages are not intersubjective leads to the conclusion that protocols can never be used as evidence for the existence of physical states of affairs, thus contradicting the obvious fact that protocols express the empirical evidence on which scientific knowledge rests. Since supposing otherwise leads to insoluble contradictions, it follows that every protocol sentence has the same content as some physical sentence.

Central to Carnap's argument is the presumption that, if two sentences are inferentially related, they possess a common content, however diverse their vocabularies might seem. (Ayer 1959, p. 91) So a phenomenalistic protocol, that seems to describe the contents of immediate experience, is really concerned with material bodies and their physical properties, since it is inferentially related to statements of the physical language. Further, Carnap construed these inferential relations with sufficient breadth to include translation schemes between distinct languages, so that the translatibility of a phenomenal protocol establishes its equivalence to its translation. Finally, Carnap regarded two sentences as inferentially related whenever current theory vindicates the conclusion of one from the other.[2] Since the meanings of statements are given by their inferential relations to other statements, this last conclusion is tantamount to the Pragmatic Theory's notion that observation reports are to be translated or interpreted in light of contemporary scientific thought. Unlike Feyerabend, who argued for the Pragmatic Theory on epistemological grounds, Carnap drew the conclusion that observation reports are theory-laden from his conception of language.

Eventually, the syntactical insights first presented in Carnap's early protocol papers were developed into the powerful doctrines of his *Logical Syntax of Language*.

In particular, his Thesis of Metalogic and the consequent distinction between the material and formal modes of speech evolved into the *Syntax's* analysis of pseudo-object sentences. (1937, Sec. 74) Likewise, other important theses of his philosophy, like his rejection of the concept of truth and related semantical notions, were founded on the deepest insights of his syntactic conception of language.[3] And though discussion of these topics must be postponed until some other time, it is essential to see how Carnap's further development of his conception of protocols depended on a further precept of his syntactic analysis of language, later known as 'The Principle of Tolerance'.

The spirit of the Principle of Tolerance was first introduced in Carnap's essay "On Protocol Sentences" to mediate a disagreement with Otto Neurath. Neurath had disclaimed Carnap's treatment of protocols as statements of a language separate from the language of systematic science. Neurath apparently felt that, despite Carnap's repeated insistence on the translatibility of protocols of every form into physical statements, treating observation reports as elements of a separate language vitiated the theses of Physicalism and the Unity of Science.[4] To arbitrate this disagreement, Carnap argued that the difference between alternative viewpoints in the protocol sentence controversy is one that is to be settled by a conventional decision. Thus the difference between the proposal to construe the protocol language as independent of the system language, and the proposal to regard protocols as physical statements, is not dependent on empirical or philosophical matters.

> My opinion here is that this is a question, not of two mutually inconsistent views, but rather of *two different methods for structuring the language of science both of which are possible and legitimate.* (1987, p. 457)

Thus the choice of the form of the protocol language comes to rest on questions of the simplicity, economy, unity, and so forth of the resulting system. Consequently, there is no real issue between Carnap's earlier view of protocols and Neurath's, but only varying preferences concerning the form of the protocol language. (1987, pp. 458, 464-5)

Carnap's point turns on the Principle of Tolerance, which later became the central thesis of *The Logical Syntax of Language*. It asserts that, roughly speaking, the choice of a language is independent of objective matters of fact so that only pragmatic considerations guide the decision to adopt a given form of language. (1937, pp. 51-2) This is not to deny that claims about the structure of the language of science serve an important, and even indispensable philosophical purpose. Rather, the thrust of the Tolerance Principle is directed towards a new assessment of the cognitive status of philosophical contentions. In short, it denies that philosophical theses are on a par with everyday assertions or typical scientific claims — contingent assertions that concern matters of fact which might or might not be the case. Rather, philosophical contentions are to be regarded as proposals to adopt a certain specific language-form. According to the Principle of Tolerance, such a recommendation must not be dogmatic: it may not *demand* the employment of the constructed language nor may it refer to any extra-linguistic reasons why the language must be adopted. There simply are no objective matters of fact that categorically compel the adoption of a particular language form. Rather, the selection of a specific language is arbitrary or conventional in the strict sense that it is never validated by reference to anything outside language. (Coffa 1976, p. 207)

When applied to the question posed by alternative constructions of protocols, the conventionalism of the Tolerance precept is a major innovation which banishes the last remnant of absolutism implicit in all previous theories of knowledge. For realists defend an absolutism of objects and idealists an absolutism of 'the given', 'phenomena', or 'immediate experience'. Even the earlier epistemology of the Vienna Circle embodies a somewhat refined absolutism of basic sentences, finally overthrown by the insight that the form of the language of science, and the question of the nature and role of protocols, is ultimately a matter of arbitrary convention. (Carnap 1987, pp. 469-70)

Carnap's banishment of absolutism from epistemology is probably the most misunderstood point of his Thirties philosophy. Indeed, it seemed to Neurath and Schlick that, in abandoning the notion that protocols have a specific syntactic form, Carnap also deserted the last vestige of the idea that certain sentences are, by their very nature, epistemologically privileged.[5] The epistemological privilege attached to observation reports derives from their special role in warranting other scientific statements even though they are not themselves warranted by any other statements. But what the other Circle members failed to realize is that Carnap banished *absolutism*, not epistemological privilege. The statements that are distinguished as observational, and endowed with epistemological privilege, are not absolutely fixed, once and for all. Instead, the distinction of certain statements as observational, and thus the determination of which statements are honored with epistemological privilege, is relativized to the language in question. Like all matters in the choice of a language, the selection of which statements are to count as observational is a conventional matter, independent of objective matters of fact. Consequently, there is no absolute class of observational sentences, fixed once and for all time, as earlier discussions had presumed. Rather, distinguishing certain statements as observational is part of determining the specifics of a language. Since these matters are conventional, they ultimately turn on pragmatic considerations and, in this limited sense, Carnap's account may be termed 'pragmatic'. But the salient point, typically misunderstood by Carnap's critics, is that even though protocols are not 'fixed' absolutely, they are 'fixed' relative to a given choice of language.

Here, too, Carnap advances the salient epistemological point —that the distinction of observation reports is relativized to a specific language— on the grounds of a precept drawn from his conception of language. On this point, as with the question of the theory-ladenness of observation, his apparent agreement with Feyerabend is vitiated by a fundamental difference with respect to the grounds on which the thesis at stake is established.

The confusion surrounding this point is compounded by the fact that, in his essay "On Protocol Sentences", Carnap discussed protocols at two different levels. At the first level, he treated protocols in general, relative to any choice of language whatsoever. It was at this level that he made his major point, to wit, that the choice among alternative schemes for construing protocols is, like any choice of language, a conventional one. Having established this point, he proceeded to compare different specific proposals for constructing the protocol language. One such proposal was his own earlier treatment of phenomenal protocols as statements of a language distinct from the system language. Contrasted with this proposal is Neurath's idea that protocols belong to the same language as all other scientific statements.

Actually, the difference in these proposals does not fully determine two specific languages, but only creates two large sub-divisions in the class of all possible treatments of protocols, for further options may be designated within each of these large categories. That is, there are any number of further conventional requirements that might be specified for protocols either outside, or within, the system language. Included in the latter class, for instance, is Neurath's idea that protocols contain the personal name of the observer, the exact words in which he originally reported his observation, as well as the details of the time and place of his report. Another method of constructing protocols within the system language is Karl Popper's view that, any statement of physical language can serve as a protocol, depending on the context of inquiry. Popper's idea was that protocol sentences are composed from any of the linguistic resources generally available in the system language, and may vary from case to case depending, in each particular situation, on an explicit decision made on the merits of the case.[6] And it may be decided to stop testing at sentences which are not 'observational' in any of the usual senses. These sentences are to be regarded as protocols, for the purposes of the particular context of testing. Then the term 'protocol' no longer exclusively denotes observational sentences, but designates any statement at which the testing process has been halted. To select this option, as Carnap was inclined to do at the time, is not to invalidate other choices of language form in which protocols are differently construed, for there is no fact of the matter concerning which of the various options represents the 'correct' method of constructing protocols.

In tracing the heredity of the Pragmatic Theory, Feyerabend emphasized Carnap's endorsement of the Popper option, and neglected the underlying conventionalism on which it was premised. Thus Feyerabend fails to appreciate the compelling link between Carnap's analysis of language and his treatment of observation. If Carnap's account of observation is pragmatic, its pragmatism is nonetheless conditioned by the principles of his philosophy of language. If the conventional choice of a language is broadly construed as the choice of a framework of justification, and only pragmatic considerations can influence this choice, then the decision to count certain sentences as observational is likewise a pragmatic issue. But Feyerabend attributes Carnap the further conclusion that observation reports are not distinguished by their contents, but by their causes; certain utterances are observation reports because their etiology conforms to certain patterns of behavior. Thus Feyerabend further misconstrues Carnap's conventionalism with respect to protocols as if it implied that

> An observation sentence is distinguished from other sentences of a theory, not, as was the case in earlier positivism, by its content; but by the cause of its production, or by the fact that its production conforms to certain behavioral patterns. (Feyerabend 1981, p. 50)

But this last notion, the idea that observation reports are pragmatically distinguished in the sense that their demarcation depends on their causal features, is a far cry from what Carnap had in mind when he granted pragmatic factors a role in the isolation of observation reports from other scientific statements.

This is especially obvious once it is noted that Feyerabend wishes to completely divorce the interpretation of observation reports from their distinction as such. Like Carnap, Feyerabend would gladly concede that the interpretation of observation is determined by the language to which they belong and is thus relativized to that language. But things are wholly otherwise with respect to the question of how observation reports are distinguished. For Feyerabend specified pragmatic (including causal) criteria for observation reports that are independent of any interpretation they might receive in a given language. Then the criterion of observationality transcends the limits of any given theoretical framework, yielding universal criteria for distinguishing observation reports.[7] Of course, the very notion of universal criteria that transcend the limits of a given theoretical context is alien to Carnap's conception. For Carnap, criteria of observationality are relativized to particular linguistic frameworks. These frameworks in turn determine the conceptual resources mobilized in the formulation of scientific theories. So the particular conventions of a language determining which of its sentences are regarded as observational would correspond, in a direct way, to the theory formulated in that language. In the rare case, where two alternative theories exploit the resources of a common language, a single criterion of observationality would naturally suffice for both. But the important point is that, for Carnap, criteria of observationality are relativized to languages, usually to theories, and never pretend to universality. Consequently, Feyerabend's pragmatic criterion for distinguishing observation reports cannot be accurately attributed to Carnap.

Then Feyerabend's historical claim, attributing the Pragmatic Theory to Carnap's early conception of protocols, is at most half-right. Carnap's account incorporated the theory-ladenness of observation and denied any fixed demarcation of observation reports. But Carnap's arguments for these points are *toto coeli* different from the epistemological and methodological considerations advanced by Feyerabend. Carnap's grounds for these ideas are based on his conception of language so that his account of observation is pervaded by logical considerations. This difference with Feyerabend's Pragmatic Theory results in radically distinct criteria for demarcating observation reports. Where Feyerabend proposes universal criteria for distinguishing observation reports, Carnap admits a plethora of tests, each relativized to a specific language, each selected on the basis of pragmatic considerations germane to the choice of that language.

It has already been mentioned that Carnap's essay "On Protocol Sentences" marked the introduction of his Principle of Tolerance, the dominant precept of all his subsequent philosophic work. Perhaps more important is the fact that from the time of Carnap's essay, observational reports have generally been thought to concern the publicly discernible properties of intersubjective material objects.[8] (Creath 1987, p. 474) And while Carnap's views on these matters crystallized into a position that he would later modify in only the most insignificant respects, his views on related issues soon underwent profound revision. Perhaps the most dramatic development in his thinking occurred toward the end of the dispute over protocols, when he acknowledged that the concept of truth had a legitimate role in scientific thinking. Impressed by Alfred Tarski's landmark paper on truth, and persuaded by Moritz Schlick's insistence that it was metaphysically harmless to speak of 'comparing' propositions with facts to see if they are true, Carnap endorsed not only the semantic explication of the concept of truth, but the idea that the truth of certain statements is determined by observational procedures.[9] (1949, pp. 119-20, 124-6) And while these developments constituted significant advances in Carnap's later thinking, they did not compromise the leading themes of his protocol era philosophy of language; rather, the definitive precepts of Carnap's early Thirties' conception of language, the Metalogic Thesis and the Principle of Tolerance, survived his development of semantics. (Coffa 1976, p. 206) True, the Thesis of Metalogic was liberalized so that philosophical contentions were construed as semantic as well as syntactic claims, and the Principle of Tolerance was thereafter called 'the principle of the conventionality of language-forms'. (Carnap 1942, Sec. 39) Thus broadened, the ideas that guided his thinking in the early Thirties continued to play a central role in his thought throughout his philosophical career. (Carnap 1963a, pp. 55-6)

Indeed, it was not long before Carnap was exploring what he called 'General Semantics', developing new logical tools to address pressing questions in the epistemology of science. (1942, Sec. 16) Chief among these was his concern with the relation between the scientific concepts in general and strictly observational ones. Carnap's initial attempt in this direction was his essay "Testability and Meaning", in which he introduced 'reduction sentences' to explicate the relation between observational concepts and dispositional notions, like "soluble" or "malleable". For the first time, Carnap publicly acknowledged that some scientific concepts could not be explicitly defined in terms of the observational notions of the everyday thing-language. (1936, Sec. 15) Further, he granted that the conventional stipulations of the meanings of scientific terms by means of so-called 'reduction pairs' have empirical content, despite their definitional role. (1936, p. 444)[10] Unfortunately, the method of reduction sentences developed in "Testability and Meaning" proved useful only for the introduction of dispositional concepts and it was not long before Carnap recognized that less restrictive means were required for the introduction of theoretical concepts, like "gene" in biology, or "electron" in modern physics. (1939, Sec. 24)

To introduce such terms, Carnap suggested a postulational approach, a method in which theoretical concepts were interrelated by postulates, and then correlated with observational notions by means of correspondence rules. This later conception of Carnap's, presented in "The Methodological Character of Theoretical Concepts", was the chief object of Feyerabend's criticisms. Specifically, it was the observational component of Carnap's well-known 'dual language' analysis of the linguistic framework of science that evoked Feyerabend's most strenuous objections. According to Carnap's conception, scientific theories are framed in a language comprised of two fragments. The observational sub-language is fully interpreted, in the sense that it serves as a means of communication for a given community of speakers. (1956, Sec. II) In contrast, the theoretical fragment is only partially interpreted via the connections of its terms with observational ones. These connections are established by means of two devices. First, theoretical and observational terms are linked by means of so-called 'correspondence rules', postulates containing terms from both fragments of the language. Further, theoretical terms are inter-related by means of the postulates that comprise the theory and which contain only expressions from the theoretical portion of the vocabulary. (1956, Sec. V) Then the whole language of science consists of a completely interpreted observational sub-language, linked by correspondence rules to a theoretical component, in which the meanings of theoretical expressions are further stipulated by the postulates of the theory.

Empirical content, assumed to be unproblematic for the observational sub-language, is conveyed to the theoretical fragment by correspondence rules. Observational meaning is thus acquired by the theoretical expressions occurring in correspondence rules and is further transmitted to the remaining theoretical terms by the postulates of the theory. Then the distribution of observational content throughout the theoretical component depends entirely on the correspondence rules and postulates. As in the case of reduction sentences, the theoretical postulates and correspondence rules play a dual role in the determination of meanings and the expression of cognitive claims. (1963b, p. 964) Of course, whether and to what extent a given combination of postulates and rules succeeds in transferring observational content to all the theoretical expressions of the language is an important question. And the goal of Carnap's "Theoretical Concepts" essay was to determine the conditions that must be satisfied by the elements of a scientific language system to ensure the (partial) observational interpretation of each theoretical expression.

Feyerabend's primary concern is that Carnap "presupposes that the meaning of observational terms is fixed independently of their connection with theoretical systems". (1981, p. 53) This presumption, Feyerabend claims, is obvious from Carnap's characterization of the observational sub-language as 'completely interpreted' in virtue of its use as a community's means of communication. Feyerabend regards this feature of the observational fragment as confirming his suspicion that "for Carnap, incorporation of a sentence into a complicated behavioral pattern has implications for its meaning". (1981, p. 54) That is just to say, Carnap assumes that the observational fragment of the scientific language is interpreted independently of its connections with the theoretical component so that the meanings of observational expressions are fixed independently of any relations between the observational and theoretical components of the language.

The consequence to be drawn is that, on the model presented in Carnap's "Theoretical Concepts" paper, linguistic behavior is the sole determinant of an expression's meaning, explicitly excluding its connections to theoretical contexts. The conclusion Feyerabend then urges is that Carnap "has silently dropped the pragmatic theory". (1981, p. 53)

Of course, Carnap never held the Pragmatic Theory as Feyerabend understands it, so the worry that it has been abandoned is irrelevant. However, it was established earlier that Carnap did recognize that the meanings of observational expressions are partially determined by their inferential connections to elements of the non-observational fragment of the scientific language. It was even granted that, since the meanings of non-observational terms are typically altered by changes in theory, observational expressions are theory-laden, though in a somewhat attenuated sense. When changes in theory result in revisions of the non-observational component, the meanings of observational expressions are likewise affected because of their inferential connections to non-observational ones. Accordingly, Feyerabend's criticism can be interpreted as the worry that Carnap's later dual language model implicitly denies the influence of theory on the meanings of observational expressions. Thus Feyerabend reports that Carnap's implication is that "the meaning of an observational statement is already fixed by the way in which the sentence expressing it is handled in the immediate observational situation". (1981, p. 54) This implies that the meanings of observational statements are fixed independently of the semantic structure of the theoretical component so that revisions of the theoretical fragment cannot have any effect on the meanings of observational expressions.

Implicit in Feyerabend's critique is the presumption that Carnap's view of the language of science is a dynamic model of theoretical concept-formation, rather than a static analysis of the empirical content of theoretical concepts. The presumption then would be that the observational language is established and interpreted prior to the introduction of theoretical expressions. The resulting image of theoretical concept-formation presents the theoretical fragment as somehow grafted onto a pre-existing, fully autonomous observation language. This image is a far cry from the one Carnap actually presents, but it makes little difference to the effectiveness of Feyerabend's critique. For Feyerabend's conclusion that the meanings of observational expressions are independent of theoretical influence still does not follow. If an autonomous observation language is given, and is subsequently expanded by a theoretical fragment including postulates and correspondence rules, establishing inferential relations between observational and theoretical expressions, the meanings of the terms of the original observational language must be affected. If the combined postulates and correspondence rules determine the meanings of the theoretical terms by relating them to observational ones, then they thereby endow observational terms with theoretical significance. The introduction of the theoretical component establishes inferential relations involving observational expressions and, to the extent that these relations involve conventional stipulations of the meanings of theoretical expressions, they also influence the meanings of the observational terms concerned. Consequently, even when Carnap's dual language analysis is construed as a dynamic model of theoretical concept-formation, Feyerabend's criticism ignores the importance of the inferential relations between observational and theoretical languages supplied by correspondence rules.

Of course, Feyerabend's criticism fares even worse when Carnap's account is regarded as a static analysis of the empirical content of theoretical concepts. Then Feyerabend's claim of an autonomous observational framework is tantamount to the denial of any inferential connections between the observational and theoretical sublanguages. And no observational statement would imply or be implied by any theoretical statement. But that is just to deny any connection between the meanings of theoretical expressions and observational ones, so that theoretical expressions are totally devoid of empirical content. Of course, Carnap's "Theoretical Concepts" essay presumed that theoretical concepts possessed empirical content, spelled out in terms of their inferential relations to observational expressions. Indeed, the point of Carnap's essay was just to analyze these relations in order to explicate the empirical content of theoretical concepts. To maintain, as Feyerabend must, that Carnap denies any connection between the theoretical and observational fragments of scientific language is palpably absurd.

Of course, Carnap further clarified his understanding of the relations between the two languages in his subsequent explications of analyticity in the theoretical language. (1963b, pp. 963-6) Probably this later work would provide an even stronger case to show the futility of Feyerabend's criticisms. But such an investigation is hardly necessary. It has been shown that, at the time of the protocol sentence controversy, Carnap admitted the influence of theoretical considerations on observation in a sense that distinguishes his view from Feyerabend's Pragmatic Theory. Further, it has been argued that Carnap rested his conception on the most fundamental precepts of his philosophy of language. Despite subsequent developments, these principles continued to guide his epistemology in the ensuing years. Indeed, it has even been maintained that his understanding of the theoretical influence on observation was essentially unchanged as late as his "Theoretical Concepts" essay. To consider, at this juncture, further pertinent developments in his thought would merely belabor what should already be obvious.

Notes

[1]This thesis was the first result of Carnap's "sleepless night" of January 1931; its discovery required the complete re-writing of *The Unity of Science*, on which he had already been working for approximately six months. (1963, p. 53; UP, item no. 029-12-70)

[2]Carnap's understanding of translations from protocol to physical language was more explicitly articulated in his essay "On Protocol Sentences". (1987, pp. 460-1) As the following discussion shows, these translations became more exactly clarified in Carnap's subsequent writings.

[3]The evidence of Carnap's June 1931 lectures on "Metalogik" indicates that his repudiation of the concept of truth in *The Logical Syntax of Language* was founded on his Thesis of Metalogic and his rejection of external meta-languages, rather than on a verificationist prejudice. (UP, item no. 081-07-17-9; Coffa 1987 pp. 566-8)

[4]Neurath's rejection of a distinct language for protocols is explicitly directed against phenomenalistic, solipsistic, or autopsychological conceptions of protocols. Several times in their correspondence, Neurath pressed Carnap to explain what he meant by "methodological solipsism", accusing him of retaining an "idealistic" element in his thought. (Carnap UP, 029-12-49, 029-12-20, 029-12-17)

[5]Neurath's objections were spelled out in two letters, in which he decried Carnap's desertion of empiricism. (Carnap UP, item nos. 029-12-38, p. 1; 029-10-71, p. 1] Carnap replied in three subsequent letters reassuring Neurath that they were in agreement over the issue of empiricism and the role of experience in science. (Carnap UP, item no. 029-10-70, p. 1; 029-09-86, pp. 1-2; 029-09-80,p. 1] Moritz Schlick's objections were presented in the form of his classic essay "On the Foundations of Knowledge" and a subsequent clarificatory note "On 'Affirmations'". (Schlick 1979, pp.370-387; 407-413] On May 17, 1934, Carnap reassured Schlick that he recognized the role of experience, as evidenced in the passage in his essay "On Protocol Sentences" where he said that all testing involves the experiences of an individual, and there was a germ of truth in 'methodological solipsism'. (Carnap 1987, pp. 469)

[6]In addition, Popper claims the decision to accept a 'basic sentence' is conventional. (1968, pp. 104-5, fn.)

[7]Feyerabend's criterion is stated in terms of the pragmatic conditions on "the relation between observation-sentences (not statements) and human beings without making any stipulations as to what those sentences are supposed to assert." (1981, p. 145) The result was his conception of 'quickly decidable sentences'. (1981, pp. 144-7)

[8]Even the obvious counterexample, Schlick's concept of a "Konstatierung", was to be understood as possessing the same content (though distinct epistemological properties) as a physicalistic protocol. (1979, pp. 409-410)

[9]Schlick's role is especially attested in his correspondence with Carnap in 1934 and 1935. (Cf. esp. Carnap UP, 029-28-14)

36

[10]It was in "Testability and Meaning" that Carnap came closest to endorsing something like Feyerabend's notion of a 'quickly decidable' sentence. (1936, pp. 454-5)

Bibliography

Ayer, A.J., (ed.) (1959) *Logical Positivism*. New York: The Free Press.

Carnap, R. (1934) *The Unity of Science*. M. Black, trans. London: Kegan Paul, Trench, Trubner and Co. Ltd.

_____. (1987) "On Protocol Sentences," R. Creath and R. Nollan, trans. *Nous* 21, 457-470.

_____. (1936) "Testability and Meaning," *Philosophy of Science*. Vol. III, pp. 420-471 and Vol. IV, pp. 1-40.

_____. (1937) *The Logical Syntax of Language*. A. Smeaton, trans. London: Routledge and Kegan Paul.

_____. (1939) *Foundations of Logic and Mathematics*. Chicago: University of Chicago Press.

_____. (1942) Introduction to Semantics. Cambridge: Harvard University Press.

_____. (1949) "Truth and Confirmation," H. Feigl, tr. in Feigl, H. and Sellars, W. (eds.), *Readings in Philosophical Analysis*. New York: Appleton-Century-Crofts pp. 119-127.

_____. (1956) "The Methodological Character of Theoretical Concepts," in Feigl, H. and Maxwell, G., eds. *Minnesota Studies in the Philosophy of Science*, Volume I. (Minneapolis: The University of Minnesota Press), pp. 38-76.

_____. (1963a) "Intellectual Autobiography," in *The Philosophy of Rudolf Carnap*. P. Schillp (ed.). LaSalle: Open Court, pp. 3-84.

_____. (1963b) "Replies and Systematic Expositions," in *The Philosophy of Rudolf Carnap*. P. Schillp (ed.). LaSalle: Open Court, pp. 859-1013.

_____. (UP) Unpublished Papers. Archives for Scientific Philosophy, University of Pittsburgh Libraries.

Coffa, J.A., (1976) "Carnap's Sprachanschauung Circa 1932," in *PSA 1976*, vol. 2, P. Asquith, and F. Suppe (eds.). Lansing: Philosophy of Science Association, , pp. 205-241.

_____. (1987) "Carnap, Tarski, and the Search for Truth," *Nous* 21: 547-572.

Creath, R. (1987) "Some Remarks on 'Protocol Sentences'," *Nous* 21: 471-5.

Feyerabend, P. K (1981). *Realism, Rationalism, and Scientific Method.* Cambridge: Cambridge University Press.

Popper, K. (1968) *The Logic of Scientific Discovery.* New York: Harper Torchbooks.

M. Schlick. (1979) *Philosophical Papers.* Vol. II. H. Mulder and B. v. d. Velde-Schlick (eds.) Dordrecht: Reidel.

THOMAS E. UEBEL*
NEURATH'S PROGRAMME FOR
NATURALISTIC EPISTEMOLOGY

Abstract — I examine the thesis that Otto Neurath anticipated the programme of
naturalised epistemology already at the time of the Vienna Circle and consider
the relation between Neurath's proposals and those of two contemporary theorists whose research programmes he would thus have broadly anticipated. The
thesis is confirmed by reference to Neurath's own writings. The connection
between Neurath's programme and the programmes of his two successors
considered here, however, is found to be highly indirect in one case and
nonexistent in the other — despite their undeniable overlap.

1.

WITH THE increasing historical and doctrinal distance from the Vienna Circle
— closely associated with the now rejected 'received view' of scientific theories
— the views of its members have begun to receive a more dispassionate
reading. To judge by the results reached so far we are in for a surprise. The
conceptual ruffians of the popular philosophical imagination turn out to have
been highly sophisticated thinkers whose ideas still bear centrally on contemporary concerns. Read aright, some of their 'mistakes' may not have been
mistakes at all, but even if they were, they cast a light on the deeper
significance of contemporary views which remains hidden to the latter's facile,
unhistorical acceptance.[1]

This revaluation of the Vienna Circle in English-speaking analytic philosophy has so far extended mainly to Rudolf Carnap, less so to Moritz Schlick,
but hardly yet to Otto Neurath, its most notorious ruffian. Some subterranean
interest, however, has been stirred by word from deepest Europe that Neurath
anticipated the programme of naturalised epistemology.[2] It is this claim that I
wish to consider here. Setting Neurath in direct opposition to Carnap's
reconstructionist programme, it is striking enough, I believe, to merit a closer
investigation of its grounds. In substantiating it, I shall also defend its

*Center for the Philosophy of Science, Cathedral of Learning, University of Pittsburgh,
Pittsburgh, PA 15260, U.S.A.

Received 9 *April* 1991.

[1]For readings of e.g. Carnap as an 'analytic neo-Kantian', see Coffa (1977), (1984), (1985),
(1986), (1987), and a forthcoming, now sadly posthumous volume; Friedman (1987), (1988),
(1991); Proust (1986), (1987), and several articles in French; Richardson (1990); Sauer (1985). For
other recent non-traditionalist readings of Carnap — and Schlick — see note 37.
[2]Haller (1979a), (1982b); Rutte (1982a), (1982b); Heidelberger (1985); Koppelberg (1987),
(1990); compare also Hempel (1982). (For a closer description of these attributions see note 30.)

Stud. Hist. Phil. Sci., Vol. 22, No. 4, pp. 623–646, 1991.
Printed in Great Britain

0039-3681/91 $3.00 + 0.00
© 1991. Pergamon Press plc.

significance against, as it were, under- and over-interpretation, but I must leave many questions open. As my title indicates, I can only lay out Neurath's programme here; his motivation and argument for, and his implementation of, this programme must be considered separately. My point is that its surprising topicality renders these further questions of more than merely historical interest.

2.

To fix the relevant ideas, it is necessary to begin by stating what shall count as naturalistic epistemology. I shall adopt as a distinctive criterion of naturalistic epistemology a point of methodology, for it seems that the best way of framing its working definition is not in terms of a particular doctrine, but in how it squares off against the way traditional epistemology is done.[3] Naturalistic epistemology rejects traditional, say Cartesian, epistemology; but does this mean that traditional epistemology is to be supplemented or replaced by another sort of inquiry, or that epistemology is to be outright abandoned?

First, what is its object of inquiry? Against traditional First Philosophy, naturalism requires that the *practice of science* be attended to. (Call this the "primacy of practice claim" (NE1).) The basic idea is to take the sciences as paradigmatic candidates for justifiable knowledge claims, if anything is. Needless to say, greater specificity is needed: most epistemologists would claim their theory to have a more than accidental bearing on our estimation of the actuality of scientific knowledge. Still, the primacy of practice claim is not without any bite: it sets limits to the standards by which knowledge candidates are assessed. Since scientific knowledge lacks the property of certainty, there is no epistemological point in trying to provide self-evident foundations.

The naturalistic constraint extends this constraint to the rejection of resort to supra-empirical principles of explanation or reconstruction. (Not all principles can be called into question at once, of course, but all can be questioned sooner or later.) Naturalistic epistemologists claim that the threat of circularity — incurred by rejecting foundationalism — cannot be met by recourse to *a priori* meta-principles, but only by disowning the threatening character of the

---

[3]The pitfall in the latter approach is that important conceptions of knowledge and its justification may be held by naturalistic and non-naturalistic epistemologists alike and particular doctrines may be controversial among naturalistic epistemologists themselves. An example of the former is the anti-foundationalist view that there are no indubitable resting points for scientific knowledge, held e.g. by Quine (1969) as a naturalist and BonJour (1985) as a non-naturalist; an example of the latter is the view that naturalized epistemology must remain descriptive, outlined as one alternative by Kornblith (1985a) as a descriptivist and as normativists by the authors cited in note 23.

circularity involved and acquiescing in it: epistemology must become a science broadly like the sciences it investigates. The 'philosophical' moral of naturalism is that the rationality of science can be shown only *from inside* of science. Traditional philosophical epistemology is thus not just to be reined in somewhat, or supplemented somehow, but it is to be *supplanted* by a many-fronted, interdisciplinary scientific inquiry. (Call this the "replacement thesis" (NE2).[4])

But what does it mean to 'explain science from within'? The anti-aprioricism of the replacement thesis forbids that human cognition be pressed into a framework of what would provide justification, if that framework is alien to scientific practice. It will not do to distinguish a context of discovery from one of justification and focus on the latter if it is riddled with in principle counterfactual assumptions about epistemic subjects. Moreover, 'explaining science from within' means that the *epistemological notions employed be explicable in terms of science.* That is to say, epistemological terms like justification or epistemic priority have to be cashed in terms applicable to and successful in the explanation of processes we antecedently deem natural, namely, processes occurring in the spatio-temporal world.[5] (Call this the "explanatory austerity demand" (NE3).)[6] Typically, explanatory austerity means the employment of causal or correlational concepts, but it should be noted that naturalistic epistemology may help itself to all the explanatory kinds used in the sciences.[7]

Turning back from the method to its object again, we may now ask what kind of practice naturalistic epistemology explains. Distinguishing sharply between a theory of truth conditions and a theory of truth claim acceptance, replacement theorists will conclude that it is not the justification of a claim to truth, but the justification of the *acceptance* of a knowledge claim that is to be

---

[4] No endorsement of the non-normativity claim associated with this moniker by Kornblith (1985a) is intended (see also note 24): the 'replacement' rather consists in the rejection of *a priori* strictures and the insistance that knowledge claims be investigated empirically (as in Boyd's definition of naturalism: (1981), p. 615).

[5] I am indebted for this formulation to Paul Pietroski. Since the formal sciences are typically involved in explanations of natural processes, the second part of E. Nagel's and R. Brandt's classic formulation of epistemological naturalism — "epistemological terms can be explained by way of empirical and logical concepts exclusively, without using the other terms in the process" (1965, p. xiii) — is upheld.

[6] Feigl (1939) already settled on something like this as the defining characteristic of *naturalism* — "the explanatory constructs of all sciences will not need to go beyond the spatio-temporal frame" — albeit *without including epistemology as comprehended thereby.*

[7] In going beyond Lehrer's definition of naturalistic epistemology (1990, p. 154) here, I admit the Achilles-heel of naturalism: what if science should use 'metaphysical' concepts? It must be noted then that naturalism itself is a historically developed doctrine: it may well be that what was once considered scientifically legitimate is no longer so (i.e. the *synthetic a priori*). It is significant, however, that the continued deployment of such now discarded concepts and principles was challenged as unwarranted by developments *within* science.

provided.[8] The primacy of practice thus becomes the *primacy of theory acceptance* (NE4). Naturalistic epistemology accordingly has recourse to the scientific investigation of epistemic subjects: linguistics, biology, psychology and the neurosciences.

The programme of naturalistic epistemology can still be further extended beyond NE1–4 (I shall do so later), but for now we may stop here. The resultant working definition of naturalistic epistemology ("austere explanation of the practice of theory acceptance") provides a minimalist conception of a consistent naturalism. Showing that Neurath endorsed NE1–4 would already suffice to establish that he was a *bona fide* naturalistic epistemologist, but leaves important question about his programme open.

Note also that our working definition clearly accords with Quine's proposals (and is not merely tailored for the task of interpreting Neurath). For Quine, epistemology is an "enterprise within natural science" (1975, p. 68), namely, to "understand the link between observation and science . . . by the very science whose link with observation we are seeking to understand" (1969a, p. 76). Accordingly, Quine suggested neurophysiology, behaviouristic psychology and evolutionary biology to furnish the required explications of epistemological vocabulary. Talk of "epistemic priority" became talk of "causal proximity to sensory receptors" (1969a, p. 84, p. 85), and "innate quality spaces" take the place of "subjective (or remembered) similarity" (1969b, p. 123); similarly, talk of the "given" is replaced by talk of "observation sentences" (1975, pp. 69–71), which are occasion sentences "whose occasion is not only intersubjectively observable, but generally adequate, moreover, to elicit assent to the sentence from any present witness conversant with the language" (1975, p. 73). The elements basic to the process which defines science — the testing of theories for confirmation or disconfirmation — are here individuated wholly within the object domain which science itself speaks of.[9] Clearly, Quine endorses NE1–3.

[8]To pursue the former — encouraged by a verificationist theory of meaning or not — would be not so much to beg the question of what knowledge is as to beg the question of what truth is. Of course, keeping our assumptions about the nature of knowledge to a minimum compatible with historical scientific practice, our practice is better described as one of theory acceptance, not of making true statements. More importantly, however, *staying within the bounds of what scientific tools allow us to establish*, justifying claims by establishing their truth, simply could not amount to more than either applying the Tarskian disquotation schema to the claim at issue or checking whether the criteria of scientific methodology (empirical adequacy, simplicity, etc. culminating in 'best explanation') have been followed in arriving at it. The former activity is better counted as philosophy of language and/or logic, and by itself hardly qualifies as epistemologically informative (though it would figure with other considerations as part of a theory of theory acceptance); the latter does not amount to more than investigating theory acceptance itself, so it might as well go under this name. Whether we accept knowledge claims as true or as instrumentally reliable is another matter still, but having begun to travel the replacement path (NE2–3) the dispute between realists and anti-realists can simply be bypassed without as yet having to argue for the deflationism which most 'naturally' goes with the envisaged position. (See Fine, 1984.)

[9]It may of course be doubted whether behavioristic psychology is the proper discipline in which to undertake the explication of what an observation sentence is, but that is a minor matter for the question of the mere naturalisability of the epistemic notion concerned, important as it may be otherwise. (This caveat also holds of Neurath, unless his strident 'behaviorism' be reinterpreted.)

The resultant circularity of justification received its acquiescence from Quine, with frequent reference to Neurath's parable of the sailors repairing their boat at sea. Given this acquiescence — and his development of the 'semantic' into the 'disquotational' conception of truth (1970, p. 12) — it is also clear that Quine takes his naturalistic programme not to be concerned with the substantiation of truth claims: his coherential reasoning aims to motivate the acceptance of knowledge claims (NE4).[10]

3.

Did Neurath propose a programme as outlined above? Discounting the outright sceptic for the moment, I also envisage another unenthusiastic reaction to this question, a short answer that is bound to lead the questioner astray: yes, Quine said so himself. The word from the Continent is but a storm in a Vienna Circle memorial mug — was it not Quine who made Neurath's Boat the virtual motto of this programme? To clear up this misapprehension, I consider how Neurath fares in the narrative of the inception of the programme in Quine's 'Epistemology Naturalized'.

In his reconstruction of the inception of naturalistic epistemology, Quine points to Carnap's *Aufbau* and its strategy of providing a rational reconstruction of empirical knowledge in terms of phenomenalist observational and logico-mathematical terms as the last hold-out of traditional epistemology. Hume had ended hopes for certainty of scientific knowledge in general, but there remained at least the hope of clarifying the basis for such knowledge. According to Quine, Carnap's *Aufbau* was to show that it was possible to accomplish "everything in terms of phenomena and set theory that we now accomplish in terms of bodies. It would have been a true reduction by translation, a legitimization by elimination". (1969a, p. 78). Alas, Quine points out, it was not to be. Carnap did not provide eliminative definitions of positions in physical space and time in terms of sense qualities: the reduction of the physical to the phenomenal failed (1969a, p. 76; 1951, p. 40). As Quine sees it, there are two kinds of consequences to be drawn from this irreducibility. First, to declare epistemology bankrupt and meaningless. As examples of this Quine points to "Carnap and the other logical positivists of the Vienna Circle" (and the later Wittgenstein). Second, to hold that "epistemology, or something like it, simply falls into place as a chapter of psychology, and hence natural science". (1969a, p. 82.) By implication, Neurath is here cast as one who drew

---

[10]Naturalistic epistemology may be seen as spelling out the terms of the pragmatism of his early views that "the considerations which guide him in warping his scientific heritage to fit his continuing sensory promptings are, where rational, pragmatic" (Quine, 1951, p. 46), or that '[o]ur standard for appraising basic changes of conceptual scheme must not be a realistic standard of correspondence to reality, but a pragmatic stand' (Quine, 1950, p. 79).

287

the first type of consequence, not as one who, like Quine himself, drew the second!

As it happens, Quine's pointer to the Vienna Circle as an example of theorists holding epistemology bankrupt and meaningless may well occasion general puzzlement. Surely Schlick did not think his grappling with "the foundations of knowledge" (1934, part-title) a meaningless pursuit: Schlick, after all, held that "philosophy must supply the ultimate support of knowledge" (1930, p. 58). Neither did Carnap think epistemology worthless. For him, epistemology as of 1930 was "applied logic"; as of 1934 it was comprehended by "the syntax of language", and later more broadly the "logic of science" in the sense of encompassing semantics.[11] His preferred method of rationally reconstructing scientific knowledge claims challenged some traditional forms of epistemology, but did not abolish the need for epistemology in the first place, in fact sought to satisfy it within the realm of formal reasoning. Long after Quine first criticised the *Aufbau*'s failure of eliminative reduction, Carnap wrote: "I still agree with . . . the possibility of the rational reconstruction of the concepts of all fields of knowledge on the basis of concepts that *refer to the immediate given*". (1961, p. v; my italics). So it cannot be Carnap either who declared epistemology meaningless. Quine's remarks apparently point to Neurath, the most militant anti-metaphysician of the Vienna Circle, who wrote in his 'Sociology in the Framework of Physicalism':

> The physicalistic language, unified language, is the Alpha and Omega of all science; there is no "phenomenal" language besides the "physical" language, no "methodological solipsism" beside some other position, no "philosophy", no "theory of knowledge", no new "Weltanschauung" beside the others: there is only Unified Science with its laws and predictions. [1932a, p. 68.]

This clearly sounds like the embrace of the first alternative of declaring, in the wake of the *Aufbau*, epistemology to be meaningless. If anybody in the Vienna Circle declared it so, Neurath did, and it must be him who Quine had in mind, convicting Carnap by association.

My reading gains support from Quine's portrayal of the Vienna Circle's so-called 'protocol sentence debate'.

> Around 1932 there was debate in the Vienna Circle over what to count as observation sentences, or *Protokollsätze*. . . . One position was that they had the form of reports of sense impressions. Another was that they were statements of an elementary sort about the external world, e.g. "A red cube is standing on the table". Another, Neurath's, was that they had the form of reports of relations between percipients and external things: "Otto now sees a red cube on the table." The worst of it was that there seemed to be no objective way of settling the matter: no way of making real sense of the question. Let us now try to view the matter unreservedly in the context of the external world . . . [1969a, p. 85.]

---

[11]Carnap (1930), p. 143; (1934/7), pp. 280–281; (1936/7), p. 2; (1938), pp. 42–43.

The main trouble with this view of the protocol sentence debate is that Quine's characterisation is not correct in its implications.[12] In the last two sentences quoted Quine alleges that neither Carnap *nor* Neurath had "an objective way" of deciding the matter of which was to be the preferred form of observation statements, that neither Carnap *nor* Neurath had a "real sense" of the issue, and that neither Carnap *nor* Neurath "unreservedly" considered the matter of scientific evidence statements in the context of the external world.

If Neurath did propose to naturalise epistemology — however darkly — then Quine is mistaken. Proposing to do so would surely count, in Quine's own terms, as having a "real sense" of the issue and as seeing the matter "unreservedly" in terms of the external world. And supposing there to have been an argument for this proposal, it is likely that he did therewith have an "objective way" of deciding the issue. That Quine twice misrepresents Neurath's position in his portrayal of the inception of the programme of naturalistic epistemology strongly suggests that he did *not* see that Neurath argued for the naturalisation of epistemology.

I stress this point not only to enhance the newsworthiness of the claim I investigate, or to guard against holding the interest of Neurath's proposal exhausted by anticipating Quine.[13] The parallels between Neurath and Quine (to be shown presently) are striking enough also to warrant a warning against yielding, on this occasion, to the lure of conspiracy theses. Given Quine's own frequent invocation of Neurath's Boat, and a recent suggestive (yet only *rational*) reconstruction of Quine's philosophy as an "attempt to synthesize the different theoretical orientations of Rudolf Carnap and Otto Neurath,"[14] this temptation may be hard to resist. It should be resisted, however. Unaware of Neurath's actual conception, Quine independently arrived at his programme.[15]

4.

Let me now turn to what I claim Quine missed and note that this is not to be taken as a reproach — clearly, he was not the only one to do so. In fact, the quotation I presented from a paper of Neurath's from the protocol sentence debate shows why the claim that he naturalised epistemology may instead be

[12]It would also be correct to object to the periodization of the debate: the particular exchange referred to (Quine mentions Neurath's 'Protocol Sentences' and Carnap's 'On Protocol Sentences') was but the end of but one stage of a debate raging since at least 1928 (though I can only assert but not show this here). Moreover, it would be correct to object that this portrayal misses the scope of the debate: at issue between Carnap and Neurath (and Schlick) was the very nature of epistemology (as is obvious, I trust, from the thesis investigated in this paper).

[13]Though that would be mistaken anyway: just think of what Quine disregards in Dewey, whom he *does* note as naturalistic precursor (Quine, 1969c, pp. 26–29).

[14]Koppelberg (1987), p. 9; see also p. 13, p. 14, and Koppelberg (1990), p. 201.

[15]So Quine himself: (1990), p. 212; compare also his remarks on the 'influence' of Duhem and Dewey.

thought rather preposterous. Yet in this *same* paper Neurath had still more to say about epistemology than is immediately suggested by that seemingly dogmatic (though not uncharacteristically blustering) claim.

First note that, whatever else he may have been up to, Neurath endorsed the primacy of practice claim (NE1) by asserting that "there is only unified science . . .". This claim becomes significant when we note that he also remarked:

> Within a consistent physicalism there can be no "theory of knowledge", *at least not in the traditional form*. It could only consist of defense actions against metaphysics, that is to say, of the unmasking of meaningless terms. [1932a, p. 67; my italics.]

This remark clearly sounds somewhat more temperate, and more akin to Carnap's conception of epistemology as applied logic in the service of "the elimination of metaphysics". (Carnap 1932, part-title.) Still, no outright positive task seems assigned to the epistemological enterprise. (For Quine, who'd count the 'ideology' in criticism of ideology' one of the bad senses of the word, it may still not amount to something 'meaningful'.) But note that Neurath continues:

> Many of the problems of the theory of knowledge will perhaps be transformable into *empirical* questions so that they can find a place in unified science. [1932a, p. 66; my italics.]

Far from emulating Carnap's perspective, this remark not only sounds like, but, I submit, *is* a demand for the naturalisation of epistemology.

Neurath here endorsed the replacement thesis (NE2). (As he also put it, less carefully, about half a year earlier: "The work on unified science replaces all former philosphy". (1931b, p. 56.)) Neurath's view is that *traditional* epistemology is meaningless. Whatever investigation there *may* remain to be pursued as part of the theory of knowledge would have to be such that it finds its place in the totality of the sciences, as an *empirical* investigation alongside others (with all the tools provided by formal science). Neurath recognised that the concepts and statements employed by a de-metaphysicalised epistemology must accordingly be capable of explications in scientific terms. Discussing varieties of 'physicalism', he notes: "The most radical formulation subjects all sentences, even the protocol sentences, to scientific criticism". (1932b, p. 568.) With this remark Neurath expressed his adherence to the principle of *explanatory austerity* (NE3): in science, after all, non-scientific concepts will not be tolerated.

But *is* there such epistemological life after metaphysics? Neurath's staunch opposition to the 'metaphysical' conceptions of Wittgenstein's *Tractatus* leaves little doubt that he was convinced of it.

> It is . . . impossible to contrast language as a whole with 'experiences' or with the 'world' or with something 'given'. Every statement of the kind: "The possibility of science rests on an orderly constitution of the world", is therefore meaningless. . . . *The possibility of science becomes apparent in science itself.* [1932a, p. 61; my italics.][16]

Here Neurath indicated the positive complement of his denial of traditional epistemology: concern with the possiblity of science is not meaningless if it means to exhibit the rationality of science from within. His conviction that the rationality of science can, indeed must, be so shown, is evident also from his opposition to Carnap's 'methodological solipsism' in the protocol sentence debate: the "most radical formulation" of physicalism was his *own*. It should be clear then that Neurath's adherence to NE3 is not merely proclaimed in the passages cited above, or implied by others like "It is impossible to separate the 'clarification of concepts' from the 'pursuit of science, to which it belongs'". (1932a, p. 59.) Neurath also practiced what he preached: his physicalistic theory of protocol sentences and the critique of its competitors constitutes precisely such an investigation of the empirical basis of science by scientific means.[17]

So far, so good. The sceptic about the word from the Continent whom I so far disregarded may well remain unimpressed, however. What of naturalistic epistemology's concern with the justification not of truth but of the acceptance of knowledge claims (NE4)? How does that sit with Neurath's notorious 'coherence theory of truth'? Consider:

> If a statement is made, it is to be confronted with the totality of existing statements. If it agrees with them, it is joined to them; if it does not agree, its called 'untrue' and rejected; or the existing complex of statements of science is modified so that the new statement is incorporated; the later decision is mostly taken with hesitation. There can be no other concept of 'truth' for science. [1931b, p. 53.]

Surely such a statement contradicts the parallel with Quine and plainly undermines the empiricism to which naturalistic epistemology lays claim; at best, the objector would conclude, Neurath grappled with some aspects of naturalism, but his 'programme' ultimately falls into *in*coherence.

---

[16]How much of Neurath's *naturalism* was engaged in his rejection of residual Wittgensteinian metaphysics is shown by the sentences preceeding the last quote: "It is . . . not possible to make pronouncements about language as whole from a 'not yet linguistic' standpoint, as Wittgenstein and some individual representatives of the Vienna Circle seek to do. A part of these attempts may perhaps, after suitable transformation, find a place within the sphere of science; while another part would have to be dropped." (1932a, p. 60–1).

[17]It might be added with Haller that his much ridiculed *index verborum prohibitorum* — which discouraged the use of much familiar philosophical terminology, e.g. 'true,' 'false,' 'fact', 'reality' (Neurath, 1944, p. 51) — is best read not as an *idée fixe* of *Gründlichkeit an und für sich* in the service of para-inquistional scientism — however much it may have overshot the mark — but as mere propaedeutics for constructive work in whatever meaningful inquiry remained of epistemology.

SHIPS 22:4-G

Well, let's consider this 'counter-evidence'. Similar passages are indeed strewn throughout Neurath's entire work from the early 1930s onward, and the passage quoted is nearly literally repeated in 'Sociology in the Framework of Physicalism' from which I drew most of my evidence so far. But note that both show concern with the rejection — and, implicitly, the acceptance — of statements, and that now Neurath speaks of "correctness" instead of "truth" (as he still later will talk of "validity" (1936b, p. 159, p. 161, p. 165,): "The definition of 'correct' and 'incorrect' as proposed here abandons the definition that is usually accepted in the Vienna Circle and recurs to 'meaning' and 'verification' ''. (1932a, p. 66.) What Neurath rejected here was the notion of correspondence truth which supposedly had been brought within the ken of empiricists by the verificationist theory of meaning. But instead of proposing an alternative theory of truth, Neurath meant to shift the Vienna Circle's debate about the 'foundations' of scientific theories towards a theory of theory acceptance.[18]

A *modicum* of interpretational charity should convince us.[19] In 1934 Neurath remarked:

> With certain justification one could propose to ban use of the terms 'induction', as well as 'true' and 'false', in the scientific language altogether. But in the interest of linguistic continuity one could also propose redefinitions which allow for the continued use of these terms. [1935, p. 631.]

Neurath's use of 'true' amounted to just such a 'redefinition'.[20] In his rebuttal of Schlick's and others' criticisms of his rejection of the correspondence theory, Neurath made clear that his concern lay with claim acceptance, not truth substantiation. It is worth quoting this passage in full for the explicit connection it draws between the naturalisitic principles NE1 and NE4:

[18]What then of Neurath's resistance to the Tarskian paradigm in later years? I hasten to add that it does *not* find its *justification*, but *only* its *explanation* in his suspicion of its supposed resuscitation of the correspondence conception (a misidentification in fact encouraged by Tarski, not wholly put to rest until Quine's 'disquotational' interpretation of the truth predicate) and of Carnap's subsequent semantic 'metaphysics'. (On the later point see Hegselmann's 1985 report of their later correspondence.)

[19]Note also his own (admittedly somewhat bumbling) disclaimers in Neurath (1934), p. 101; (1935c), p. 666; (1944), note 16. For independent corroboration, Carnap (1963), p. 864, which spells out one of the addresses of his somewhat misleading claim in his (1936/49, p. 19, p. 120) that the discussions in the Vienna Circle about 'truth' prior to this date (Paris Congress 1935) had been marred by insufficient attention to the distinction between 'truth' and 'confirmation'. But Neurath was not merely inattentive to this distinction. (For a contrasting attempt to reconstruct Neurath's 'theory of truth', see Hofmann-Grüneberg, 1988.)

[20]Note also that in 'Sociology . . .' he had remarked: "The arguments of this paper link up best with a *behaviourist* basic attitude. We speak not of "thinking" but straight away of 'speech-thinking', this is, of *statements as physical events*" (1932a, p. 67.) Now, what would behaviourism have to contribute to a theory of truth? Behaviourism bears only on the problem at hand if it is understood to explicate the concept of theory acceptance empirically. In later papers he reaffirmed this behaviourist attitude alongside his non-standard use of 'truth': " 'Certain' is to be defined as a term within the doctrine of human 'behaviour' ''. (1934, p. 104.)

However, if Vogel . . . is of the opinion that one has to have the system axiomatized in order to have contradiction revealed, he overlooks the fact that in *practice* one proceeds much more clumsily and is mostly glad to have some contradiction pointed out or a greater number of conformities. It is precisely the history of physics that shows that our procedures are often quite consciously defective. It happens that occasionally two contradictory hypotheses about the same subject are used at two places with some degree of success. And still, one knows that in a more complete system only *one* hypothesis should be used throughout. We just resign ourselves to a moderate clarification in order to delete or *accept* statements later. [1934, p. 109; first and third italics mine.]

I trust I need not turn to Neurath's later writings to exhibit similar and further confirmation of his concern with theory acceptance rather than truth claims.[21]

So Neurath's rejection of a First Philosophy does not only betoken the rejection of traditional epistemology as meaningless, but also the proposal to naturalise epistemology. Moreover, Neurath did not only anticipate one or another aspect of naturalistic epistemology, even less so haphazardly; there is a system in Neurath's anti-epistemological (and anti-metaphysical) tantrums. Neurath's programme meets our narrower working definition as the "austere explanation of the practice of theory acceptance" (NE1–4).

5.

As laid out so far, the outlines of Neurath's programme agree with Quine's. Yet Neurath's program anticipated not only Quine, but also other 'anti-positivists', indeed ones whose theories Quine seems not to feel wholly comfortable with and who, in turn, might want to dissociate themselves from the naturalistic project.[22]

Naturalistic epistemology may be extended beyond NE1–4. Understood as concerned with the human practice of theory acceptance, it may demand that it take account *in full* of the context in which this kind of practice is exercised. Doing so leads — via the recognition of the historical nature of concepts of knowledge themselves and of the general social determinants of this practice — to the inclusion of the history and sociology of inquiry. Theory acceptance is thus viewed as determined not only by methodological 'canons' and linguistic and psychological facts, but also by factors concerning the internal organisation of science, by the potential applicability of results and availability of sponsorship for research programmes, and by general historical conditions

[21]Particularly Neurath (1936b), *passim*, esp. p. 169; cf. (1935b), p. 124, and (1944), p. 12.
[22]See Quine's scepticism of Kuhn's (and Hanson's) "cultural relativism" (1969a, p. 87) and Kuhn's recent embrace of the *Ding an sich* (1991), but note that Giere (1985) drafts Kuhn for the project of a naturalistic philosophy of science.

from *Zeitgeist* to the development of productive forces. Naturalistic epistemology may, in short, also recognize *extra-cognitive factors in theory acceptance* (NE5).

The 'externalist' conception may of course be endorsed in a stronger or weaker form, setting scientific knowledge in the context of the socio-economic development of society as a whole or merely in the context of the so-called 'scientific community'. Kuhn, for one, followed the narrower reading (but that still proved controversial enough):

> The new paradigm implies a new and more rigid definition of the field. Those unwilling or unable to accommodate their work to it must proceed in isolation or attach themselves to some other group. . . . one of the things a scientific community acquires with a paradigm is a criterion for chosing problems that, while the paradigm is taken for granted, can be presumed to have solutions. To a great extent these are the only problems that a community will admit as scientific or encourage its members to undertake. [1962, p. 19, p. 37.]

These considerations hold for 'normal' science; in times of revolutionary change such (narrow) extra-cognitive influences can be expected to increase in importance.

Now, can the replacement of traditional epistemology stop short of the abandonment of epistemology altogether?[23] Only if the question of the *legitimation* of scientific knowledge claims still remains broachable in this successor discipline, if the envisaged inquiry is not condemned to remain merely descriptive, but possesses the resources to criticise claims as inadequate. (Call this the "legitimation" claim (NE6).) It is, of course, a highly contentious hope that science may not only explain but also legitimise itself, but it would be rash to equate naturalism with non-normativity.[24] The question of the possibility of a normative naturalistic epistemology rather calls forth the very question whether objectivity as a universal claim to epistemic allegiance is itself a scientifically respectable notion or needs to be replaced.[25] (Naturalising epistemology may effect not only a shift in the concept of knowledge, but also in the concept of rationality.) Fortunately, I need not pursue this question here, for I shall claim merely that Neurath also endorsed NE5 and is driven to endorse

[23]Such doubts are raised by Siegel (1980), (1984); Stroud (1981); Putnam (1982); Kim (1988).

[24]For this reason I added NE3 to NE2 and thus spelt out 'replacement' — in addition to anti-aprioricism — as the 'principle of explanatory austerity'. Whether NE2 ultimately leads to non-normativity is thus left open. Zolo's (1986/9) reading of Neurath as a non- even anti-naturalist depends (in part) on presuming naturalism to amount to non-normativity. That naturalistic epistemology does have a normative aspect is claimed e.g. by Roth (1983), (1991); Giere (1985); Antony (1987); Laudan (1987), (1990); Brown (1989); and Quine (1990b), pp. 19–21.

[25]It may appear that the task becomes more difficult when NE5 is added to NE1–4: the possibility that the descriptive adequacy of naturalistic epistemology is bought for the price of its normative authority looms rather large. On the other hand, once 'normativity, is understood as an essentially 'social' concept, the addition of NE5 promises to provide the very tools needed for its naturalistic reconstitution; whether relativism could be held at bay sufficiently remains an open question.

NE6, and it suffices that it be noted what problems loom on Neurath's horizon.

To see that a Neurathian de-metaphysicalised epistemology would also recognise the existence of external, non-cognitive influences and constraints on the pursuit of science, consider his following remarks about theory acceptance:

> We select one of the systems of statements that are in competition with each other. The system thus selected is not, however, logically distinguished. . . . The practice of living reduces the multiplicity quickly. . . . Furthermore we are restrained by the views of our environment. [1934, p. 105.]
>
> And how soon one senses the weakening effect of isolation. Thus one deserts the lonely, though perhaps auspicious, notions of an outsider to join in the work in a way of thought that enjoys more support and has therefore better chances of greater scientific achievement . . . not to speak of the cases in which certain trains of thought are anathema, persecuted and suppressed. [1935a, p. 117.]

Now these are hardly earth-shaking pronouncements indeed they strike me as rather common-sensical. (I shall not here explore Neurath's own stronger extension of NE5.[26].) Nonetheless, they are not without consequence. Russell once ridiculed Neurath's views as allowing that "empirical truth can be determined by the police". (1940, p. 186.) Take away the misinterpretation of Neurath as a truth theorist and substitute his concern with the acceptance of knowledge claims and it is clear that the denial of the alleged absurdity is itself empirically inadequate. Neurath's dry remark in exile: "Scholars, like other human beings, fear their neighbors, the police, and other people" (1944, p. 43) need not be amplified.

But what do such external constraints have to do with epistemology? it may be asked. Neurath's answer:

> Who knows the truth? Perhaps the imagined 'solitary thinker of absolute constancy of personality' [imagined by Russell; T.E.U.]? I do not think that we can describe a fight between 'error and truth', but only between different groups of thinkers. [1941, p. 229.]

It need not be stressed that Neurath did not mean to give *carte blanche* to unrestrained relativism here, given his strenuous efforts over the years to pit the 'scientific world-conception' against metaphysics and defend it against 'pseudo-rationalist' misuse. What cannot be stressed enough, however, is that Neurath's conception of knowledge, whether common sense or science, was one of communal achievement: there existed no isolated epistemic subjects, nay, not even language users.[27] "Our proposals lead to history and sociology of

---

[26]"Though today we can observe a growth of metaphysics there are also many indications that doctrines free from metaphysics are also spreading and are gaining more and more ground as a new 'superstructure' of the changing economic 'substructure' of our time". (Neurath 1932a, p. 90.)
[27]This touches on the question of Neurath's arguments for his programme, to be explored elsewhere.

the sciences and to a stressing of the social implications of language". (1941, p. 229.) If knowledge is viewed as a social enterprise, then it follows that a theory of knowledge must reflect this and, as a naturalistic theory, take recourse to the social sciences.

So Neurath's naturalistic epistemology also anticipated the socio-historical turn of the post-'received view' philosophy of science. What makes this particular turn of his, unlike that of some of his successors, a naturalistic one? On this central point, note Neurath's equation of "naturalistic" and "scientific" (1931c, p. 361): it is Neurath's adherence to the 'explanatory austerity demand' (NE3). External influences on hypothesis formation and theory choice have to be established in empirically respectable terms: Weberian *Zeitgeists* are out. But this of course also raises the question what precisely Neurath's understanding of the 'unity of science' amounted to. The implementation of Neurath's naturalistic epistemology must find a path between the reductionism associated with the 'received view' and the traditional understanding of 'interpretative' sociology.

Did Neurath's epistemology then remain purely descriptive? Clearly, taking account of extra-cognitive factors on knowledge claim acceptance does not mean that all of them must be accepted as legitimate influences. Yet neither does, on the other hand, mere observance of the rules of the scientific language (e.g. the testability of its statements) make for good science. Aware of this, Neurath noted: "Books on racialism may be written in plain empiricist language but nevertheless lack a critical attitude." (1944, p. 19.)[28] To encourage a critical attitude was precisely Neurath's intention in promoting the 'scientific world-conception'; thus he exhorted scientists to "take responsibility for their concepts" (1935c, p. 699). It may be noted, then, that in order to engage in the critique of metaphysics' — and so "serve life"[29] — Neurath's socio-historical naturalistic epistemology also required some semblance of normative authority (NE6).

6.

My exposition has systematised and partly amplified the various insights of Haller, Rutte, Heidelberger and Koppelberg (in terminology not identical to the one used here and not always with the references cited here). The

[28]The "racialist" books in question are the numerous publications on "racial hygiene" circulating since the beginning of the twentieth century and officially approbated in Nazi Germany. (For an earlier comment in a related vein, with a different addressee, see also his (1932a), p. 89.)

[29]"The scientific world-conception serves life, and life receives it." (Carnap, Hahn, Neurath, 1929, p. 318.) On this intended 'intervention' of theory in practice, see Nemeth (1981).

remarkable word from the Continent thus stands confirmed.[30] My exposition makes plain that — for all the charity they need — Neurath's own various *aperçus* are not mere throw away remarks that individually invite deep interpretation, but that their ensemble points to a systematic programme. What Neurath later on called "the behaviouristic of scholars" (1936b) amounts to naturalistic epistemology. Furthermore, Neurath's programme constitutes not only a very early, but also a very radical version of it, perhaps even an unusual one: the historical and sociological turn in the philosophy of science and epistemology is, in any case, not always associated with the programme of naturalism (nor *vice versa*). Reconstructing Neurath's proposals in the framework of a radical, extended naturalistic epistemology makes plain that he was no *mere* precursor of Quine on the one hand, and Kuhn and the 'anti-positivists' on the other: it invites us to join his ideas to contemporary debates.

My reconstruction also suggests part of the explanation which the long neglect of Neurath's proposals surely requires, even given the somewhat obscure manner in which they were put forward. What seems to have happened is that Neurath's original proposals not only had to struggle (mostly in vain) to make themselves heard amidst the din of the Vienna Circle party line propaganda with which Neurath chose to lace his articles, but that his proposals also did not 'pan out' into something easily recognisable when readers actually got hold of one of their strands. This holds for readers in overall sympathy the Vienna Circle philosophy. Yet others, potentially at least sympathetic to Neurath's own distinct programme, but not to that of the Circle at large, may not have bothered even to look, given his Vienna Circle

---

[30]Neurath's refusal to engage with the traditional epistemological project has of course been noted widely — but does this attitude amount to naturalistic epistemology? Hempel (1982), pp. 15–16, concluded that Neurath would have found such a programme congenial (and changed his mind from his (1935) reading of Neurath as a coherence theorist of truth to one of justification). This conclusion is sharpened into the attribution of its *actual anticipation* by Haller, Rutte and Koppelberg. Their attributions can be roughly coordinated to my expository schema as follows. Neurath's adherence to the primacy of practice thesis (NE1) was diagnosed by Haller (1979a, p. 29) as a "hypothetical naturalism", by Koppelberg (1987, p. 23) as naturalistic epistemology simpliciter. Neurath's adherence to the replacement demand (NE2) was diagnosed by Haller (1982b, p. 195) and Rutte (1982, p. 92) as Neurath's anticipation of Quinean natural epistemo-logy, by Rutte (1982b, p. 185) as his refusal to engage with the traditional problem of justification; Neurath's view of such traditional treatments as impossible and unnecessary is regarded by Koppelberg (1990, p. 203) as constituting his anticipation of naturalistic epistemology. Neurath's adherence to the principle of explanatory austerity (NE3) has not been diagnosed explicitly (though it is implicit in Haller 1982b, p. 195, and also follows once we regard his naturalism *vis-à-vis* scientific concepts in general (Koppelberg 1987, pp. 53–55) and social, mental and linguistic terminology in particular (Heidelberger, 1985, p. 182; on the latter also Koppelberg, 1990, p. 203, as also pertaining to epistemological notions). Rutte (1982a) and Koppelberg (1990), however, also read Neurath as a coherence theorist of truth; by contrast, Neurath's concern with the practice of theory acceptance (NE4) and the externalist consequences drawn therefrom (NE5) had previously been noted by Haller (1979b, pp. 35–36) (without explicitly associating it with naturalistic epistemology). The recognition of both NE4 and NE5 is also basic to Rutte (1982b) who, like Haller (1982b), investigated in what sense Neurath could have sustained the legitimation claim (NE6).

association. In two of the theorists whose positions Neurath anticipated we can, I believe, see this process at work.

First I return once more to the objection that Quine did announce Neurath as his naturalistic precursor — in so many words. We saw that the very protocol sentence debate which Quine pointed to as a morass of confusion in fact provided the forum for Neurath's proposal that epistemology be naturalized. Still, the objector might hold that Quine portrayed Neurath as a naturalistic epistemologist in another passage — one in the light of which all his other invocations of the parable of the mariners would also amount to just such an attribution:

> We are after an understanding of science as an institution or process in the world, and we do not intend that understanding to be any better than the science which is its object. This attitude is indeed one that Neurath was already urging in Vienna Circle days, with his parable of the mariner . . . [1969a, p. 84.]

But is far from clear that Quine here credits Neurath with the naturalisation of epistemology: he credits him with an *attitude*, not a programme. In the paragraph at issue, Quine was concerned to defend his acquiescence in the kind of circularity which according to traditional epistemology could only be avoided by recurring to certain foundations: how could, so this objection would go, scientific knowledge be justified with reference to the results of scientific inquiry? Neurath, according to Quine in my reading, simply urged the anti-foundationalist rejection of the dream of a certain base and the adoption, in its stead, of a coherentist strategy of justification. (After all, is it not this bravely cheerful circularity which the boat simile depicts most prominently?[31]) The scientistic attitude so applauded by Quine apparently yielded as yet only parables.

Once again it is confirmed that Quine did not credit Neurath with the proposal to naturalise epistemology.[32] The attitude which Quine credited to Neurath amounts to crediting him with the primacy of practice claim (NE1), to be sure, but not with the principle of explanatory austerity. Without the latter, however, so brilliantly developed and deployed by Quine himself, concern with scientific practice is hardly a distinctively naturalistic doctrine. Quine then is a

---

[31]Thus Koppelberg (1987), p. 298–299, takes Neurath's "fallibilistic holism" to be the meaning of the Boat and declares this to be Neurath's most important influence on Quine.

[32]Compare also Quine's remark that "The Vienna Circle espoused the verification theory of meaning but did not take it seriously enough. If we recognize with Pierce that the meaning of a sentence turns purely on what would count as evidence for its truth, and if we recognise with Duhem that theoretical sentences have their evidence not as single sentences but only as larger blocks of theory, then the indeterminacy of translation of theoretical sentences is the natural conclusion. And most sentences, apart from observation sentences are theoretical". (Quine, 1969a, pp. 79–80.) For Neurath's endorsement of the so-called 'Duhem-Quine thesis' as of 1913, redubbed the 'Neurath principle', see Haller (1982c). For Neurath endorsement of Duhem's conclusions and their extension to all of science still three years earlier, see Neurath (1910), pp. 44–45.

case of recognising some, but hardly all of Neurath's non-standard Vienna Circle philosophy.[33]

Next I turn to consider why the anticipation of the socio-historical turn of metascientific thought by Neurath was not noticed as such by the theorists we nowadays mostly associate with it, for example Kuhn. As in the case of Quine, I must warn against the lure of conspiracy theses — and stress that I intend no reproach, nor claim that Neurath anticipated all of Kuhn's arguments.

Kuhn's *Structure of Scientific Revolutions* was published as one of the last instalments of the *International Encyclopedia of Unified Science*. It is common to find it considered a 'marvellous irony' that the book which gave such impetus to the overcoming of the 'received view,' should have been published in this series so closely associated with logical empiricism — indeed, in which the 'received view' found a first canonical expression (Carnap 1939, pt III). What is overlooked in such a judgement — quite apart from obvious consent of the surviving editors Carnap and Morris — is that the founder of this series, Otto Neurath, would not at all have found it ironic, but only fitting.[34] However monumental his plans for it, Neurath never intended the *Encyclopedia* to provide an orthodox party line.[35] Moreover, the very historicism 'introduced' by Kuhn was nothing new to Neurath. Consider the opening sentence of a paper of his published in 1915:

> History of science, if seen as more than a chronicle of findings and biographies, is a young discipline. It can aim much higher: like the history of any field of enquiry it may try to shed light on the psychology of the enquirer; besides, it may exhibit the logical structure of theories and from it derive how they may develop. [1915, p. 101.]

The reader may compare the first sentence of *Structure of Scientific Revolutions*;[36] quite clearly, Neurath would only have applauded Kuhn's enterprise.

So was there a direct influence? Not if we believe Kuhn, and there is no reason not to believe him. Important for the development of his thought was the work of Ludwik Fleck (Kuhn 1962, pp. vi–vii; 1979). When we turn to

[33]In a conversation on 13 May 1987, Quine recalled reading only Neurath (1931d) and (1932c) during his 1932–33 visit to Vienna, Prague and Warsaw; copies of these were given to him by Carnap (cf. his letter to Carnap, 26 November 1935, also acknowledging receipt of some further unnamed offprints (Creath, 1991a, p. 192)). Having "come around to naturalism" Neurath's metaphor provided a welcome expository device, but it did not precipitate Quine's naturalism.

[34]From the beginning of its publication in 1937 onwards, a volume on the 'sociology of science' was planned, with varying authors over the years: up to the late '40s Louis Wirth, up to the mid-'50s I. B. Cohen, until Kuhn was commissioned (cf. the series announcements on the inside back covers of the various impressions of the single monographs). On the history of the *Encyclopedia* in general see Morris (1960).

[35]Indeed Neurath (1944), his own volume in this series, did not exactly sit well with the then emerging post-Vienna orthodoxy and only thinly veiled his dissent from the developments Carnap's thought had taken (e.g. note 19). Carnap in turn insisted on a note that absolved him from co-'editorial responsibility for this monograph' (p. ii)!

[36]I thank Paul Hoyningen-Huene for pointing out this parallel when I cited the Neurath passage in a discussion of his early metascientific thought.

Fleck's long-neglected book we find that he discussed the Vienna Circle at two points. Once he remarked that its members construed thinking as "something fixed and absolute", but an empirical fact as "relative" (Fleck 1935, p. 50); later on, he stated that

> [i]t is all but impossible to make any protocol statements based on direct observation and from which the results should follow as logical conclusion. . . . Observation without assumption, which psychologically is nonsense and logically a game, can be dismissed. [1935, p. 89, p. 92.]

In the attached footnote Fleck briefly reviewed Carnap's retreat on this position from the *Aufbau* via (what in English became) the *Unity of Science* to 'On Protocol Sentences', and notes about Carnap's position in the latter — "according to which the absolute character of protocol statements is already rejected" — that

> one would hope that eventually he might discover the social conditioning of thought. This would liberate him from any absolutism in the standards of thought, but of course he would also have to renounce the concept of 'unified science'. [1935, p. 177 (ch. 4, fn. 3)]

Of course, it was criticism by Neurath (not mentioned by Fleck) which prompted Carnap's 'On Protocol Sentences'; but, given this review, who, enamoured of Fleck's position, would study the debate referred to? It would only be natural to identify with the "thought collective" represented by Fleck, count his opponents as one's own — and leave the Vienna Circle alone. Kuhn then is a case of not recognising Neurath's anticipation for reasons of mistaking his views from a distance as uncongenial.[37]

7.

I have argued that Neurath anticipated naturalistic epistemology and the socio-historical turn of post-positivist analytical philosophy. This claim is not without interest for historians. I have also argued that his anticipation went virtually unnoticed. But is it not the case that "unappreciated precursors are of antiquarian interest only" (Hull 1988, p. 72)? It is true in the sciences if the cognitive content of their theories has been exhausted by posterity. That, however, seems not to be the case with Neurath. If my reconstruction of Neurath's programme of 'scientific philosophy' is correct, he proposed a type

---

[37]The case of Feyerabend, whose dissertation concerned Popper's 'basic sentences,' may be more complicated: in his papers from the early 1960s Feyerabend portrays Neurath together with Carnap as holding the 'pragmatic theory of observation' (for criticism of this reading of Carnap see Oberdan (1990)); by 1978, in any case, Feyerabend lauded Neurath as the only neo-positivist who "had a clear conception of the properties of scientific research (as opposed to philosophical analysis)" (1978, p. 86).

of naturalism that has only very recently become topical. We may still have to learn from his mistakes, if nothing else.

Moreover, in a discipline like philosophy that rather thrives on self-reflection (and in this sense largely manufactures its data for its own consumption), history and historical figures play perhaps a somewhat larger role than in the sciences, even in 'scientific philosophy'. In this respect, the claim investigated and supported here raises new and interesting questions about the Viennese prehistory of analytic philosophy and about the self-image of post-positivist analytical philosophy, complementary to those raised by the rediscovery of the 'analytic neo-Kantian' or other variants of the non-foundationalists Carnap and Schlick.[38]

Either of these benefits of the Neurath rediscovery are contingent, of course, on also understanding Neurath's motivation and argument for his programme and his implementation of it. I have not advanced these latter tasks here, but I hope I have shown some of the interest that lies in them.[39]

References

Antony, L. M. (1987) 'Naturalistic Epistemology and the Study of Language', in A. Shimony and D. Nails (eds), *Naturalistic Epistemology* (Dordrecht: Reidel), pp. 235–257.

Ayer, A. J. (ed.) (1959) *Logical Positivism* (New York: Free Press).

Barrett, R. B. and Gibson, R. F. (eds) (1990) *Perspectives on Quine* (Oxford: Blackwell).

BonJour, L. (1985) *The Structure of Empirical Knowledge* (Cambridge, Mass.: Harvard).

Boyd, R. (1981) 'Scientific Realism and Naturalistic Epistemology', in *PSA 1980 vol. 2* (East Lansing: Philosophy of Science Association), pp. 613–662.

Brown, H. I. (1989) 'Normative Epistemology and Naturalized Epistemology', *Inquiry* **31**, pp. 53–78.

Carnap, R. (1930) 'Die alte und die neue Logik', *Erkenntnis* **1**, pp. 12–26; transl. I. Levi, 'The Old and the New Logic', in Ayer (1959), pp. 133–146.

Carnap, R. (1932) Überwindung der Metaphysik durch logische Analyse der Sprache', *Erkenntnis* **2**, pp. 219–241; transl. A. Pap, 'The Elimination of Metaphysics through Logical Analysis of Language', in Ayer (1959), pp. 60–81.

[38]For non-traditionalist dissent from the Carnap interpretation given by the authors mentioned in note 1 from the perspective of conventionalism, see Ryckman (1991); also Creath (1990), (1991b); from that of philosophical minimalism, see Dreben (1991). Note that, if Neurath is read as a *naturalist*, his dispute with Carnap (*whichever* of these 'non-foundationalist readings of the latter we adopt) and Schlick (for 'non-foundationalist' readings see e.g. Daum (1982), Friedman (1983), Haller (1985), Lewis (1991), Sheldon (1989), Zhai (1990)) becomes decidedly 'post-modern'.

[39]The preparation of this paper was supported by a Postdoctoral Fellowship in History and Philosophy of Science at Northwestern University. My understanding of Neurath was greatly helped by numerous discussions with Sylvian Bromberger, Robert S. Cohen, James Higginbotham and Paul Horwich — and the study of the authors discussed in this paper. I also profited from an instructive interview with W. V. O. Quine. Helpful comments by Arthur Fine, Michael Friedman, David L. Hull, Thomas S. Kuhn and Thomas A. Ryckman improved the present paper. I thank them all.

Carnap, R. (1934/7) *Die logische Syntax der Sprache* (Springer, 1934); transl. of rev edn A. Smeaton, *The Logical Syntax of Language* (London: Kegan, Paul, Trench Trubner & Co, 1937; repr. Patterson, NJ: Littlefield, Adams & Co., 1959).

Carnap, R. (1936/7) 'Testability and Meaning', *Philosophy of Science* **3**, pp. 419–471; **4**, pp. 1–40.

Carnap, R. (1936/49) 'Wahrheit und Bewährung', in *Actes du Congrès International de Philosophie Scientifique, Sorbonne, Paris, 1935, Facs. IV: Induction et Probabilité* (Paris: Hermann & Cie), pp. 60–70; partial transl. H. Feigl, 'Truth and Confirmation', in H. Feigl and W. Sellars (eds), *Readings in Philosophical Analysis* (New York: Appleton-Century-Crofts, 1949), pp. 119–127.

Carnap, R. (1938) 'Logical Foundations of the Unity of Science', in International Encyclopedia of Unified Science, vol. 1, no. 1 (Chicago: University of Chicago Press).

Carnap, R. (1939) *Foundations of Logic and Mathematics*, International Encyclopedia of Unified Science, vol. 1, no. 3 (Chicago: University of Chicago Press).

Carnap, R. (1961) 'Vorwort zur zweiten Auflage', in his *Der logische Aufbau der Welt: Scheinprobleme in der Philosophie* (Hamburg: Meiner); transl. R. A. George, 'Preface to 2nd edition', in *Logical Structure of the World: Pseudo-problems of Philosophy* (Berkeley: University of California Press, 1967), pp. v.–xi.

Carnap, R., Hahn, H. and Neurath, O. (1929) *Wissenschaftliche Weltanschauung: Der Wiener Kreis* (Wien: Gerold); transl. 'The Scientific World-Conception: The Vienna Circle', in Neurath (1973), pp. 299–318.

Coffa, A. (1977) 'Carnap's *Sprachanschauung* ca. 1932', *PSA 1976 vol. 2* (East Lansing: Philosophy of Science Association), pp. 205–241.

Coffa, A. (1984) 'Logical Positivism, the Semantic Tradition and the A Priori', *Fundamenta Scientiae* **5**, pp. 237–254.

Coffa, A. (1985) 'Idealism and the Aufbau', in N. Rescher (ed.), *The Heritage of Logical Positivism* (Lanham: University Press of America), pp. 133–156.

Coffa, A. (1986) 'From Geometry to Tolerance: Sources of Conventionalism in Nineteenth-Century Geometry', in R. G. Colodny (ed.), *From Quarks to Quasars* (Pittsburgh: University of Pittsburgh Press), pp. 3–70.

Coffa, A. (1987) 'Carnap, Tarski and the Search for Truth', *Nous* **21**, pp. 547–572.

Creath, R. (1990) 'Carnap, Quine and the Rejection of Intuition', in Barrett and Gibson (1990), pp. 55–66.

Creath, R. (ed.) (1991a) *Dear Carnap, Dear Van: The Quine-Carnap Correspondence and Related Work* (Berkeley/Los Angeles: University of California Press).

Creath, R. (1991b) 'Contribution to the Carnap Centenary', in Fine, Forbes, Wessels (1990–91), vol. 2.

Dahms, H.-J. (ed) (1985) *Philosophie, Wissenschaft, Aufklärung* (Berlin: de Gruyter).

Daum, A. (1982) 'Schlick's Empiricist Critical Realism', *Synthese* **52**, pp. 449–493.

Dreben, B. (1991) 'Contribution to the Carnap Centenary', in Fine, Forbes, Wessels (1990–91), vol. 2.

Feigl, H. (1939) 'Unity of Science and Unitary Science', *Journal of Unified Science (Erkenntnis)* **9**, repr. in H. Feigl and M. Brodbeck (eds), *Readings in the Philosophy of Science* (New York: Appleton-Century-Crofts, 1953), pp. 382–384.

Feyerabend, P. K. (1978) 'From Incompetent Professionalism to Professional Incompetence', *Philosophy of Social Science* **8**, pp. 37ff.; partly repr. as 'Philosophy of Science versus Scientific Practice: Observations on Mach, his Followers and Opponents', in his *Philosophical Papers vol. 2: Problems of Empiricism* (Cambridge: Cambridge University Press, 1981).

Fine, A. (1984) 'The Natural Ontological Attitude', in J. Leplin (ed.), *Scientific Realism* (Berkeley: University of California Press), pp. 83–107; repr. in Fine, *The Shaky Game* (Chicago: Chicago University Press, 1986), pp. 112–135.

Fine, A., Forbes, M. and Wessels, L. (eds) (1990–91) *PSA 1990*, 2 vols (East Lansing: Philosophy of Science Association).

Fleck, L., (1935) *Entstehung und Entwicklung einer wissenschaftlichen Tatsache: Einführung in die Lehre vom Denkstil und Denkkollektiv.* (Basel: Schwabe & Co.); transl. F. Bradley and T. J. Trenn, ed. T. J. Trenn and R. K. Merton, *Development of a Scientific Fact* (Chicago: University of Chicago Press, 1979).

Friedman, M. (1983) 'Critical Notice: Moritz Schlick, Philosophical Papers', *Philosophy of Science* **50**, pp. 498–514.

Friedman, M. (1987) 'Carnap's *Aufbau* Reconsidered', *Nous* **21**, pp 521–545.

Friedman, M. (1988) 'Logical Truth and Analyticity in Carnap's *Logical Syntax of Language'*, in W. Aspray and P. Kitcher (eds), *Essays in the History and Philosophy of Mathematics* (Minneapolis: University of Minnesota Press).

Friedman, M. (1991) 'Contribution to the Carnap Centenary', in Fine, Forbes, Wessels (1990–91), vol. 2.

Giere, R. N. (1985) 'Philosophy of Science Naturalized', *Philosophy of Science* **52**, pp. 331–357.

Haller, R. (1979a) 'Über Otto Neurath', in his *Studien zur Österreichischen Philosophie* (Rodopi), pp. 99–106; transl. T. E. Uebel, 'On Otto Neurath', in Uebel (1991), pp. 25–32.

Haller, R. (1979b) 'Geschichte und wissenschaftliches System bei Otto Neurath', in Berghel, Hübner and Köhler (eds), *Wittgenstein, der Wiener Kreis und der Kritische Rationalismus* (Wien: Hölder-Pichler-Tempsky), pp. 302–307; transl. T. E. Uebel, 'History and System of Science in Otto Neurath', in Uebel (1991), pp. 33–40.

Haller, R. (ed.) (1982a) *Schlick und Neurath: Ein Symposion*, Grazer Philosophische Studien vol 16/17 (Amsterdam: Rodopi).

Haller, R. (1982b) 'Zwei Arten der Erfahrungsbegründung', in Haller (1982a), pp. 19–33; transl. T. E. Uebel, 'Two Ways of Experiential Justification', in Uebel (1991), pp. 191–203.

Haller, R. (1982c) 'Das Neurath–Prinzip: Grundlagen und Folgerungen', in Stadler (1982), pp. 79–87, and in Dahms (1985), pp. 205–220, pp. 108–124; transl. T. E. Uebel, 'The Neurath Principle: Its Grounds and Consequences', in Uebel (1991), pp. 117–130.

Haller, R. (1985) 'Problems of Knowledge in Moritz Schlick', *Synthese* **64**, pp. 283–296.

Hegselmann, R. (1979) 'Otto Neurath, empiristischer Aufklärer und zozialreformer', in Otto Neurath, *Wissenschaftliche Weltauffassung, Sozialismus und Logischer Empirismus*, ed. R. Hegselmann (Frankfurt: Suhrkamp).

Hegselmann, R. (1985) 'Die Korrespondenz zwischen Otto Neurath und Rudolf Carnap aus den Jahren 1934–1945', in Dahms (1985), pp. 276–290.

Heidelberger, M. (1985) 'Zerspaltung und Einheit: vom logischen Aufbau der Welt zum Physikalismus', in Dahms (1985), pp. 144–189.

Hempel, C. G. (1935) 'On the Logical Positivists' Theory of Truth', *Analysis* **2**, pp. 49–59.

Hempel, C. G. (1982) 'Schlick und Neurath: Fundierung vs. Kohaerenz in der wissenschaftlichen Erkenntnis', in Haller (1982a), pp. 1–18.

Hofmann-Grüneberg, F. (1988) *Radikal-empiristische Wahrheitstheorie. Eine Studie über Otto Neurath, den Wiener Kreis und das Wahrheitsproblem* (Wien: Hölder-Pichler-Tempsky).

Hoyningen-Huene, P. (1989) *Die Wissenschaftsphilosophie Thomas S. Kuhns* (Braunschweig: Vieweg).

Hull, D. L. (1988) *The Process of Science* (Chicago: University of Chicago Press).

Kim, J. (1988) 'What Is "Naturalized Epistemology"?', in J. Tomberlin (ed.) *Philosophical Perspectives: 2, Epistemology* (Atascadero, Cal.: Ridgeview), pp. 381–405.

Koppelberg, D. (1987) *Die Aufhebung der analytischen Philosophie* (Frankfurt: Suhrkamp).

Koppelberg, D. (1990) 'Why and How to Naturalize Epistemology', in Barrett and Gibson (1990), pp. 200–211.

Kornblith, H. (1985a) 'Introducton: What is Naturalistic Epistemology', in Kornblith (1985b), pp. 1–14.

Kornblith, H. (ed.) (1985b) *Naturalizing Epistemology* (Cambridge, Mass.: MIT Press).

Kuhn, T. S. (1962) *The Structure of Scientific Revolutions*, International Encyclopedia of Unified Science, vol. 2, no. 2; 2nd edn. (1970, Chicago: University of Chicago Press).

Kuhn, T. S. (1979) 'Forword', in transl. of Fleck (1935), pp. vii–xi.

Kuhn, T. S. (1991) 'Presidential Address', in Fine, Forbes, Wessels (1990–91), vol. 2.

Laudan, L. (1987) 'Progress or Rationality? The Prospects for Normative Naturalism', *American Philosophical Quarterly* **24**, pp. 19–31.

Laudan, L. (1990) 'Normative Naturalism', *Philosophy of Science* **57**, pp. 44–59.

Lehrer, K. (1990) *Theory of Knowledge* (Boulder: Westview Press).

Lewis, J. (1991) 'Hidden Agendas: Knowledge and Verification', in Fine, Forbes, Wessels (1990–91), vol 2.

Morris, C. (1960) 'On the History of the International Encyclopedia of Unified Science', *Synthese* **12**, pp. 517–521; repr. in: Neurath (1973), pp. 64–68.

Nagel, E. and Brandt, R. B. (eds) (1965) *Meaning and Knowledge* (New York: Harcourt, Brace & Jovanovich).

Nemeth, E. (1981) *Otto Neurath und der Wiener Kreis: Wissenschaftlichkeit als revolutionärer politischer Anspruch* (Frankfurt: Campus).

Neurath, O. (1910) 'Zur Theorie der Sozialwissenschaften', *Jahrbuch für Gesetzgebung, Verwaltung und Volkswirtschaft im deutschen Reich* **34**, pp. 37–67; repr. in: Neurath (1981), pp. 23–46.

Neurath, O. (1915) 'Prinzipielles zur Geschichte der Optik', *Archive für die Geschichte der Naturwissenschaften und Technik* **5**, pp. 371–389; transl. 'On the Foundations of the History of Optics', in Neurath (1973), pp. 101–112.

Neurath, O. (1929) 'Bertrand Russell, der Sozialist', *Der Kampf* **22**, pp. 234–238; repr. in Neurath (1981), pp. 337–343.

Neurath, O. (1931a) 'B. Russell, Wissen und Wahn', *Der Kampf* **24**, pp. 187–189; repr. in Neurath (1981), pp. 397–400.

Neurath, O. (1931b) 'Physikalismus', *Scientia* (Nov. 1931), pp. 297–303; transl. 'Physicalism', in Neurath (1983), pp. 52–57.

Neurath, O. (1931c) *Empirische Soziologie: Der wissenschaftliche Gehalt der Geschichte und Nationalökonomie* (Wien: Springer); repr. in Neurath (1981), pp. 423–527; excerpts transl. 'Empirical Sociology', in Neurath (1973), pp. 391–421.

Neurath, O., (1931d) 'Physicalism: The Philosophy of the Vienna Circle', *The Monist* **41**, pp. 618–623; repr. in Neurath (1983), pp. 48–51.

Neurath, O. (1932a) 'Soziologie im Physikalismus', *Erkenntnis* **2**, pp. 393–431; transl. 'Sociology in the Framework of Physicalism', in Neurath (1983), pp. 58–90.

Neurath, O. (1932b) 'Socialbehaviourismus', *Sociologicus* **8**, pp. 281–288; repr. in Neurath (1983), pp. 563–570.

Neurath, O. (1932c) 'Protokollsätze', *Erkenntnis* **3**, pp. 204–214; transl. 'Protocol Statements', in Neurath (1983), pp. 91–99.

Neurath, O. (1934) 'Radikaler Physikalismus und "wirkliche Welt" ', *Erkenntnis* **4**, pp. 346–362; transl. 'Radical Physicalism and "the Real World" ', in Neurath (1983), pp. 100–114.

Neurath, O. (1935a) 'Einheit der Wissenschaft als Aufgabe', *Erkenntnis* **5**, pp. 16–22; transl. 'The Unity of Science as a Task', in Neurath (1983), pp. 115–120.

Neurath, O. (1935b) 'Pseudorationalismus der Falsifikation', *Erkenntnis* **5**, pp. 353–365; transl. 'Pseudorationalism of Falsification', in Neurath (1983), pp. 121–131.

Neurath, O. (1935c) '1. Internationaler Kongress für Einheit der Wissenschaft in Paris 1935', *Erkenntnis* **5**, pp. 377–406; repr. in Neurath (1981), pp. 649–672.

Neurath, O. (1935d) 'Zur Induktionsfrage', *Erkenntnis* **5**, pp. 173–174; repr. in Neurath (1981), pp. 631–632.

Neurath, O. (1936a) *Le devéloppement du Cercle de Vienne et l'avenir de l'Empiréisme logique* (Paris: Hermann & Cie); transl. B. Treschmitzer and H. G. Zilian, 'Die Entwicklung des Wiener Kreises und die Zukunft des logischen Empirismus', in Neurath (1981), pp. 637–703.

Neurath, O. (1936b) 'Physikalismus und Erkenntnisforschung', *Theoria* **2**, pp. 97–105, pp. 234–237; transl. 'Physicalism and the Investigation of Knowledge', in Neurath (1983), pp. 159–167.

Neurath, O. (1941) 'Universal Jargon and Terminology', *Proceedings of the Aristotelian Society* n.s. **41**, pp. 127–148; repr. in Neurath (1983), pp. 213–229.

Neurath, O. (1944) *Foundations of the Social Sciences* International Encyclopedia of Unified Science, vol. 2, no. 1 (Chicago: University of Chicago Press).

Neurath, O. (1973) *Empiricism and Sociology*, eds M. Neurath and R. S. Cohen, transl. P. Foulkes and M. Neurath (Dordrecht: Reidel).

Neurath, O. (1981) *Gesammelte philosophische und methodologische Schriften*, eds R. Haller and H. Rutte (Wien: Hölder-Pichler-Tempsky).

Neurath, O. (1983) *Philosophical Papers 1913–1946*, eds and transl. R. S. Cohen and M. Neurath (Dordrecht: Reidel).

Oberdan, T. (1990) 'Positivism and the Theory of Observation', in Fine, Forbes, Wessels (1990–91), vol. 1, pp. 25–37.

Proust, J. (1986) *Questions de Forme: Logique et proposition analytique de Kant à Carnap* (Paris: Fayard), transl. A. A. Brenner, *Questions of Form: Logic and the Analytic Proposition from Kant to Carnap* (Minneapolis: University of Minnesota Press, 1989).

Proust, J. (1987) 'Formal Logic as Transcendental in Wittgenstein and Carnap', *Nous* **21**, pp. 501–520.

Putnam, H. (1982) 'Why Reason Can't be Naturalized', *Synthese* **52**, pp. 3–24; repr. in Putnam, *Realism and Reason: Philosophical Papers, vol. 3* (Cambridge: Cambridge University Press, 1983), pp. 229–247.

Quine, W. V. O. (1950) 'Identity, Ostension, and Hypostasis', *Journal of Philosophy* **47**, pp. 621–633; repr. in Quine (1953), pp. 65–79.

Quine, W. V. O. (1951) 'Two Dogmas of Empiricism', *Philosophical Review* **60**, pp. 20–43; repr. in Quine (1953), pp. 20–46.

Quine, W. V. O. (1953) *From a Logical Point of View*, rev. edn 1980 (Cambridge, Mass.: Harvard University Press).

Quine, W. V. O. (1969a) 'Epistemology Naturalized', in Quine (1969c), pp. 69–90.

Quine, W. V. O. (1969b) 'Natural Kinds', in N. Rescher et al. (eds), *Essays in Honor Carl G. Hempel* (Dordrecht: Reidel), pp. 5–23; repr. in Quine (1969c), pp. 114–138.

Quine, W. V. O. (1969c) *Ontological Relativity and Other Essays* (New York: Columbia University Press).

Quine, W. V. O. (1970) *Philosophy of Logic* (Englewood Cliffs: Prentice-Hall; 2nd edn, Cambridge, Mass.: Harvard University Press, 1986).

Quine, W. V. O. (1975) 'The Nature of Natural Knowledge', in S. Guttenplan (ed.), *Mind and Language* (Oxford: Oxford University Press), pp. 67–82.

Quine, W. V. O. (1990a) 'Comment on Koppelberg', in Barrett and Gibson (1990), p. 212.

Quine, W. V. O. (1990b) *Pursuit of Truth* (Cambridge, Mass.: Harvard University Press).

Richardson, A. (1990) 'How not to Russell Carnap's Aufbau', in Fine, Forbes, Wessels (1990–91), vol. 1, pp. 3–14.

Roth, P. A. (1983) 'Siegel on Naturalistic Epistemology and Natural Science', *Philosophy of Science* **50**, pp. 482–493.

Roth, P. A. (1991) 'Doing What Comes Naturalistically', forthcoming.

Russell, B. (1940) *An Inquiry into Meaning and Truth* (London: Allen & Unwin).

Rutte, H. (1982a) 'Der Philosoph Otto Neurath', in Stadler (1982), pp. 70–78; transl. T. E. Uebel, 'The Philosopher Otto Neurath', in Uebel (1991), pp. 81–94.

Rutte, H. (1982b) 'Über Neuraths Empirismus und seine Kritik am Empirismus', in

Haller (1982a), pp. 366–384; transl. T. E. Uebel, 'On Neurath's Empiricism and his Critique of Empiricism', in Uebel (1991), pp. 175–190.

Ryckman, T. A. (1991) 'Coordination and Convention: A Chapter of Early Logical Empiricism', in Fine, Forbes, Wessels (1990–91), vol. 2.

Sauer, W. (1985) 'Carnaps Aufbau in kantischer Sicht', *Grazer Philosophische Studien* **23**, pp. 19–35.

Schlick, M. (1930) 'Die Wende in der Philosophie', *Erkenntnis* **1**, pp. 4–11; transl. D. Rynin, 'The Turning Point in Philosophy', in Ayer (1959), pp 53–59.

Schlick, M. (1934) 'Über das Fundament der Erkenntnis', *Erkenntnis* **4**, pp. 79–99; transl. D. Rynin, 'The Foundation of Knowledge', in Ayer (1959), pp. 209–227.

Sheldon, J. (1989) 'Schlick's Theory of Knowledge', *Synthese* **79**, pp. 305–317.

Siegel, H. (1980) 'Justification, Discovery and the Naturalizing of Epistemology', *Philosophy of Science* **47**, pp. 297–321.

Siegel, H. (1984) 'Empirical Psychology, Naturalized Epistemology, and First Philosophy', *Philosophy of Science* **51**, pp. 667–676.

Stadler, F. ed. (1982) *Arbeiterbildung in der Zwischenkriegszeit: Otto Neurath–Gerd Arntz* (Wien: Österreichisches Gesellschafts- und Wirtschaftsmuseum).

Stroud, B. (1981) 'The Significance of Naturalized Epistemology,' in *Midwest Studies in Philosophy, vol. IV* (Minneapolis: University of Minnesota Press), pp. 455–471; repr. in Kornblith (1985b), pp. 71–90.

Uebel, T. E. (ed.) (1991) *Rediscovering the Forgotten Vienna Circle: Austrian Studies on Otto Neurath and the Vienna Circle*, Boston Studies in the Philosophy of Science (Dordrecht: Kluwer).

Zhai, Z. (1990) 'The Problem of Protocol Statements and Schlick's Concept of "Konstatierungen" ', in Fine, Forbes, Wessels (1990–91), vol. 1, pp. 15–23.

Zolo, D. (1986/9) *Scienza e politica in Otto Neurath. Una prospettiva post-empiristica* (Milan: Feltrinelli, 1986); rev. edn transl. D. McKie, *Reflexive Epistemology: The Philosophical Legacy of Otto Neurath*, Boston Studies in the Philosophy of Science, vol. 118 (Dordrecht: Kluwer, 1989).

WESLEY C. SALMON

HANS REICHENBACH'S VINDICATION
OF INDUCTION

Hans Reichenbach believed that he had solved Hume's problem of the justification of induction, but his arguments have not proved persuasive to most other philosophers. The majority of those who addressed the problem held, for one reason or another, that it is a pseudo-problem. In a number of articles during the 1950s and 1960s I tried to refute this position.[1] I still believe it is incorrect – that the problem of justification of induction is a genuine and profoundly important problem – but I shall not rehearse that issue here. In this article I shall first discuss Reichenbach's justification and the problems confronting it. I shall then consider other attempts at vindication that in one way or another pursue a similar goal. In the end, I shall maintain, Reichenbach's program can succeed, given certain additional considerations, a crucial one of which is found in his own writings.

1. REICHENBACH'S JUSTIFICATION

One of the key steps in Reichenbach's solution to the problem of induction was his recognition that what is needed is the justification of a rule, not the proof of a factual proposition such as the uniformity of nature.[2] He realized, in addition, that it is impossible to justify the rule in question by proving that it will always, or even sometimes, yield true conclusions, given true premises. He argued, nevertheless, that his *rule of induction*[3] should be adopted because one has everything to gain and nothing to lose by employing it. Although successful prediction cannot be guaranteed, he argued, if any method works the rule of induction will work. The structure of his argument is rather similar to Pascal's wager; it seeks not to justify belief in a proposition but rather to justify a practice. For that reason, Reichenbach characterized his justification as *pragmatic*.

Herbert Feigl, who was strongly sympathetic to Reichenbach's approach, drew a basic distinction between two kinds of justification, namely, *validation* and *vindication* (1950). One validates a rule or proposition by deriving it from a more fundamental principle. In deductive

Erkenntnis **35**: 99–122, 1991.
© 1991 *Kluwer Academic Publishers. Printed in the Netherlands.*

TABLE I

	Nature is uniform	Nature is not uniform
Induction is used	Success	Failure
Induction is not used	Success or failure	Failure

logic, for example, the rule of conditional proof is validated by means of the deduction theorem, which shows that any conclusion deduced through use of conditional proof can be deduced without appeal to that rule. Obviously, the most fundamental rules cannot be validated. If they can be justified at all, it must be by vindication. One vindicates a rule by showing that its use is well suited to the achievement of some aim we have. The rules of propositional logic can be vindicated by showing them to be truth-preserving. Their use fulfills our desire to avoid deriving false conclusions from true premises.

Since Reichenbach's justification of induction does not consist in the derivation of the rule of induction from a more fundamental rule, his pragmatic justification qualifies as a vindication. Our goal is to make correct predictions of future events (or more generally, to make correct inferences from observed phenomena to as yet unobserved phenomena). He argues roughly as follows (see Table I).[4] As Hume has shown, we cannot know whether nature is uniform or not. If we are fortunate and nature is uniform then the use of induction will fulfill our goal. This does not mean that every prediction will be correct, but we will be successful on the whole. If we are unlucky and nature turns out not to be uniform we may fail miserably. Perhaps there will be a few lucky guesses, but overall we will suffer failure in our attempts at prediction.

Suppose, instead, that we do not use induction. This might happen in either of two ways. In the first place, we might simply refuse to make any inferences at all. This alternative obviously fails whether nature is uniform or not. Nothing ventured, nothing gained. In the second place, we might try some different method for making predictions, for example, making wild guesses, consulting a crystal gazer, or believing what is found in Chinese fortune cookies. If nature exhibits uniformities, any of these methods might work, but there is no guarantee of success. If nature is uniform, then, it seems clear that induction is the best method, for it is bound to work on the whole, whereas the others may or may not be successful.

But what if nature is not uniform? In that case, no method can yield consistent success, for the *consistent* success of any noninductive method would be a uniformity, contary to the hypothesis that nature is not uniform. Moreover, if such a uniformity did transpire, that regularity could be exploited inductively. If, for example, the crystal gazer achieves a good predictive record – consistently predicting the outcomes of future horse races – we could use induction to predict that such predictions will continue to be accurate. It would *not* be foolish to use such information for the placing of bets. Therefore, Reichenbach concludes, if nature is not uniform the user of inductive reasoning is no worse off than anyone who uses any other method. Consequently, we have everything to gain and nothing to lose by employing inductive methods.

Although the foregoing argument gives some of the flavor of Reichenbach's vindication, it suffers from excessive vagueness. How uniform must nature be to qualify as uniform? What sorts of uniformities are important? What exactly do we mean by "success" and "failure"? How are the various kinds of methods to be precisely characterized?

Reichenbach was well aware of these difficulties, and he offered a more precise version of the foregoing pragmatic argument (1949, Sec. 91). Whether he knew of Bishop Butler's famous aphorism or not, he concurred fully that *probability is the very guide of life*. The fundamental goal in terms of which induction is to be vindicated is the acquiring of knowledge of probabilities. Inasmuch as he advocated the limiting frequency interpretation of probability, the goal is the ascertainment of the values of limits of relative frequencies in potentially infinite sequences of events. Given any such sequence of events, and any attribute that can meaningfully be predicated of its members, we can attempt to ascertain its limiting frequency in that sequence. There is, of course, no a priori guarantee that the limit in question exists. If it does, in fact, exist, then nature is uniform in that respect. The probability constitutes a statistical regularity. As we try to establish the value of the probability we are concerned with precisely that uniformity. If the frequency with which the attribute occurs in the sequence does not converge to any limit, nature fails to be uniform in the pertinent respect.

In order to make the notion of using induction more precise, Reichenbach offers his *rule of induction*. which may be formulated as follows:

> If an observed initial section consisting of n members of a sequence of As contains m elements with the attribute B

TABLE II

	Sequence has a limit	Sequence has no limit
Induction is used	Success	Failure
Induction is not used	Success or failure	Failure

> POSIT THAT the limit of the relative frequency of B in A lies within the interval $m/n \pm \delta$.

It is to be understood that nothing is known about the probability of B within A beyond the observed frequency of B in the specified initial section of the sequence. Moreover, the rule is to be used repeatedly as larger and larger initial sections of the sequence are observed. The size of δ is determined by pragmatic aspects of the context – the degree of precision required in that situation. Looking at the problem in terms of these more precise concepts, we may offer a refined version of Table I (see Table II).

Given the precise formulation of the rule of induction, Reichenbach points out, it is an immediate consequence of the mathematical definition of the limit of a sequence that, if the limit exists, repeated application of that rule will lead sooner or later to posits that are accurate to any desired degree of approximation; moreover, further posits, based upon larger and larger observed initial sections of that sequence, will continue to be at least that accurate. Thus, he argues, it is an analytic truth that *if success in ascertaining a limit is possible his rule of induction will yield success.* If no limit exists, obviously no method will succeed. This is his argument to establish the first row of Table II.

Various objections have been raised against this part of Reichenbach's argument. Some authors have questioned the presumption that ascertainment of limits of relative frequencies can seriously be considered a goal of human inquiry.[5] The first question to ask, it seems to me, is whether knowledge of the objective physical probability relations that obtain in the world is an aim of our endeavor. With Bishop Butler's aphorism in mind, I think the answer must be affirmative. The next question concerns the appropriate interpretation of physical probabilities. Reichenbach, of course, adopted the limiting frequency interpretation. Nowadays the so-called propensity interpretation is more popular, but it has severe drawbacks, including the fact that it is not an admissible interpretation of the probability calculus.[6] Since the controversy over these interpretations is beyond the scope of this paper, I shall adopt Reichenbach's position for purposes of argument herein.

Another objection arises from the fact that, with respect to any given sequence, we cannot predict how far we must go before reaching posits that are accurate to any particular degree. Moreover, even if we have arrived at that point, we have no way of knowing that we are there. For all we can ever know, a sequence might not begin to converge until an initial section containing billions upon billions of members had elapsed. One can only hope that convergence occurs reasonably rapidly, for convergence that occurs too slowly for the benefit of human investigators is no better than complete lack of convergence. Reichenbach, and John Venn[7] before him, stated clearly that the infinite sequence is a mathematical idealization of a very large finite class, much as the Euclidean plane is an idealization of a large surface that is approximately flat. Reichenbach (1949, 347–48, 447–48) called specific attention to this issue by introducing the concept of the *practical limit*.

There are obviously many ways of attempting to ascertain the value of the limiting frequency of a sequence – assuming it exists – that may or may not yield approximately correct results. One could, for instance, write a lot of rational fractions between zero and one on slips of paper, place them in a hat, mix them well, and draw one out. One could then posit that the number drawn is within δ of the actual limit. Still assuming that the sequence has a limit, there is no proof that this method will not work, but there is no proof that it will. Another example is the *counter-inductive rule* – discussed, but not advocated, by Max Black (1954) – according to which one posits that the relative frequency of non-B in the observed sample, $(n - m)/n$, is approximately equal to the limiting frequency. In any case in which the limiting frequency is not near 1/2, the persistent use of this method is guaranteed to yield posits that do not approximate the limiting frequency. In my view, Reichenbach's convergence argument is sufficient to show the superiority of his rule of induction to either of these noninductive methods. The counter-inductive rule can, incidentally, be rejected on other grounds as well, for it yields radically incoherent sets of probability values.[8]

Unfortunately for Reichenbach's attempted pragmatic justification, there exists a nondenumerably infinite set of rules each of which shares

WESLEY C. SALMON

TABLE III

	Sequence has a limit	Sequence has no limit
Rule of induction used	Success	Failure
Other asymptotic rule used	Success	Failure
Nonasymptotic method used	Success or failure	Failure

the convergence property of his rule of induction. He was fully aware of this set; he called them *asymptotic rules*. These rules can be characterized as follows:

> If an observed initial section consisting of n members of a sequence of As contains m elements with the attribute B **POSIT THAT** the limit of the relative frequency of B in A lies within the interval $(m/n + c_n) \pm \delta$, where $c_n \to 0$ as $n \to \infty$.

We may think of c_n as a 'corrective term' that modifies the observed frequency for the sake of a 'better posit'. Because the sequence of posits endorsed by any rule of this type converges to the sequence of posits endorsed by the rule of induction, both sequences will converge to the actual limiting frequency provided that such a limit exists. The situation is shown in Table III.

In view of this circumstance, Reichenbach cannot claim that his rule of induction is the *only* rule that is guaranteed to succeed if any method can succeed; any of the asymptotic rules will succeed if success is possible. He realized, moreover, that it is impossible to show that his rule of induction will yield faster convergence than any others of the asymptotic rules.[9] If his rule of induction is to be justified, some adequate reason must be given to prefer it to the other asymptotic rules. He was fully aware of that fact, and he offered a justification that he considered sufficient. From the entire class of asymptotic rules, he said, we select the rule of induction on grounds of descriptive simplicity (1949, p. 447).

Reichenbach had distinguished two types of simplicity: inductive and descriptive (1938, Sec. 42). Suppose we have two hypotheses, one simpler, the other more complex, both of which are compatible with all of our observations up to the present. They are, however, factually distinct; indeed, they are mutually incompatible. Faced with a choice between them, we sometimes select the simpler because we believe it is more likely to be true. This is *inductive simplicity*. In such cases there is evidence, not in our possession as yet, to undermine at least one of them. They are *not* observationally equivalent.

Descriptive simplicity comes into play when we have two or more theories that *are* observationally equivalent. The most vivid example arises in his theory of space and geometry. He maintained that the physical space of our universe can be described equally adequately by Euclidean geometry augmented by a suitable set of universal forces or by a non-Euclidean geometry without universal forces. Given such a pair of descriptions, either both are true or both are false. There is no empirical or factual difference between them. In company with Einstein, he maintains, we choose the description that eschews universal forces, but it is a matter of aesthetics or intellectual economy. Truth or falsity does not enter into the choice.

Reichenbach claimed that, since all of the asymptotic rules, including his rule of induction, converge to the same limits in the long run, they are empirically equivalent. We are free to choose the rule of induction because it is the descriptively simplest rule in the set of asymptotic rules. It appears, however, that a serious lapse has occurred in his argument. We noted at the outset that he placed great emphasis upon the fact that what stands in need of justification is not a statement but, rather, a rule. Clearly, both inductive and descriptive simplicity apply only to selections among statements, not to selections among rules. If Reichenbach's claim about the equivalence of all asymptotic rules has any merit at all, it must refer to equivalence "in the long run". According to a famous aphorism of J. M. Keynes, *in the long run we will all be dead*. If we look at the asymptotic rules in terms of human application, they are as radically nonequivalent as any rules could be.

Reichenbach's attempt to vindicate his rule of induction cannot be considered successful.

2. SALMON'S ATTEMPT TO FILL THE GAP

When I realized that the counter-inductive rule runs into incoherence, it occurred to me that the same consideration imposed a serious constraint on Reichenbach's 'corrective term' c_n. Any asymptotic rule in which c_n is a function of n alone will lead to the same sort of incoherence (Salmon, 1956). After imposing a suitable coherence condition – which I first called *regularity*, but later referred to as *normalizing conditions*

– on asymptotic rules, I was able to show that, even after disqualifying the irregular asymptotic rules, an infinity remained. This collection is so broad that it contains rules to license any posit whatever regarding limiting frequencies. More explicitly, one may arbitrarily choose any positive integer n as the size of the observed sample (initial section) of the sequence A, any nonnegative integer $m \leq n$ to represent the number of elements of the sample having the property B, and any real number $p(0 \leq p \leq 1)$ as the value of the limit of the frequency of B in A. Then, there exists among the regular asymptotic rules some rule that directs one to posit p as the limit on the basis of the observed frequency m/n (Salmon, 1957a). If, for example, a million members of A have been observed, all of which have possessed the attribute B, there is an asymptotic rule that licenses the posit that the limiting frequency of B within A is zero. Although the set of regular asymptotic rules is convergent, it is *nonuniformly* convergent. That means that there is no finite integer N representing a sample size at which all of the regular asymptotic rules begin to converge. For purposes of human prediction, this set of rules is as divergent – as empirically nonequivalent – as it could possibly be. Descriptive simplicity is not a suitable criterion for making a selection from that class. If simplicity is to be invoked for purposes of justifying an inductive rule, it must be a different sort of simplicity. Since it would be applicable to rules, I suggest that it be called *methodological simplicity*. I shall return to that concept in Section 4.

Having noted the foregoing difficulty regarding regular asymptotic rules, and having taken into account the fact that Reichenbach's 'corrective term' c_n cannot be a function of n alone, I began looking at other variables on which it might depend. The search was facilitated by consideration of Carnap's continuum of inductive methods (1952). All of these methods, with the exception of the straight rule, are dependent upon the language in which the evidence and hypotheses are stated. It seemed to me at the time – and it still does – that our inductive rules should be invariant across languages. For example, a switch from metric units to English units should make no difference to the inductive relationship between the hypothesis and the experiment. Similarly, if a hypothesis can be articulated in German and in English, and if the evidence can also be described in both languages, then the degree to which the hypothesis is supported or undermined should be the same in both cases. It would be strange, indeed, if a native speaker of English – displeased with the outcome of an experiment with respect to a hypothesis – could find a more pleasing inductive result by translating the experimental outcome and the hypothesis into German.

In an attempt to bring these considerations to bear on the problem at hand, I imposed – in addition to the regularity requirement or normalizing conditions – a *requirement of linguistic invariance*. I then offered a rather tedious mathematical argument to show that these two requirements constrained c_n to be identically zero, thus establishing Reichenbach's rule of induction as the only acceptable rule (1963a). Regrettably – at least from my standpoint – the argument was flawed.[10] As Ian Hacking pointed out, I had failed to take into account the fact that we often know, not only the relative frequency of the attribute in the sample, but also the order in which the Bs occur within the sample. It is possible to construct asymptotic rules that violate neither of my requirements but are not identical with Reichenbach's rule of induction (Hacking, 1968, esp. pp. 57–59).

Hacking did more; he proved a general theorem, showing that the fulfillment of three conditions – (1) *consistency*, (2) *symmetry*, and (3) *invariance* – is necessary and sufficient to select the Reichenbach rule of induction. The first condition, consistency, is unproblematic from my standpoint. It is closely related to the normalizing conditions (regularity requirement). It is somewhat stronger, but it is demonstrably satisfied by limits of relative frequencies. The third condition, invariance, is related to my criterion of linguistic invariance, but is considerably stronger. It includes a condition of *statistical invariance* over and above linguistic invariance. In a commentary on Carnap's inductive logic, I had already argued for Hacking's stronger condition (1967). As we shall see, it still appears to be defensible.

Hacking's second condition, symmetry, is closely related to what personalists call *exchangeability*. It says, in effect, that for a given relative frequency in an initial section of a sequence, the posited value for the limiting frequency must be the same regardless of the order in which the members of the sample occur. I shall return to this requirement in Sections 6–7. For 25 years, it has seemed to me an insurmountable obstacle to the kind of vindication I had hoped to provide.

My attempt to vindicate Reichenbach's rule of induction cannot be considered successful.

3. SELLARS: INDUCTION AS VINDICATION

More than a quarter of a century ago, Wilfrid Sellars published "Induction as Vindication" (1964) in which he offered his resolution of the problem of justifying induction. At the very heart of his argument, and of the article (paragraph 55), there occurs a brief subsidiary argument that deserves careful scrutiny.

Let me begin by setting the stage. The context in which the crucial argument occurs is one in which Sellars is concerned with inferences from a finite observed sample K of population X to an unobserved finite sample ΔK of that population with respect to an attribute Y. The practice to be justified is to infer that the relative frequency of Y in ΔK is approximately the same as it is in K. Sellars maintains quite explicitly that he is trying to justify a practice rather than attempting to establish any sort of lawful statistical generalization. More precisely, he wants to establish "the state of being able to draw inferences concerning the composition with respect to a given property Y of unexamined finite samples (ΔK) of a kind, X, in a way which also provides an explanatory account of the composition with respect to Y of the total examined sample, K, of X" (1964, p. 215). As I understand it, Sellars is making a distinction between (1) simply drawing an inference from a frequency in an observed sample to the frequency in an unobserved sample of the same population and (2) drawing an inference from something that *explains* the frequency in the observed sample to the frequency in an unobserved sample of the same population. We can now consider the crucial argument; I shall quote it in full:

. . . to give an explanatory account of the composition of the class K of *examined* Xs one must, logically, assert that the composition in question is the most statistically probable composition on the basis of the finite population (P) of Xs which are known to exist but of which only the members of K have been examined. If we take the finite unexamined remainder of P as ΔK, so that

$$P = K + \Delta K$$

then, since the statistically probable composition of a random sample approximates that of the population, the above condition logically requires the acceptance of

$$rf(Y, P) \approx rf(Y, \Delta K) \quad ['\approx' \text{ means 'approximates to'}]$$

which, in turn, logically requires the acceptance of

$$rf(Y, \Delta K) \approx rf(Y, K)$$

and hence that the proportion [mentioned above] be specified as (approximately) the proportion of Ys in the examined sample K. [p. 216]

This argument involves, I think, a serious ambiguity.

When Sellars refers to "the statistically probable composition of a random sample", he seems to be referring to the combinatorial fact that, *given a finite population* P, in most subsets K (of reasonable size *m*) of P the relative frequency of Y in K is approximately equal to the relative frequency of Y in P. This fact is not in dispute; it is a consequence of the Bernoulli theorem. If you pick *randomly* one subset K from the set of all subsets of specified size of P, you will probably get a representative sample. (We shall consider the meaning of the concept of randomness and its role in Sellars's arguments below.)

It is crucial to realize, however, that the foregoing fact does not imply that, *given an observed set* K *of size m containing n elements with property* Y, it very probably comes from a population P in which the relative frequency is approximately *n/m*. This probability must be computed by means of Bayes's theorem, and to do so requires prior probabilities. No conclusion can be drawn concerning the relative frequency of Y in P unless we are given a distribution of prior probabilities for the logically possible relative frequencies of that attribute in that population.

This point can be illustrated by a simple example. Suppose we have a sack containing a million pennies, all but one of which are standard fair coins. The exceptional one is two-headed. Someone draws a coin at random from this sack and proceeds to flip it ten thousand times. We have no opportunity to examine the coin, nor to witness all of the tosses, but we do have the opportunity to learn the outcomes of ten tosses selected at random. All of the ten tosses resulted in heads. It would obviously be unwarranted to infer that the relative frequency of heads in the set of ten thousand tosses is approximately one. It would obviously be unwarranted to infer that the relative frequency of heads in another (nonoverlapping) randomly selected subset of ten would be approximately one.

Now Sellars has, in the foregoing argument, insisted on the importance of being able to *explain* the frequency in the observed sample, and the principle of explanation he adopts is essentially of a maximum likelihood type. Recall his assertion that "to give an explanatory account of the composition of the class K of *examined* Xs one must, logically, assert that the composition in question is the most statistically probable composition on the basis of the finite population (P)". In spite of Sellars's use of such expressions as "must, logically", maximum likelihood explanations are not uncontroversial. For instance, it would be rash in the extreme to offer as an explanation of the ten heads in the foregoing example that the coin being tossed has two heads.[11] A much more plausible explanation of the set of ten heads in this context would be that it was a fairly improbable (1/1024) chance occurrence that came about as a result of flipping a fair coin. And, from a practical standpoint, it would surely be unwise to take a number close to one as a betting quotient for purposes of wagering that the next toss will be a head.

In the above-quoted argument, Sellars appears to make use of two transitivity relations, both of which are illegitimate. In the first place, he claims that we are logically required to accept

(1)　　　　　$rf(Y, \Delta K) \approx rf(Y, P)$

"which, in turn, logically requires the acceptance of

(2)　　　　　$rf(Y, \Delta K) \approx rf(Y, K)$".

How could (1) logically require the acceptance of (2)? As far as I can see, the justification for accepting the second approximation, given the first, is the assumption that

(3)　　　　　$rf(Y, P) \approx rf(Y, K)$.[12]

If "$\approx$" designated a transitive relation, (2) would follow from (1) and (3); however, it is well-known that relations of this sort (approximate matching relations) are not transitive (Salmon, 1984, pp. 78–79). But this misuse of transitivity is relatively innocuous compared to the second.[13]

Sellars is not claiming, after all, that the frequency makeup of the sample *must* approximate that of the population, even that of population P as he defines it. In order to make his argument go through, Sellars is apparently relying on something like the transitivity of probabilistic support. The argument would seem to go as follows: Given that the frequency of Y in K = r, it is highly probable that the frequency of Y in P = $r \pm \delta$, and given that the frequency of Y in P = $r \pm \delta$, it is highly probable that the frequency in $\Delta K = r \pm \epsilon$ (for suitably chosen δ and ϵ).[14] My favorite counterexample to probabilistic transitivity at the time

Sellars's article was published was this: Given that x is a scientist, it is highly probable that x is alive,[15] and given that x is alive, it is highly probable that x is a micro-organism. The reader can complete the syllogism.

In subsequent paragraphs (81–82) Sellars addresses the issue of inferences from samples to populations: ". . . given an identificatory ordering of the classes K_i which are members of ^2K [the class of all subclasses of P having m members]

$$K_1, K_2, \ldots, K_\mu$$

formulating the class of questions

Does K_i match the B composition of P within ϵ?

and answering them all in the affirmative, a majority of the answers, for properly chosen ϵ, will be true". This much is certainly correct (at least for samples and populations of reasonable size). But Sellars continues, ". . . we can argue

I shall accept all the affirmative answers to the question 'Does K_i match P in B within ϵ?'
Therefore since S is a random sample of P having m members, it is identical with one of the members K_i of ^2K
Therefore I shall accept 'S matches [P] in B within ϵ'
But the B composition of S is n/m
Therefore I shall accept 'the B composition of P is within ϵ of m/n'".

Seductive as this reasoning may be, it is unsound. Sellars begins with the true claim that, if we examine *each and every* m-size sample of P *exactly once*, we shall be right in the majority of cases if we assert that the sample approximately matches the population.[16] But that is not the situation in which we find ourselves for the most part. We ordinarily observe one or a few samples of the given population. If we are to have any basis for claiming that these observed frequencies match the population, we must, as Sellars seems to realize, assume that the observed samples are random samples.

A good deal of trouble is caused in Sellars's arguments by his use of the term "random". Unfortunately, it is seriously ambiguous. The meaning Sellars adopts, in a footnote to the quoted passage, is "a sample concerning which *nothing else is known* relevant to its matching o[r] not matching the composition of P". Another standard meaning is *a sample selected by a method under which all possible samples have an equal probability of being drawn.* Now Sellars's discussion of the numbers of samples of a given composition from a given population is pertinent if the proportion of such samples in the population reflects the frequency with which they are drawn. Ignorance of bias is not, however, an adequate basis for concluding that there is absence of bias. For example, if the members of the population are presented sequentially, and if the probability that an element possesses an attribute is *not* independent of the possession of that attribute by its predecessor, then sequential sampling is not random. Consider the probability that a day is warm and sunny. In Pittsburgh, for example, it is far more probable that a warm sunny day follows a warm sunny day than that it follows a cold snowy day. You will not get a random sample with respect to warmth and sunshine by visiting Pittsburgh for a month, whichever month you choose. The samples nature gives us are often far from random; frequently we must contrive cleverly to find random samples.

The randomness issue arises again near the conclusion of Sellars's essay. In paragraph 86 he deals with inferences regarding an attribute B from the composition of a finite population P to the composition of a *random sample* S drawn from that population. Citing the combinatorial facts discussed above, he concludes, "I shall accept 'S matches the B composition of P within ϵ'". The problem is, how are we to know whether an actual sample drawn from an actual population is random or not? Sellars's answer seems to be that we are entitled to consider any sample random if we have no knowledge to the contrary (1964, p. 225, n. 16). This answer constitutes an appeal to Laplace's *principle of indifference*: two events are equally probable if we have no reason to prefer one to the other. In this context Sellars is saying that any particular *m*-member sample of the population in question is just as likely to be drawn as any other sample of the same size. The principle of indifference has been widely criticized, and I am convinced that it cannot be sustained in the face of the well-known objections, the chief among which is that its unbridled use leads to outright logical contradiction (Salmon, 1967a, pp. 65–68). In his theory of confirmation, Carnap attempted to preserve what he considered "the valid core" of the principle, but even a limited adoption seems to me to entail intolerable a priorism. I cannot see that Sellars has escaped these well-known difficulties.

Sellars's attempt to show that induction is vindication cannot be considered successful.

4. CLENDINNEN'S APPEAL TO SIMPLICITY

In *Experience and Prediction* (1938, p. 355) Reichenbach remarked that the adoption of any asymptotic method in which the 'corrective term' c_n is not identically zero would be *arbitrary*, but he did not elaborate.[17] Instead, he argued that the adoption of any such rule would carry a greater risk of error than would adoption of his rule of induction. Later, in his *Theory of Probability*, he saw that this argument is unfounded, and he abandoned it in favor of an appeal to descriptive simplicity.

In his (1982), F. John Clendinnen develops an argument for a vindication of induction that hinges crucially on the notion of arbitrariness. The point of departure is a claim that *it is irrational to believe in a proposition that we have arrived at by guessing*. He readily concedes that there are circumstances in which guessing is perfectly appropriate. Suppose I am on my way to Damascus, and I come to a fork in the road. I have no idea which fork leads to Damascus – no evidence to which I can appeal. Unfortunately, there are no inveterate liars, truth-tellers, or anyone else for me to interrogate. If I stay at the fork in the road I am likely to die of thirst before anyone shows up. I must choose one way or the other. I could flip a coin or just make a wild guess. That would be better than no choice at all, for the arbitrary choice gives me some chance of getting to Damascus, while the absence of choice will surely prevent me from reaching my goal. It would be rational, in these circumstances, to make an arbitrary choice, and to *hope* that it is the right choice. It would, however, be irrational to *believe* that the road I choose will lead to Damascus, or even to believe it *more likely* than the road I reject to lead to Damascus.[18]

How does all of this apply to the problem of induction? Clendinnen offers the following suggestion:

In outline the argument is that non-inductively based prediction is, if not itself a mere guess, based on a procedure which includes a purely arbitrary step. As such, non-inductive predictions are no better than guesses. And it is irrational to place any more reliance on one guess than any of the other guesses which might have been made (p. 3).

Suppose that we have observed an initial section consisting of n members of a sequence A and that we have found m of them to be B. Suppose also that we have no further evidence concerning the probability that an A is a B. If we use Reichenbach's rule of induction we will posit that the limit of the relative frequency is approximately equal to m/n. If, instead, we use a different asymptotic rule, we must add a (possibly negative) quantity c_n to the observed frequency. As we have already noticed, there is a vast plethora of asymptotic rules which furnish a superabundance of 'corrective terms'; hence, the decision to use one value instead of another is simply a blind guess. To guess when you don't have to is irrational.

It might be objected that even the use of Reichenbach's rule of induction is arbitrary. As Hume's skeptical arguments show, it might be said, any prediction we happen to make is just as arbitrary as any other. Clendinnen would deny this claim. Reichenbach's convergence argument shows that the frequency in the observed initial section of a sequence is relevant evidence regarding the probability of a given attribute in that sequence.[19] We are, after all, investigating the behavior of a sequence of relative frequencies, and the observed frequencies are constitutive of that very sequence. Rationality requires that we utilize all of the available relevant evidence; it requires, further, that we not introduce unnecessary arbitrary elements. The quantities we get by adding a 'corrective term' to the observed frequency may or may not occur anywhere in the sequence of relative frequencies. Hence, to use Reichenbach's rule of induction constitutes use of the available evidence. However, to modify that rule by adding a 'corrective term' is to add an element that is arbitrary. To a sound inductive procedure it adds an irrational guess.

The outcome of Clendinnen's argument is a principle that warrants designation as *the principle of methodological simplicity*:

Adopt the simplest system of predicting rules which are compatible with, and exemplified in, the set of known facts (p. 20).

It is manifestly more appropriate to appeal to this concept of simplicity, which pertains to the selection of rules, than to appeal either to Reichenbach's concept of descriptive simplicity or to his concept of inductive simplicity, both of which pertain to the selection of statements or propositions. Moreover, Clendinnen's principle, prohibiting reliance on guesses, seems to me eminently sound.

Although Clendinnen's essay does not, in my opinion, furnish a fully articulated vindication of induction, it does provide a significant step in the right direction.

5. CARNAP'S REJECTION OF THE STRAIGHT RULE

Given what strikes many philosophers as the reasonableness of the straight rule, it is incumbent upon us to consider Carnap's reasons for rejecting it. His argument is actually quite simple. In any case in which all observed As are B, or in which no observed As are B, Reichenbach's rule of induction would have us posit that the limit of the relative frequency, and hence the probability, is one or zero respectively. In Carnap's view, a major function of probabilities is to serve as betting quotients, and it is easy to see that zero and one are unsuitable values for them. A betting quotient of one would mandate a wager of a million dollars to a penny on the next event in the sequence. Such bets would obviously be absurd.

The source of Carnap's difficulty is his claim that statements of probability$_1$ (degree of confirmation) are, if true, analytic, or, if false, contradictory. One can be as certain of the correctness of the value as one is of any other truths of arithmetic. When one has a probability value of unity, one can be certain that it is the correct betting quotient. In consequence, to protect against foolhardy betting quotients, Carnap builds in what might be called a "safety factor" which keeps various probabilities from assuming the offending values. It is the safety factor rather than an arbitrary guess that makes Carnap's probabilities differ from those the straight rule would give. Users of the straight rule – Reichenbach's rule of induction – are willing to posit probabilities having the extreme values, but they recognize full well that they *cannot* be confident that the posited values are truly the limiting values of the relative frequencies. A posit is *by definition* something about which we cannot be certain. Rather than building the safety factor into their inductive rules, they would offer such practical advice as to avoid making large bets at unfavorable odds on the basis of probabilities whose values are not known with great confidence.

6. TWO REICHENBACHIAN DISTINCTIONS

In Section 2 I mentioned the three criteria – consistency, symmetry, and invariance – shown by Hacking to be necessary and sufficient for justification of Reichenbach's rule of induction. Consistency poses no problems; it amounts to probabilistic coherence and is demonstrably satisfied by limits of relative frequencies. The invariance requirement, as I remarked above, corresponds to two invariance criteria I have proposed, namely, linguistic invariance and statistical invariance.[20] These two invariance requirements are, I think, fully justified by Clendinnen's principle of methodological simplicity; a violation of either would be a case of proscribed arbitrariness. The criterion of symmetry is another story.

The reason symmetry is required, according to Hacking, is that we often have knowledge, not only of the relative frequency of a given attribute in a given sample, but also of the order in which the elements occur. Symmetry requires that the order have nothing to do with our posit regarding the limiting frequency. One can devise asymptotic rules that satisfy both consistency and invariance, as Hacking has shown (1968, pp. 50–51), but that differ from Reichenbach's rule of induction. This particular example can be ruled out by Clendinnen's methodological simplicity principle, but I do not think all rules that violate symmetry can be handled quite that easily.

Depending on one's philosophical proclivities, there are various ways of dealing with the symmetry issue. A subjective Bayesian can simply invoke exchangeability, meaning that his or her subjective probability with respect to an A being a B would not be any different regardless of the order of B and non-B in the sample. Reichenbach, as a dedicated frequentist, would not have admitted subjective probabilities.

Another approach would be to apply statistical estimation methods. In stating his rule of induction, Reichenbach consistently used the term "posit", meaning something like a bet or a wager. It is not entirely clear whether he intended this term in the sense of an inference or in the sense of an estimate, but the fact that he included a margin of error δ in the formulation of the rule suggests that it can properly be construed as an estimate. On this construal, we can circumvent the symmetry requirement by adopting the observed frequency S_n as our estimator. It has the sorts of virtues we would seek; it is convergent, additive, unbiased, and has minimal variance. The observed frequency provides a sufficient statistic; the order of occurrence of the constituents is irrelevant.

Reichenbach's objection to this approach would be, I believe, to call attention to the fact that it requires factual assumptions about the nature of the population under investigation.[21] In his treatment of induction, he distinguished *primitive knowledge* from *advanced knowledge*. In primitive knowledge we have no previous inductive results upon which to depend; in advanced knowledge the results of previous inductive inferences are available. His rule of induction is a method for primitive knowledge, and this is what he was attempting to justify. Thus, he would argue, since we have no results of previous inductions to establish these factual assumptions, we are not entitled to make them.

From a *psychological* standpoint, the distinction between primitive and advanced knowledge is probably unfounded; it is doubtful that any such psychological state as primitive knowledge exists. We can, nevertheless, make a useful *logical* distinction between primitive inductive rules and advanced inductive methods. Primitive rules are employed where no previous inductive results are available – the permissible inputs for such rules are statements of observed facts, but no inductive generalizations are allowed. According to Reichenbach, all inductions, primitive or advanced, are carried out by means of the rule of induction and the probability calculus. The theorems of Bernoulli and Bayes play prominent parts in the development of advanced induction. On the frequency interpretation of probability, he argues, all of the axioms and theorems are logically necessary. The rule of induction provides the only nondeductive element in inductive logic. Relying on no inductive assumptions, it provides the probability values to plug into the formulas of the mathematical calculus in order to get the whole inductive enterprise off the ground. It should be noted, by the way, that Reichenbach (1949, Sec. 86) offers a treatment of induction by enumeration in advanced knowledge. In this context he is free to make use of any of the tools of mathematical statistics, provided there is inductive evidence to support the assumptions demanded by these methods.

Reichenbach made another fundamental distinction, namely, between the *context of discovery* and the *context of justification*. Although he often emphasized its importance,[22] he did not, to the best of my knowledge, appeal to it in his arguments concerning the justification of induction. We might begin by asking whether his rule of induction belongs to the former or the latter context. Clearly any decision to examine As to find out whether or how often they are Bs belongs in the context of discovery. What about the inductive posit itself? The use of the rule of induction to *arrive at* a value to posit is also part of the context of discovery; at the same time, it looks like part of the context of justification as well, for the posit is justified by virtue of the rule of induction. Although, as I have argued in detail (1970), it is possible for a given item to belong to both contexts, let us consider it, in the first instance, as a method of discovery only.[23]

Suppose, then, that somehow our curiosity has been aroused regarding the probability that an A is a B. We look at n of the As and find m to be B. We might posit – i.e., guess – that the probability of an A being a B is about m/n. We might further hypothesize – i.e., guess – that the distribution is Bernoullian.[24] On that assumption about the distribution, we can calculate the size n of the sample required to have a given degree of confidence that the actual probability is within a *specified* δ of the observed frequency S_n.[25] Our hypothesis may, of course, be false; the distribution may be far from Bernoullian. This does not vitiate our investigation, however, since the assumption is itself subject to statistical testing by standard means. Thus, the assumption that was introduced as a hypothesis in the context of discovery can be confirmed or disconfirmed in the context of justification.

The Bayesian approach can also employ the distinction between discovery and justification. What the subjective Bayesian takes as a personal probability the objective Bayesian can regard as a guess in the context of discovery. For an objective Bayesian, the guess might take the form of an assumption that the sampling procedure is random. This, too, can be subjected to statistical tests.

It may seem that we run a risk of violating the requirement of total evidence if we decide to ignore the order of items in the sample, but this is not necessarily the case. The requirement of total evidence requires us to take account of all available *relevant* evidence. We normally have a great deal of evidence about our samples that is not taken into account in making various inferences or estimates – e.g., whether the evidence was collected at night or during the day, whether the collector's hair was blond or not, whether the study was conducted in winter or some other season. Notice that such information would clearly be relevant to certain investigations, but for many others it would not. Standard practice, in general, ignores much available information; otherwise statistical studies would be too complicated to be practical. In estimating or inferring values of limiting frequencies, under many circumstances, the order in the sample can also be ignored as irrelevant.[26]

7. CONCLUSION

Reichenbach sought to resolve Hume's problem of the justification of induction by means of a pragmatic vindication that relies heavily on the convergence properties of his rule of induction. His attempt to rule out all other asymptotic methods by an appeal to descriptive simplicity was unavailing. We found that important progress in that direction could be made by invoking normalizing conditions (consistency) and methodological simplicity (as a basis for invariance), but that they did not do the whole job. I am proposing that, in the end, Reichenbach's own distinction between discovery and justification holds the key to the solution.

8. ACKNOWLEDGMENT

I should like to express my sincere gratitude to Deborah Mayo for critical comments and constructive suggestions on an earlier draft of this paper.

NOTES

[1] See, for example, Salmon (1957), (1965), (1967a), and (1968a).
[2] Or any functional equivalent thereof, such as Keynes's principle of limited independent variety or Russell's postulates of scientific inference. Herbert Feigl (1949) had urged the same point.
[3] It is the counterpart of what Carnap called "the straight rule".
[4] Reichenbach gives a very informal sketch of this argument in (1951), chap. 14.
[5] See, for example, Sellars (1964), Sec. XII, pp. 212–14.
[6] This point was first made by Paul Humphreys (1985). I have discussed the propensity 'interpretation' at length in (1979).
[7] Although many authors, including Aristotle, had hinted at a frequency interpretation of probability, John Venn, in *The Logic of Chance* (1866) was the first to spell it out fully; see Salmon (1980).
[8] See Salmon (1956).
[9] We may as well classify the rule of induction as an asymptotic rule, namely, that for which c_n identically zero for all values of n.

[10] This point was first made by the statistician I. Richard Savage in discussion at the Minnesota Center for the Philosophy of Science.

[11] Using Bayes's theorem, we can calculate that, in the light of the evidence that the ten tosses resulted in heads, the probability that the coin is two-headed is about 0.001.

[12] The key principle seems to be that the frequency composition of any randomly selected sample very probably nearly matches the frequency composition of the population.

[13] By suitably specifying precise degrees of approximation in the three formulas, something akin to transitivity can be salvaged, though it is not actually transitivity, for "$\approx$" does not remain univocal throughout the argument.

[14] The fallacy in arguments of this type was pointed out in my (1961) and (1965).

[15] At that time it had been estimated that 90% of all scientists that ever lived were then alive. I do not know what the current percentage would be – still fairly high I should imagine.

[16] We shall also be easy victims of Kyburg's lottery paradox.

[17] Herbert Feigl made a similar point, and elaborated it somewhat more fully, in (1949).

[18] Actually, I would have preferred to go to Rome, but since, as the saying goes, all roads lead to Rome, that would not have constituted a suitable example.

[19] It has often been noted, as I did above, that any observed frequency is *deductively* compatible with any limit. The issue here is not, however, *deductive relevance* but rather *inductive relevance*.

[20] Linguistic invariance requires invariance under permutations of predicates; statistical invariance requires invariance under permutations of properties.

[21] At any rate, this was my longstanding objection to attempts to justify induction by appealing to standard statistical methods.

[22] See Reichenbach (1938, pp. 6–7), (1947, p. 2), and (1949, pp. 433–34). Robert McLaughlin (1982) has argued persuasively that it would be preferable to call these the *context of invention* and the *context of appraisal*; however, in this historical discussion I shall retain Reichenbach's terminology.

[23] In an unpublished manuscript, Deborah Mayo has pointed out that an excellent method for discovering and justifying a statement about the mean score on a test is to add up all of the scores and divide by the number of students in the class taking the test.

[24] One may, of course, hypothesize a different sort of distribution, but it, too, can be tested statistically.

[25] Reichenbach has often been criticized for failure to provide any way of establishing specific values for δ.

[26] In his (1980) Hacking relinquishes his earlier position on foundations of statistics, and adopts the Neyman–Pearson approach. It seems to me likely that his position regarding the symmetry requirement and the status of information about order in the sample would be revised in consequence.

REFERENCES

Black, Max: 1954, *Problems of Analysis*. Ithaca, NY: Cornell University Press.
Carnap, Rudolf: 1952, *The Continuum of Inductive Methods*. Chicago: University of Chicago Press.

Clendinnen, F. John: 1982, 'Rational Expectation and Simplicity', in Robert McLaughlin (ed.), *What? Where? When? Why?* (Dordrecht: D. Reidel Publishing Co.), pp. 1–25.

Feigl, Herbert: 1949, 'The Logical Character of the Principle of Induction', in Herbert Feigl and Wilfrid Sellars (eds.), *Readings in Philosophical Analysis* (New York: Appleton-Century-Crofts), pp. 297–304. Originally published in *Philosophy of Science* I (1934).

Feigl, Herbert: 1950, 'De Principiis Non Disputandum. . .?' in Max Black, ed., *Philosophical Analysis* (Ithaca, NY: Cornell University Press), pp. 119–56.

Hacking, Ian: 1968, 'One problem about induction',' in Imre Lakatos, ed., *The Problem of Inductive Logic* (Amsterdam: North-Holland Publishing Co.), pp. 44–59.

Hacking, Ian: 1980, 'The theory of probable inference: Neyman, Peirce and Braithwaite', in D. H. Mellor (ed.), *Science, Belief and Behaviour* (Cambridge: Cambridge University Press), pp. 141–60.

Humphreys, Paul: 1985, 'Why Propensities Cannot Be Probabilities', *Philosophical Review* XCIV, 557–70.

McLaughlin, Robert: 1982, 'Invention and Appraisal', in Robert McLaughlin (ed.), *What? Where? When? Why?* (Dordrecht: D. Reidel Publishing Co.), pp. 69–100.

Popper, Karl R.: 1959, *The Logic of Scientific Discovery*. New York: Basic Books.

Reichenbach, Hans: 1938, *Experience and Prediction*. Chicago: University of Chicago Press.

Reichenbach, Hans: 1947, *Elements of Symbolic Logic*. New York: Macmillan.

Reichenbach, Hans: 1949, *The Theory of Probability*, 2nd ed. Berkeley & Los Angeles: University of California Press.

Reichenbach, Hans: 1951, *The Rise of Scientific Philosophy*. Berkeley & Los Angeles: University of California Press.

Reichenbach, Hans: 1954, *Nomological Statements and Admissible Operations*. Amsterdam: North-Holland Publishing Co. Reissued by the University of Califomia Press under the title, *Laws, Modalities, and Counterfactuals* (1976, same pagination).

Salmon, Wesley C.: 1956, 'Regular Rules of Induction', *Philosophical Review* LXV, 385–88.

Salmon, Wesley C.: 1957, 'Should We Attempt to Justify Induction?' *Philosophical Studies* **8**, 33–48.

Salmon, Wesley C.: 1957a, 'The Predictive Inference', *Philosophy of Science* 24, pp. 180–90.

Salmon, Wesley C.: 1963, Review of John Patrick Day, *Inductive Probability*, in *Philosophical Review* LXXII, 392–96.

Salmon, Wesley C.: 1963a 'On Vindicating Induction', *Philosophy of Science* 30, 252–61.

Salmon, Wesley C.: 1965, 'Consistency, Transitivity, and Inductive Support', *Ratio* VII, 164–69.

Salmon, Wesley C.: 1967, 'Carnap's Inductive Logic', *Journal of Philosophy* LXIV, pp. 725–39.

Salmon, Wesley C.: 1967a, *The Foundations of Scientific Inference*. Pittsburgh: University of Pittsburgh Press.

Salmon, Wesley C.: 1968, 'Reply', in Imre Lakatos (ed.), *The Problem of Inductive Logic* (Amsterdam: North-Holland Publishing Co.), pp. 84–85.

Salmon, Wesley C.: 1968a, 'The Justification of Inductive Rules of Inference', in Imre Lakatos (ed.), *The Problem of Inductive Logic* (Amsterdam: North-Holland Publishing Co.), pp. 24–43.

Salmon, Wesley C.: 1970, 'Bayes's Theorem and the History of Science', in Roger H. Stuewer (ed.), *Minnesota Studies in the Philosophy of Science*, vol. V (Minneapolis: University of Minnesota Press), pp. 68–86.

Salmon, Wesley C.: 1979, 'Propensities: A Discussion Review', *Erkenntnis* 14, 183–216.

Salmon, Wesley C.: 1980, 'John Venn's Logic of Chance', in J. Hintikka and David Gruender (eds.), *Probabilistic Thinking, Thermodynamics, and the Interaction of History and Philosophy of Science*, vol. II (Dordrecht: D. Reidel Publishing Co.), pp. 125–38.

Salmon, Wesley C.: 1984, *Logic*, 3rd ed. Englewood Cliffs, NJ: Prentice-Hall.

Salmon, Wesley C., *et al.*: 1965, 'Symposium on Inductive Evidence', *American Philosophical Quarterly* 2, 1–16.

Sellars, Wilfrid: 1964, 'Induction as Vindication', *Philosophy of Science* 31, 197–231.

Venn, John: 1866, *The Logic of Chance*, 1st ed. London & Cambridge: Macmillan & Co.

Department of Philosophy
University of Pittsburgh
Pittsburgh, PA 15260
U.S.A.

BRIAN SKYRMS

CARNAPIAN INDUCTIVE LOGIC FOR MARKOV CHAINS

"Then the criticism of the so-far-constructed 'c-func-
tions' is that they correspond to 'learning machines'
of very low power. They can extrapolate the simplest
possible empirical generalizations, for example: 'approximately nine-tenths of the balls are red', but they
cannot extrapolate so simple a regularity as 'every
other ball is red'."

Hilary Putnam
'Probability and Confirmation'

ABSTRACT. Carnap's Inductive Logic, like most philosophical discussions of induction,
is designed for the case of independent trials. To take account of periodicities, and more
generally of order, the account must be extended. From both a physical and a probabilistic
point of view, the first and fundamental step is to extend Carnap's inductive logic to the
case of finite Markov chains. Kuipers (1988) and Martin (1967) suggest a natural way in
which this can be done. The probabilistic character of Carnapian inductive logic(s) for
Markov chains and their relationship to Carnap's inductive logic(s) is discussed at various
levels of Bayesian analysis.

1. INDEPENDENCE AND MARKOV DEPENDENCE

Suppose that one is investigating a sequence of tosses of a coin with
unknown bias, or a sequence of random drawings with replacement
from an urn filled with balls of a finite number of different colors. Then
Rudolf Carnap's inductive logic – his original inductive logic, c^* – will
turn in a creditable performance. In the case of the biased coin, Carnap's advice is that before the first toss – in the absence of any prior
knowledge – you should consider heads and tails as equiprobable, and
in general – on the evidence of n heads in N tosses – you should take
the probability of heads to depend on the relative frequency of heads
and the number of tosses as follows:

$$pr(\text{Heads}) = \frac{n+1}{N+2}$$

Erkenntnis **35**: 439–460, 1991.
© 1991 *Kluwer Academic Publishers. Printed in the Netherlands.*

In the case of the urn with balls of k colors, Carnap uses the natural generalization from 2 to k properties. In N trials in which color C comes up n times we have:

$$\text{pr}(C) = \frac{n+1}{N+k}$$

If we take $\alpha = k/(N + k)$, then we can rewrite Carnap's rule as a weighted average of the a priori probability and the relative frequency:

$$\alpha \frac{1}{k} + (1 - \alpha) \frac{n}{N}$$

This rule gives reasonable results when the trials are – in fact – independent,[2] and empirical investigation is like sampling from the great urn of nature.[3]

But nature serves up dependence as well as independence. Indeed, any investigation of causation – deterministic or probabilistic – must deal in dependence. Deterministic laws of periodicity are a special case, and Carnap's original system was vigorously criticized by Putnam (1963a,b) and Achinstein (1963) for its inability to confirm such laws.

The first, fundamental step in dealing with dependence is to consider Markov chains – in which the probability of the state of a system at a moment of time depends only on the state of the system at the preceding moment. Most of science deals with processes which – when properly described – exhibit this Markov property.

In the simplest case, we can replace the iconic example of the coin flip with that of the Markov thumb tack of Diaconis and Freedman (1980a). A thumb tack is repeatedly flicked as it lays. It can come to rest in either of two positions: point up or point down. The chance of the next state may well depend on the previous one. Thus there are unknown transition probabilities:

	PU	PD
PU	Pr(PU|PU)	Pr(PD|PU)
PD	Pr(PU|PD)	Pr(PD|PD)

which an adequate inductive logic should allow us to learn. Here are two extreme examples:

Example 1, Independence: Independent flips of a coin biased 9 to 1 in favor of heads.

$$\begin{bmatrix} .9 & .1 \\ .9 & .1 \end{bmatrix}$$

Example 2, Putnam: This is an example of a strictly alternating sequence of the kind suggested in the epigraph.

$$\begin{bmatrix} 0 & 1 \\ 1 & 0 \end{bmatrix}$$

More generally, where the physical system has a finite number of states we can fix on the canonical urn model for a Markov chain (Feller (1966)). A Markov chain on a system with k-states is equivalent to an urn model with $k + 1$ urns. There are balls of k different colors in the urns. A ball is drawn at random from the first urn to determine the first state of the system. The remaining k urns are each labeled with a color. Depending on which of the k colors is drawn at a given time, the next ball is drawn from the next urn to determine the next state of the system and then replaced.

2. KUIPERS' CARNAPIAN INDUCTIVE LOGIC FOR MARKOV CHAINS

How should one treat Markov chains in the spirit of Carnap's original inductive logic? Carnap never addressed this question in his published work. Taking into account both his publications and conversations and correspondence with his coworkers, I believe that the question would fall somewhere on Carnap's agenda for the future development of inductive logic but it was not one which he actively tried to answer.

However, there is a natural answer, which fits into Carnap's program much more neatly than some of the generalizations that he did consider. It is put forward by Theo Kuipers in Kuipers (1988). The leading idea is this: Carnap already has an inductive logic suitable for sampling from an urn with replacement. Just apply this inductive logic to the natural urn model of a Markov chain, under the assumption that transitions originating in one state give us no information about transitions originating in a different state. Parametric Bayesians, such as Martin (1967), operating in a somewhat different tradition, have followed the same path (see Section 6).

Suppose that we have a Markov chain with unknown transition probabilities, and our inductive problem is to watch the system evolve and predict the transitions. Applying Kuipers' idea to Carnap's original logic, c^*, for a two state physical system gives the inductive rule:

$$\Pr(S_j|S_i) = \frac{1 + N[S_j, S_i]}{2 + N[S_j, S_i] + N[S_i, S_i]}$$

where $N[S_j, S_i]$ is the number of transitions from state S_i to state S_j observed.

More generally, for a physical system with states $S_1 \ldots S_k$, a Markov Chain (with stationary transition probabilities) consists of a k-place probability vector for the initial state[4] together with a k-by-k square matrix of transition probabilities. If a Carnapian is confronted with a totally new system which he wishes to treat as a Markov chain, he will assign equal probabilities for the initial state, and initially use transition probabilities which assign equal probabilities for every state given any state as predecessor, and update the transition probabilities using the transition counts in the style of c^*:

$$\Pr(S_j|S_i) = \frac{1 + N[S_j, S_i]}{k + \Sigma_{m=1}^{k} N[S_m, S_i]}$$

(If it is possible to repeatedly restart a Markov chain or to observe many independent instantiations of it, then the probabilities for the initial state could be learned by Carnap's original inductive logic.)

Carnapian inductive logic for Markov chains is a natural generalization of Carnap's inductive logic for coin flipping. It learns properly in our two examples of independence and strict periodicity. It also can be shown to be the correct generalization at a deeper level.

3. CLASSICAL BAYESIAN ANALYSIS

Carnap's inductive rule for coin flipping has a Bayesian interpretation which predates Carnap and, in fact, is due to the Reverend Thomas Bayes himself (Bayes (1765)). Zabell (1989) makes a convincing case that Bayes intended his analysis as a reply to the inductive skepticism of David Hume. I will sketch the essence of Bayes' argument.

For a coin with known probability p of heads, the probability of n heads in N independent trials is:

$$p^n(1 - p)^{N-n}$$

Where the bias is unknown he puts a prior probability on the bias. That is a prior degree-of-belief probability on the objective probability of heads, which may be anywhere in the interval $[0, 1]$. In the absence of any relevant knowledge about the coin, Bayes uses the uniform "flat" prior. The probability of n heads N trials is gotten by mixing with respect to the prior:

$$\int_0^1 p^n(1 - p)^{N-n}\, dp$$

and the conditional probability of a head on the N + 1th trial given n heads in N trials is then:

$$\frac{\int_0^1 p^{n+1}(1 - p)^{N-n}\, dp}{\int_0^1 p^n(1 - p)^{N-n}\, dp} = \frac{n + 1}{N + 2}$$

Independently of Bayes, but somewhat later, Laplace gave the same derivation of this inductive rule and it was baptized by John Venn as *Laplace's rule of succession*.

Laplace also discussed the natural generalization from the binomial to the multinomial case of rolling a die or sampling from an urn with a finite number of colors. The full analysis is carried out in Lidstone (1920). Assuming a uniform prior and integrating gives us Carnap's inductive rule.

Suppose that we treat the case of the Markov thumbtack with unknown transition probabilities in the same spirit. For each state, the transition probabilities from that state to point up can range from 0 to 1. Any point in the unit square represents a possible combination of transition probabilities. An arbitrary subjective prior could be any probability distribution over the unit square. Some priors would make the rows in the Markov matrix dependent. That is, information about the true transition probabilities from state PU might be relevant to beliefs about the transition probabilities from state PD, or *vice versa*.

But let us, following classical intuitions of symmetry as before, take the uniform distribution over the unit square as the appropriate prior. This has the uniform distributions on the unit interval as marginals for transition probabilities from PU, $p1$, and from PD, $p2$. Moreover the distribution on the unit square is the product measure, $p = (p1)(p2)$; rows are independent. Let us suppose for the moment that we simply start the thumbtack point up. Then the probability of an outcome sequence which starts with the stipulated initial state PU and which has the transition counts Tuu, Tud, Tdu, Tdd is:

$$\int_{\Omega} p_1^{\mathrm{Tuu}}(1 - p_1)^{\mathrm{Tud}} p_2^{\mathrm{Tdu}}(1 - p_2)^{\mathrm{Tdd}}\, dp$$

$$= \int_0^1 p_1^{\mathrm{Tuu}}(1 - p_1)^{\mathrm{Tud}}\, dp_1 \cdot \int_0^1 p_2^{\mathrm{Tdu}}(1 - p_2)^{\mathrm{Tdd}}\, dp_2$$

Now suppose that such a sequence ends with point up and we want to predict the result of the next toss. The conditional probability of PU on the next trial is:

$$\frac{\int_0^1 p_1^{\mathrm{Tuu}+1}(1 - p_1)^{\mathrm{Tud}}\, dp_1}{\int_0^1 p_1^{\mathrm{Tuu}}(1 - p_1)^{\mathrm{Tud}}\, dp_1} = \frac{\mathrm{Tuu} + 1}{\mathrm{Tuu} + \mathrm{Tud} + 2}$$

(The calculation is similar if the sequence ends PD or if we start by placing the thumbtack PD.) This is Carnapian inductive logic for Markov Chains.

Suppose we go to the more general case where we flip a biased coin to decide the initial position of the thumbtack. In this setting, the full specification of the Markov chain is a point in the unit cube, with one dimension for the probability of the initial state and the other two for the transition probabilities. Putting the uniform prior on the unit cube makes the probability of the initial state independent of the transition probabilities, so there is just one more term which cancels out in the calculation of the conditional probability, and the result is the same. Here again, independence is not an extra assumption but rather a consequence of the choice of the uniform prior as the appropriate quantization of ignorance.

If the physical system has some arbitrary finite number of states, the Bayesian analysis goes in much the same way. Each row in the Markov transition matrix can be thought of as sampling from an urn without replacement. Taking the uniform prior for a multinomial process gives us Carnap's inductive logic. Taking a uniform prior for Markov chains makes probability of the initial state and rows in the transition matrix independent and factors into uniform priors on each giving us the Carnapian inductive logic for Markov chains of Section 2.

It is also desirable to consider physical systems whose state corresponds to the value of some continuous physical magnitude (or a vector of values of a number of physical magnitudes), and the associated Markov chains. But Carnap's inductive logic was not developed sufficiently to deal with such systems even when the trials are independent. The extension to continuous magnitudes was also on Carnap's agenda for inductive logic. Something can be done along these lines in the same spirit as the foregoing treatment, but space does not permit going into this matter here.

4. CONSISTENCY

Suppose that we have a statistical model in which the chances are determined by a parameter whose value is unknown. An inductive rule is *consistent* (relative to the model) for a value of the parameter if when the parameter has that value, as more and more data are observed the inductive rule will (with chance equal to 1) yield degrees of belief which in the limit concentrate at a point mass on the true chance probability. The inductive rule is consistent for the model overall if it is consistent for every value of the parameter.

Laplace's rule of succession is consistent for coin tossing. By the strong law of large numbers, the limiting relative frequency coincides (almost surely in chance) with the single case chance probability. In the limit, application of Laplace's rule leads to degrees of belief which correspond to relative frequencies. So we have consistency for all values of "the chance of heads". By the same reasoning, the Laplace–Lidstone–Carnap generalization to n-states is consistent for a multinomial process. It should be clear from the argument that we could formulate many consistent alternative rules as well.

We would like to investigate the question of consistency for Carnapian inductive logic for Markov chains. There are really a number of different questions depending on what is assumed known or unknown about the Markov chain, and how the data stream is generated. With regard to the last question, can we "restart" a chain or observe arbitrarily many independent copies of it, or are we limited to observing one sequence of events.

To begin with, let us suppose that the initial state of the physical system is generated with probability 1 and known, but the (stationary) transition probabilities are unknown. The chain evolves and we observe just that one sequence of states, and update according to Carnapian inductive logic for Markov chains. When do we have consistency? A simple example shows that we do not always have it. Suppose that a two state system state in state 1 and that the transition matrix is of the following type:

$$\begin{bmatrix} 1 & 0 \\ x & (1-x) \end{bmatrix}$$

Then, with probability 1, the system will stay in state 1 and we will never get any data for transitions from state 2. Carnapian inductive logic is only consistent here for the case where $x = 1/2$ and our initial guess at the transition probabilities from state 2 happens to be right.

A state of a Markov chain is called *recurrent* if the probability that it is visited an infinite number of times is 1. The chain is recurrent if all its states are recurrent. Both the examples of Section 1, started deterministically in any state, give recurrent Markov chains. In the Putnam example the states are periodic, while in the independence example they are ergodic.

Carnapian inductive logic for Markov chains is consistent for recurrent Markov chains. Here is a quick sketch of a proof. Suppose that the true state of nature is a recurrent Markov chain (with a finite number of states). Then the set of sample sequences in which some state does not recur has probability zero. Delete these sequences and restrict the chance measure to get a new probability space. In this space define the random variables $f[nij]$ as having the value 1 if the nth occurrence of state i is followed by state j. The sequence $f[1ij], f[2ij], \ldots$ is an infinite sequence of independent and identically distributed random variables, so the strong law of large numbers applies. This means that the limiting transition relative frequencies from i to j equal the true transition probabilities with chance 1. Thus if the true state of nature is a recurrent Markov chain, then there are only a finite number of ways in which Carnapian inductive logic for Markov chains can fail to learn the true transition matrix, and each of these has a chance of zero.

What are we to think of the failure of consistency of the Carnapian method for some non-recurrent Markov chains? It hardly seems a failure of the inductive method where nature conspires not to provide the relevant data. And Carnapians can take some comfort from two considerations: (1) From the classical Bayesian standpoint the non-recurrent chains have probability zero and (2) if nature prefers not to visit some states it may not be so important to know the true transition probabilities from those states.

On the other hand, if we can observe many independent Markov chains with the same transition probabilities more learning possibilities open up. If we can arbitrarily start a process in any state we please, as seems natural in the thumbtack example, and do it independently as many times as we please, then Carnapian inductive logic can obviously learn the true transition probabilities for even non-recurrent chains. If we want to learn the probability vector for the initial state – say that a biased coin is flipped to decide the initial position of the thumbtack – then by looking at independent trials of the starting mechanism the correct chances for the initial state can be learned by Carnap's inductive rule. These sorts of possibilities are not usually discussed in the literature on Markov chains because they have nothing to do with the Markov structure. But they are nevertheless quite relevant to the application of inductive logic. If this kind of learning is allowed consistency of Carnapian inductive logic for Markov chains is unrestricted. Again, however, we should emphasize that Carnapian inductive logic for Markov chains is not unique in having this virtue. Many other inductive rules can be formulated which share it, some of which will be examined later in this essay.

5. COORDINATE LANGUAGES, STOCHASTIC PROCESSES, AND CELLULAR AUTOMATA

Carnap worked in a framework where names were mere indices, meant to convey no substantive information about the things named. This lay behind his adoption of a postulate of *symmetry*, that probability should be invariant under finite permutations of names. If the indices are a set with a natural order the principle of symmetry says that order makes no difference with respect to probability. In such a language, if one wants to express temporal or spatial order which is to make a difference, one would need to introduce relational predicates to carry the information.

It is simpler to let the indices carry information about order as is routinely done in the theory of stochastic processes. Carnap believed, early and late,[5] that a mature inductive logic should have a formulation for such *coordinate languages*. This is the sort of framework that we have implicitly been using in the foregoing discussion of Markov chains. Let us make the framework explicit. A stochastic process is a family $[X_i : i \in I]$ of random variables on a probability space (W, F, pr). The sort of Markov chains we considered are discrete, one sided stochastic processes where the index set I consists of the positive integers. The indices are usually taken to represent discrete times and their order to represent temporal order. The possible states of the system are the same at every time, so the measurable space (W, F) is a product of the measurable spaces for a single trial. The value of the random variables represent the state of the system at the time of its index. In the case we considered the system has only a finite number of possible states, so the random variables need take on only a finite number of values. The Markov property can then be written:

$$Pr(X_{n+1} = x | X_1 = x_1 \& \ldots \& X_n = x_n) = Pr(X_{n+1} = x | X_n = x_n)$$

In linguistic terms, W is a set of possible worlds and F is a boolean σ-algebra of propositions. The indices of the random variables are names of times, and that a random variable takes on a value corresponds to the proposition in F which is the inverse image of that value under that random variable.

Carnap was – of course – interested in coordinate languages with more than one dimension. There is no reason why the indices have to only be ordered in one dimension. All sorts of orderings or partial orderings of the index set are possibilities. For example, we could consider the case of a product space for a physical system with a finite number of states where we have 3 dimensions of finite discrete space, and one dimension of one-sided infinite time. The appropriate Markov property would be that the state of a spatial point at time $t + 1$ depends only on that of that point and its neighbors at time t. Possible instantiations would be Ising spin models of ferromagnets, lattice gases, or stochastic cellular automata. If a Carnapian started by observing all points at time 1 and then at each point in time observed all points again, how should she update her probabilities? I will briefly sketch an answer.

For each point in space, there is an associated transition matrix. If the physical system in question has k possible states, then the matrix has k columns. If the point has m neighbors, the number of rows is

$m_k + 1$, with each row specifying the state of that point and its neighbors at the preceding time. (The number of neighbors depends here on whether the spatial point is on a vertex, edge, face, or in the interior.) Thus the transition matrix determines an urn model where each row in the matrix corresponds to an urn containing balls of k different colors. Applying Carnap's inductive logic to the urns gives us an answer. Initially, each entry in each transition matrix will be $1/k$. A Carnapian will update the transition matrix for a point using the transition counts, so that if the number of transitions from row r to column c is Trc and the total number of transitions from row r to any column in Tr, the updated probability in the cell of the transition matrix for that point is:

$$\frac{Trc + 1}{Tr + k}$$

Perhaps the true transition probabilities are assumed stationary in space as well as time. That is, every interior point must have the same true transition probabilities – and likewise for every point on a face, on an edge, at a vertex. Then updating at a time, a Carnapian can use for any point in the interior the total transition counts for all points in the interior – and likewise for the other spatial classes of points.

It would be of some interest to explore the Carnapian inductive logic of general coordinate languages. Here the proper setting – at least initially – is the theory of Markov random fields.[6] Another direction for generalization is to consider higher-order Markov chains – where the probability of a state depends on some finite number, k, of preceding states. For a system with a finite number, n, of states the transition matrix has n^k rows and n columns, each row corresponding to an urn for the transition matrix from the natural urn model as before. One sort of application with some physical motivation would be a cellular automaton which is Markov in time but only higher order Markov in space.

6. PARAMETRIC MODELS, CONJUGATE PRIORS AND CONTINUA OF INDUCTIVE METHODS

In 1952 Carnap generalized his original inductive logic to a continuum of inductive rules. The continuum of 1952 gives us an inductive method which depends on a parameter, λ. The inductive method of 1950 is a special case. In the posthumously published 'Basic System' Carnap introduced another parameter, γ. The resulting $\lambda - \gamma$ continuum gives a treatment identical to the parametric Bayesian approach. The change in viewpoint is dramatic. Originally, inductive logic was to be for Carnap an *a priori* organon which evaluated the claims of all beliefs to support by empirical evidence. But what is the status of the parameters? Are they to be evaluated on the basis of empirical evidence and if so, how? Or can they be freely chosen, opening the doors to subjectivism and relativism? In his intellectual autobiography (1963) Carnap appears to lean in the latter direction:

As far as we can judge the situation at the present time, it seems that an observer is free to choose any of the admissible values of λ and thereby an inductive method. If we find that the person X chooses a greater value of λ than a person Y, then we recognize that X is more cautious than Y, i.e. X waits for a larger class of observational data than Y before he is willing to deviate in his estimate of relative frequency by a certain amount from its a priori value. (p. 75)

In part I of the 'Basic System' (1971) he leaves the question open:[7]

There will presumably be future axioms, justified in the same way by considerations of rationality. We do not know today whether in this future development the number of admissible M-functions will always remain infinite or will become finite and possibly even be reduced to one. Therefore, at the present time I do not assert that there is only one rational Cr_0-function. (p. 27)

The inductive methods in Carnap's (1952) λ-continuum generalize his (1950) rule by adding a parameter, λ ($\lambda > 0$), which determines how much weight is put on the a priori probability. Instead of:

$$pr(C) = \frac{n+1}{N+k}$$

we have:

$$pr(C) = \frac{n+\lambda}{N+\lambda k}$$

If λ is 1, we get the original rule. If λ is smaller the relative frequencies swamp the a priori probabilities more quickly; if λ is smaller the relative frequencies exert their effect more slowly.

In his (1971) 'Basic System of Inductive Logic', Carnap added another parameter, γ. The effect is to enlarge the λ-continuum so that the various outcomes need not be equiprobable a priori. If there are m possible outcomes, let $\gamma_1 \ldots \gamma_m$ be positive numbers summing to 1. Then the rule of succession becomes:

$$pr(C_i) = \frac{n + \lambda \gamma_i}{N + \lambda k}$$

It should be clear from the discussion of Section 4 that all of the inductive rules in this continuum are *consistent*.

The foregoing generalizations have a Bayesian interpretation in terms of *natural conjugate priors*.[8] As an example, consider once more flipping a coin with unknown bias and assume that the prior probability density for $x = $ Chance of heads on $[0, 1]$ is Beta (with parameters α and β):

$$f_{\alpha\beta}(x) = \frac{\Gamma(\alpha + \beta)}{\Gamma(\alpha) + \Gamma(\beta)} x^{\alpha-1}(1 - x)^{\beta-1}$$

The first term is simply a normalizing constant to make the integral on $[0, 1]$ come out to 1. The (degree of belief) probability of heads is then just the expectation of chance:

$$\frac{\alpha}{\alpha + \beta}$$

Suppose that a coin is flipped N times giving n heads. The probability of this result conditional on the chance of heads being x is:

$$x^n(1 - x)^{N-n}$$

Then by Bayes' Theorem the posterior density for x is again Beta with new parameters $\alpha' = \alpha + n$ and $\beta' = \beta + (N - n)$. The Beta is known as the *natural conjugate prior* for a Bernoulli process. Use of the β prior leads to a generalized rule of succession. On the evidence of a series of N trial with n heads, the probability of heads is:

$$\frac{\alpha'}{\alpha' + \beta'} = \frac{\alpha + n}{\alpha + \beta + N}$$

When $\alpha = \beta = 1$ the Beta prior is flat and we have Laplace's rule of succession.

For a multinomial model such as sampling from an urn with replacement, the story is similar. Suppose that an urn contains balls of k different colors, and let the unknown chance of color 1 be $x_1, \ldots,$

color k be x_k. Then the prior density will be on the random vector $X = \langle x_1, \ldots, x_k \rangle$. The natural conjugate prior on the appropriate simplex is the Dirichlet:

$$f_\alpha = \frac{\Gamma(\alpha_1 + \cdots + \alpha_k)}{\Gamma(\alpha_1) \ldots \Gamma(\alpha_k)} x_1^{\alpha_1 - 1} \ldots x_k^{\alpha_k - 1}$$

with parameter $\alpha = \langle \alpha_1, \ldots, \alpha_k \rangle$ (all α_i positive).

The a priori probability of color i is the expectation of its chance:

$$E(x_i) = \frac{\alpha_i}{\Sigma_{j=1}^k \alpha_j}$$

Since the likelihood is proportional to the prior density, the posterior on the evidence of a sample of N balls, n_1 of color $1 \ldots n_k$ of color k, is again with parameter $\alpha' = \langle \alpha_1 + n_1, \ldots, \alpha_k + n_k \rangle$. This yields a generalized rule of succession. The posterior probability of color i is:

$$\frac{\alpha_i + n_i}{\Sigma_{j=1}^k \alpha_j + n_j}$$

If we restrict ourselves to symmetric Dirichlet priors this gives us Carnap's λ-continuum. If we allow arbitrary Dirichlet priors we get the $\lambda - \gamma$ continuum of 'Carnap's Basic System'.

How can one get a reasonable conjugate prior analysis for Markov chains with unknown transition probabilities? The standard parametric Bayesian approach, as in Martin (1967) is to apply the foregoing analysis to the natural urn model for the transitions and take the product measure to get the appropriate prior. Here independence is simply assumed, rather than being derived from indifference as in the classical Bayesian approach of Section 3. Martin calls this prior the "Matrix Beta" prior. This approach yields a Carnapian $\lambda - \gamma$ continuum of inductive methods for transition probabilities of Markov chains:

$$(\text{Pr}S_j | S_i) = \frac{\lambda \gamma_{ij} + N[S_j, S_i]}{\lambda k + \Sigma_{m=1}^k N[S_m, S_i]}$$

The suggestion of Kuipers (1988) was, in fact, to use the part of this continuum which corresponds to Carnap's λ-continuum. All of these methods share the consistency properties of Carnapian Inductive Logic for Markov chains that were pointed out in Section 4.

7. SUBJECTIVE BAYESIAN ANALYSIS

Subjective Bayesians do not believe in chances, and will not postulate a chance statistical model such as coin flipping or multinomial sampling or a Markov chain. The role that statistical models play for parametric Bayesians is, for them, taken over by *symmetries* in degrees of belief. The fundamental example is that of what de Finetti called *exchangeability*. Carnap defended a principle of symmetry which bears a superficial resemblance to exchangeability in his work on inductive logic, that is that the probability should be invariant under permutations of names. This was done in systems for finite languages, in which names were supposed to be mere identifying tags carrying no additional information as to order. Carnap was well aware that in the context of a temporal coordinate language the assumption of symmetry cannot have a priori status.[9] Here the indices carry information about temporal order, and the principle of symmetry tells us that temporal order makes no probabilistic difference. De Finetti may be thought of as working in a coordinate language where the nth integer names the nth trial. A degree of belief probability is *exchangeable* if it is invariant under finite permutations of trials. If we know that we are flipping a biased coin with the tosses independent in the true chance probabilities but are uncertain as to the bias so that we mix the possible Bernoulli processes, the result gives us degree of belief probabilities which do not make the trials independent, but do make them exchangeable. De Finetti proved the converse. Exchangeable sequences have a unique representation as mixtures of independent ones. This is true not only for two valued random variables, but also for the random variables taking a finite number of values that we have discussed (and for much more general cases as well).[10] Thus a subjective Bayesian can take symmetry of degrees of belief here as a mark that one will act *as if* one were drawing balls from an urn of unknown composition. In other words, for all intents and purposes the symmetry in the degrees of belief here are tantamount to identifying a statistical model.

Exchangeability has an equivalent formulation which brings out an important connection with relative frequencies. Suppose that we have a physical system with a finite number of states, $S_1 \ldots S_m$, and are considering a sequence of trials on each of which the system takes one of these states. Sampling from an urn or rolling a die are examples. For a finite sequence of trials consider $C = \langle C_1, \ldots, C_n \rangle$, the vector of frequency counts for occurrences of the states $S_1, \ldots, S_n$, respectively. C will be said to be a *sufficient statistic* if any two finite sample sequences with the same value of C are equiprobable. A probability measure is exchangeable just in case it makes C a sufficient statistic. Thus we can say that if the vector of frequency counts is a sufficient statistic, relative to degrees of belief, then the beliefs are as if they came from a multinomial sampling model. The prior on the parameters is, however, so far undetermined.

A stronger sense of sufficiency gets stronger results. Suppose in addition to exchangeability that the probability of the outcome on the next trial given the sample size and relative frequency of the outcome in all previous trials is equal to the probability of that outcome on the next trial given the complete record of all previous trials. Suppose also that for any state the numerical dependence is the same, that is that the probability of that state on the next trial given a frequency count for it and number of trials is the same as the probability of any other state given the same number of trials and the same frequency count for the latter state. That is, the probability of a state, S_i, on the next trial depends only on C_i and $\Sigma_j C_j$. This is W. E. Johnson's (1932) *"sufficiency postulate"*. If the physical system has 3 or more states and if we have both exchangeability and Johnson's sufficiency postulate satisfied by degrees-of-belief, then they are as if we put a symmetric Dirichlet prior on a multinomial model. That is to say that we are operating within Carnap's λ-continuum of inductive methods.

If, for repeated trials with a finite number of outcomes, you have an initial probability which is exchangeable and which makes every relative frequency of outcomes equally likely for a fixed sample size (equiprobability of structure descriptions in Carnap's terms) then you get Carnap's (1950) c^* inductive logic. This was demonstrated by W. E. Johnson (1924), before either Carnap or de Finetti.[11] Exchangeability was Johnson's "permutation postulate". Equiprobability of structure descriptions was Johnson's "combination postulate". The permutation postulate was tantamount to identifying a statistical model of multinomial sampling. The combination postulate in addition was tantamount to assuming a flat prior on the parameters of the model. Together they give the generalization of Laplace's rule of succession that Carnap advocated in *Logical Foundations of Probability*.

In dealing with Markov chains we do not have exchangeability, but perhaps a weaker kind of symmetry condition can play a role with respect to Markov processes analogous to that played by exchangeability with respect to Bernoulli processes. Such an analysis is developed in Freedman (1962), de Finetti (1974), Diaconis and Freedman (1980b). Let us return to the Markov thumbtack. Suppose we are sure that this was a Markov process but were unsure about the transition probabilities and probability of the initial state, and mixed over the possibilities. The result would be a stochastic process which would, in general, not be Markov, but which has the property that the vector of initial state and transition counts is a sufficient statistic for all finite sequences of given length generated by the process. That is to say that sequences of the same length having the same transition counts and the same initial state, are equiprobable. When this is the case, say the process is *Markov exchangeable*.

Markov exchangeability, like ordinary exchangeability, can also be given an equivalent formulation in terms of invariance (Diaconis and Freedman (1980)). A primitive block-switch transformation of a sequence takes two disjoint blocks of the sequence with the same starting and ending states and switches them. A block switch transformation is the composition of a finite number of primitive block switch transformations. A probability is then Markov exchangeable just in case it is invariant under all block switch transformations.

Diaconis and Freedman (1980) show that *recurrent* stochastic processes of this type which are Markov exchangeable have a unique representation as a mixture of Markov chains. The leading idea of the proof involves looking at sequences of blocks. A 1-block of an infinite sequence is a finite string of states which begins with state 1 and proceeds until (but not including) the next occurrence of state 1. Consider the infinite sequence of 1-blocks, which is well defined with probability 1 because of recurrence. Permuting 1-blocks leave the transition counts in the original sequence unchanged. Then, as a consequence of Markov exchangeability, the sequence of 1-blocks is exchangeable. De Finetti's theorem is applied to this higher order sequence of 1-blocks to get the desired conclusion. The resulting representation theorem for recurrent Markov-exchangeable stochastic processes gives us the foundation for a wholly subjective approach to the theory of Markov chains.

In this spirit, Carnapian inductive logic for Markov chains can be given a purely subjective development parallel to W. E. Johnson's pathbreaking analyses of generalized rules of succession. Suppose that in degrees of belief we have a recurrent, Markov exchangeable, stochastic process with a finite number of states. Then by the Diaconis and

Freedman result, our degrees of belief are as if we had a Markov chain with unknown transition probabilities. Now suppose further that for all states, our degrees of belief for transitions from that state satisfy the appropriate form of *W. E. Johnson's sufficiency postulate*. That is that $Pr(S_j|S_i)$ depend only on $N[S_j, S_i]$ and $\Sigma_m N[S_m, S_i]$. Notice that this automatically gets us the independence that Martin and Kuipers assume. If our beliefs satisfy the postulate then transition counts from one state are not taken as giving any evidence about transitions from another state. Furthermore, our beliefs about transitions from a state must be as if we were sampling from an urn with unknown composition and symmetric Dirichlet prior, with the same prior for each state. In other words, we are in the Kuipers λ-continuum of Carnapian inductive methods for Markov chains:

$$Pr(S_j|S_i) = \frac{\lambda + N[S_j, S_i]}{\lambda k + \Sigma_{m=1}^k N[S_m, S_i]}$$

If the degrees of belief in addition satisfy the analogous form of Johnson's *combination postulate*, then they embody the Carnapian inductive logic for Markov chains of Section 2. However, from a subjective point of view none of Johnson's postulates has any *logical* status; they merely serve to characterize cases in which degrees of belief have certain interesting and computationally tractable symmetries.

8. ORDER, PERIODICITY AND INDUCTIVE LOGIC

In the 1960's Carnap was challenged by Achinstein and Putnam to extend his inductive logic to make it sensitive to questions of order and periodicity. Achinstein (1961) looked towards a higher order language as the proper framework to accomplish this task. Carnap (1963a) replies by trying to code the order information into first order predicates:

Achinstein studies chiefly laws of periodicity . . . . He believes that a solution will require a much stronger language containing predicates and quantified variables of higher order.

I do not share this belief. I will briefly indicate how I would approach the problem of a coordinate language in inductive logic. As an example, let us think of a family of five predicates for simple qualitative properties $P_1, \ldots P_5$, say colors. An m-segment is a series of m consecutive positions. I introduce Q^m-for the possible properties ("m-species") of m-segments. For example, I define:

(6) $Q^3_{5,1,4}(n) =_{df} P_5(n).P_1(n + 1).P_4(n + 2).$

Thus the sentence '$Q^3_{5,1,4}(8)$' ascribes to the 3-segment beginning with position 8 the 3-species consisting of $P_5, P_{1,4}$ in this order; but formally '$Q^3_{5,1,4}$' is a one-place predicate of positions.

These two approaches are perhaps not so far apart as Carnap seems to suppose. If we are not too particular about details we might say that the analysis in terms of blocks is consonant with the general spirit of Achinstein while the argument sketched in Section 4 is closer to the general spirit of Carnap.

At this time Carnap was then already in his 70s, and he spent the time he had left pursuing other generalizations of his system. Order was to be dealt with properly within a framework of coordinate languages whose implementation was seen as far in the future. The natural place to start such an investigation is with the question of inductive logic for finite Markov chains. Far from being impossible – or even difficult – the adaptation of Carnapian methods here is smooth and natural. Generalization to higher order Markov chains and to finite cellular automata is equally straightforward. In such a setting Carnapian inductive logic can learn the laws of chance that are responsible for the order in some very complicated phenomena.

The problem here for Carnap's original program is not that one cannot get a good inductive learning machine, but that there are so many ways in which one can do it. Any member of the full $\lambda - \gamma$ continuum of Carnapian inductive methods for Markov chains will do. All of them share the desirable consistency properties, but they are not the only methods which have such consistency properties. And there is even some reason to look outside this class of methods because in some applications the "sufficiency" postulates may be inappropriate. Perhaps transitions from one state may be evidentially relevant to our beliefs about the transitions from another state. Suppose that there are 100 states, and there have been 100,000 trials. State 100 always has gone to state 1, state 1 to state 2 . . . state 98 to state 99. Might this evidence not raise slightly the transition probability that the system will go to state 100 given that it was previously in state 99? Such analogical reasoning should not be excluded a priori. Carnap had already met the problem of non-uniqueness in inductive logic for multinomial sampling. There he moved to more and more general continua of inductive methods. Consideration of wider domains of stochastic phenomena can only reinforce this tendency to move towards an all embracing Bayesian methodology. Both Carnap (1959), (1971), (1980) and de Finetti (1938), (1972) discuss additional forms of dependence which are important for inductive reasoning. These discussions point to promising territory for further extension of inductive methods in the spirit of Carnap.

NOTES

[1] I would like to thank Haim Gaifman, Richard Jeffrey, and John Kemeny for sharing recollections of the times when they assisted Carnap, and Persi Diaconis for comments on an earlier draft of this paper. Research upon which this paper was based was partially supported by the National Science Foundation.

[2] That is, our degrees of belief make them independent conditional on the chance hypotheses. Unconditionally, our degrees of belief make them exchangeable.

[3] See Strong (1976) for an account of the historical life of this metaphor.

[4] So to make the Markov thumbtack example into a full Markov chain we can imagine flipping a (possibly biased) coin to determine the initial position of the thumbtack.

[5] See Carnap (1950) sec. 15B; (1963b) p. 226; (1971) pp. 118–120; (1980) pp. 69–70.

[6] See Preston (1976) and Georgii (1979).

[7] As he does in Carnap (1980) p. 119:

> Suppose that at some point in the context of a given problem, say, the choice of a parameter value, we find that we have a free choice within certain boundaries, and at the moment we cannot think of any additional rationality requirement which would constrict the boundaries. Then it would certainly be imprudent to assert that the present range of choice will always remain open. How could we deny the possibility that we shall find tomorrow an additional requirement? But it would also be unwise to regard it as certain that such an additional requirement will be found, or even to predict that by the discovery of further requirements the range will shrink to one point.

[8] See Raiffa and Schlaifer (1961) for analyses using natural conjugate priors. Giving a satisfactory characterization of natural conjugate priors which is not vacuously satisfied is not unproblematic. See Diaconis and Ylvisaker (1979).

[9] See Carnap (1971) pp. 118–120; (1980) pp. 69–70.

[10] See Hewitt and Savage (1955).

[11] See Zabell for details of both this and of Johnson's 1932 generalization.

REFERENCES

Achinstein, P.: 1963, 'Confirmation Theory, Order and Periodicity', *Philosophy of Science* **30**, 17–35.

Bayes, Thomas: 1765, 'An Essay Toward Solving a Problem in the Doctrine of Chances', *Philosophical Transactions of the Royal Society of London* **53**, 370–418.

Burks, A.: 1970, *Essays on Cellular Automata*, University of Illinois Press, Urbana.

Carnap, R.: 1950, *Logical Foundations of Probability*, University of Chicago Press (second edition 1962), Chicago.

Carnap, R.: 1952, *The Continuum of Inductive Methods*, University of Chicago Press, Chicago.

Carnap, R. and Stegmüller, W.: 1959, *Induktive Logik und Wahrscheinlichkeit*, Springer, Vienna.

Carnap, R.: 1963a, 'Variety, Analogy and Periodicity in Inductive Logic', *Philosophy of Science* **30**, 222–227.

Carnap, R.: 1963b, 'Replies and Systematic Expositions', in P. A. Schilpp (ed.), *The Philosophy of Rudolf Carnap*, Open Court, Lasalle, Illinois, pp. 711–737.

Carnap, R.: 1971, 'A Basic System of Inductive Logic, Part 1', in R. Carnap and R. C. Jeffrey (eds.), *Studies in Inductive Logic and Probability*, vol. I. University of California Press, Berkeley.

Carnap, R.: 1980, 'A Basic System of Inductive Logic, Part 2', in R. C. Jeffrey (ed.), *Studies in Inductive Logic and Probability*, vol. II. University of California Press, Berkeley.

De Finetti, B.: 1937, 'La Prevision: ses lois logiques, ses sources subjectives', *Annales de l'Institut Henri Poincaré* **7**, 1–68; tr. as 'Foresight: its Logical Laws, its Subjective Sources', in H. E. Kyburg, Jr. and H. Smokler (eds.), *Studies in Subjective Probability*, 1980. Kreiger, Huntington, N.Y.

De Finetti, B.: 1938, 'Sur la condition d'equivalence partielle', *Actualités scientifiques et industrielles*, No. 739 (Paris: Hermann & Cie). Translated by P. Benacerraf and R. Jeffrey as 'On the Condition of Partial Exchangeability', in R. C. Jeffrey (ed.), *Studies in Inductive Logic and Probability*, vol. II. University of California Press, Berkeley, pp. 193–205.

De Finetti, B.: 1974, *Probability, Induction and Statistics*. Wiley, New York.

Dynkin, E.: 1978, 'Sufficient Statistics and Extreme Points', *Annals of Probability* **6**, 705–730.

Diaconis, P. and Freedman, D.: 1980, 'De Finetti's Theorem for Markov Chains', *Annals of Probability* **8**, 115–130.

Diaconis, P. and Freedman, D.: 1981, 'Partial Exchangeability and Sufficiency', in *Statistics: Applications and New Directions* (Proceedings of the Indian Statistical Institute Golden Jubilee International Conference) Calcutta: Indian Statistical Institute, pp. 205–236.

Diaconis, P. and Freedman, D.: 1986, 'On the Consistency of Bayes Estimates', *Annals of Statistics* **14**, 1–26.

Diaconis, P. and Ylvisaker, D.: 1979, 'Conjugate Priors for Exponential Families', *Annals of Statistics* **7**, 269–281.

Feller, W.: 1966, *An Introduction to Probability Theory and its Applications*, vol. 1, 2nd. ed. Wiley, New York.

Freedman, D.: 1962, 'Mixtures of Markov Processes', *Annals of Mathematical Statistics* **2**, 615–629.

Georgii, H. O.: 1979, *Canonical Gibbs Measures* (Lecture Notes in Mathematics 760) Springer, Berlin.

Good, I. J.: 1965, *The Estimation of Probabilities: An Essay on Modern Bayesian Methods*, MIT Press, Cambridge, Massachusetts.

Hewitt, E. and Savage, L. J.: 1955, 'Symmetric Measures on Cartesian Products', *Transactions of the American Mathematical Society* **80**, 470–501.

Johnson, W. E.: 1924, *Logic, Part III: The Logical Foundations of Science*, Cambridge University Press, Cambridge.

Johnson, W. E.: 1932, 'Probability: the Deductive and Inductive Problems', *MIND* **49**, 409–423.

Kemeny, J.: 1964, 'Carnap on Probability and Induction" in P. A. Schilpp (ed.), *The Philosophy of Rudolf Carnap*, Open Court, Lasalle, Illinois, pp. 711–737.

Kuipers, T. A. F.: 1988, 'Inductive Analogy by Similarity and Proximity', in D. H. Helman (ed.), *Analogical Reasoning*, Kluwer, Dordrecht.

Lidstone, G. J.: 1920, 'Note on the General Case of the Bayes–Laplace Formula for Inductive or posteriori Probabilities', *Transactions of the Faculty of Actuaries* **8**, 182–192.

Martin, J. J.: 1967, *Bayesian Decision Problems and Markov Chains*, Wiley, New York.

Preston, C.: 1976, *Random Fields*, (Springer Lecture Notes in Mathematics 534) Springer, Berlin.

Putnam, H.: 1963a, 'Degree of Confirmation and Inductive Logic', in P. A. Schilpp (ed.), *The Philosophy of Rudolf Carnap*, Open Court, Lasalle, Illinois, pp. 761–783. Reprinted in Putnam, H.: 1975, *Mathematics, Matter and Method. Philosophical Papers*, vol. 1. Cambridge University Press, Cambridge, pp. 270–292.

Putnam, H.: 1963b, 'Probability and Confirmation', in *The Voice of America, Forum Philosophy of Science* 10. U.S. Information Agency. Reprinted in Putnam, H.: 1975, *Mathematics, Matter and Method. Philosophical Papers*, vol. 1. *Cambridge University Press, Cambridge, pp. 293–304*.

Raiffa, H. and Schlaiffer, R.: 1961, *Applied Statistical Decision Theory*. Division of Research, Boston, Graduate School of Business Administration Harvard University.

Strong, J. V.: 1976, 'The Infinite Ballot Box of Nature: De Morgan, Boole and Jevons on Probability and the Logic of Induction', *PSA 1976*, vol. 1 [Proceedings of the Philosophy of Science Association] East Lansing, Michigan.

Zabell, S. L.: 1982, 'W. E. Johnson's "sufficientness" postulate', *Annals of Statistics* **10**, 1091–1099.

Zabell, S. L.: 1988, 'Symmetry and its Discontents', in B. Skyrms and W. L. Harper (eds.), *Causation, Chance and Credence*. Kluwer, Dordrecht, pp. 155–190.

Zabell, S. L.: 1989, 'The Rule of Succession', forthcoming in *Erkenntnis*.

Department of Philosophy
University of California
Irvine, CA 92717
U.S.A.

HOWARD STEIN

WAS CARNAP ENTIRELY WRONG, AFTER ALL?

In suggesting that Carnap was not altogether wrong, I do not by any means intend to imply that he was in every respect right. But the criticisms brought against the logical empiricists' program, and against Carnap in particular, by Quine in the early 1950s are quite generally regarded as having been successful – not just in the historical sense (namely, that those criticisms were of great influence upon philosophical opinion, and may even be said to have initiated the decline of logical empiricism as a living enterprise), but in the substantive philosophical sense that this entirely favorable reception of Quine's criticisms was warranted. I have occasionally remarked that in my own view, something valuable was lost in this dismissal of logical empiricism; and that is the theme I wish here to elaborate upon. More exactly, there are three interrelated themes to be treated: (1) the question of the cogency of Quine's critique (and of some of his related positive views); (2) some of the weaknesses, and some of the strengths, of Carnap's position as I understand it; and (3) the character of Carnap's philosophy, as it is manifested in his later writings. I include this third topic because I believe that Carnap is a far subtler and a far more interesting philosopher than he is usually taken to be.

The attack by Quine in 1951 was directed against three points: the two famous 'dogmas of empiricism', and the views on ontology that Carnap had expressed in his paper 'Empiricism, Semantics, and Ontology' (Carnap 1950). Of these, the most celebrated is the critique of the concept (or alleged concept) of the analytic, which was renewed in Quine's contribution to the Schilpp volume on Carnap. I want first to comment on a passage in this latter source – the article 'Carnap and Logical Truth' (Quine 1960) – as the later and presumably more considered statement of Quine's criticism.

The attack upon the notion of analyticity was, of course, an integral part of Quine's general critique of the concept of meaning; and in 'Two Dogmas of Empiricism', Quine tells us that the two dogmas there exposed – that of the analytic/synthetic distinction, and that (in his terminology) of 'reductionism' – are at root identical: both depend

Synthese **93**: 275–295, 1992.
© 1992 *Kluwer Academic Publishers. Printed in the Netherlands.*

upon a specious distinction between "a linguistic component and a factual component in the truth of any individual statement" (Quine 1963, pp. 41–42). "If this view is right" Quine says (ibid., p. 43), "it is misleading to speak of the empirical content of an individual statement"; and he adds, "especially if it is a statement at all remote from the experiential periphery of the field" (that is, of 'total science', which Quine has just analogized to "a field of force whose boundary conditions are experience").

Now, here is a passage from 'Carnap and Logical Truth'; "Already the obviousness (or potential obviousness) of elementary logic can be seen to present an insuperable obstacle to our assigning any *experimental meaning* to the linguistic doctrine of elementary logical truth" (Carnap et al. 1963, p. 389; the emphasis is mine). How is one to understand this demand for 'experimental meaning', in the light of Quine's rejection of the dogma of reductionism?

It may be that the answer has to take account of another Quinean notion or rubric: that of fact of the matter. In an illuminating discussion of this notion in the Schilpp volume on Quine – one that is heartily endorsed by Quine himself in his reply – Roger Gibson tells us that Quine's use of this expression does not pertain to methodology (or epistemology); rather, "Quine's understanding of this term is decidedly *naturalistic* and *physicalistic*" (Gibson 1986, p. 143). Again: "Facts of the matter belong to the ontological phase of inquiry, not to the epistemological phase. *Ontology* is the theory of what there is . . . . Ontology is that theoretical structure that links past and present sensory stimulations to future ones – it is a theory of objects" (ibid., p. 151); and: "Ontology and epistemology are concerned with different issues. Ontology focuses on the issue of what there is; and what there is is a question of *truth*" (ibid., p. 147). The context of Gibson's discussion is Quine's doctrine that there is no fact of the matter – and Gibson particularizes, in view of Quine's physicalism, "no *physical* fact of the matter" (emphasis added) – to translation; thus, to assertions of sameness of meaning. In short, then, one might consider Quine's point to be that what he calls 'the linguistic doctrine of logical truth' (but Carnap prefers to call that of 'truth based on meanings') cannot be construed as saying anything about the physical world.

If that is really what Quine intends, he has surely not said it with precision. Indeed, his 'insuperable obstacle' was to the assignment of experimental, not of physical, meaning to Carnap's doctrine; and that, surely, suggests a methodological rather than an ontological concern.

In any case, Quine's objection seems to me to pose very serious puzzles. His demand for 'experimental meaning' not only seems on the face of it, if construed as the methodological (or epistemological) demand suggested by its wording, to rest upon some version of the dogma of 'reductionism': it stands on *either* reading in a very strange relation to the position Carnap himself takes on the status of his own doctrine. I must now attempt to explain this.

Just after posing the objection under discussion, Quine says the following:

The philosopher, like the beginner in algebra, works in danger of finding that his solution-in-progress reduces to '0 = 0' Such is the threat to the linguistic theory of elementary logical truth. For, that theory now seems to imply nothing that is not already implied by the fact that elementary logic is obvious or can be resolved into obvious steps. (Carnap et al. 1963, p. 389)

Carnap, in his reply to Quine, subjects the last sentence to a (perhaps heavy-handed, but in my opinion not entirely unmerited) satirical analysis (ibid., pp. 916–17), the gist of which is that since his theory implies itself, it follows from Quine's statement that Carnap's theory – which Quine rejects – follows from a fact that Quine asserts. But there is a little more to this than a joke at the expense of a glib Quinean aphorism.

What, after all, does reduction to '0 = 0' really mean? Quine, of course, is thinking of a student who is trying to solve an equation, and who fails to obtain an equivalent from which the value of an unknown quantity can be determined. But if the original equation can in fact be *reduced* to '0 = 0', in the sense of being shown not merely to imply, but also to be implied by this latter, then of course one has established by the reduction that the original equation is an identity – in other words, one has proved a theorem of algebra. In Carnap's language, such an identity is called 'analytic'; so in Carnap's own terms, Quine's refutation of Carnap's doctrine is to be seen as a proof that that doctrine is true, and, indeed, analytic.

The objection Quine has posed to Carnap's view of logical truth (and, more generally, of analytic truth) is one that has been raised by others against the so-called verifiability theory of meaning (one is reminded of Quine's diagnosis: that the two dogmas are at root identical). In the latter case, the allegedly unanswerable question is, "How can the principle of verifiability itself be verified?" – in other words, as one might put it: What is the experimental meaning of that principle? Carnap's answer is the same in both cases: the doctrine of analyticity, and the doctrine (in so far as he maintains one) of empirical meaning, are both – formally considered – analytic.

At this point, I have the uncomfortable feeling that my discussion may seem to the reader – and threatens to become by my own lights as well – scholastic, or talmudic, in the pejorative sense of those words. It is not my intention to pit a representation of the issue in Carnapian terms against a representation in Quinean terms, for the mere purpose of doctrinal hair-splitting. Let me try to explain what I think Carnap's view of the matter really amounts to.

A historical remark may be helpful here. (I am about to make what will probably be my only eyewitness contribution to the history of philosophy.) In 1951, Quine read a paper to the departmental colloquium of the Philosophy Department at the University of Chicago containing the criticisms prompted by Carnap's (then recent) article on questions of ontology. The first main conclusion of that paper (part of which has appeared in print, as 'On Carnap's Views on Ontology' – Quine 1951a) is that the distinction called by Quine that between 'category questions' and 'subclass questions' – his own rephrasing of Carnap's distinction between questions 'external' and 'internal' to a 'linguistic framework' – is both ill grounded and unnecessary; that is, unnecessary for Carnap himself. He expressed the hope that he could persuade Carnap of this latter point; for, he said, no more is needed for Carnap's own philosophical purpose than the distinction between analytic and synthetic. Quine then proceeded to explain that this first elimination was not the end of his dissent; that, indeed, it had just led to the basic point of contention – namely, "the distinction between analytic and synthetic itself". In the paper as published, the further discussion of this basic point is waived in favor of a reference to 'Two Dogmas'.

So far, of course, I have said of this occasion only what can be read in the public record. My original contribution to the history concerns the discussion that followed Quine's talk. For I was present at the colloquium, and I found that discussion not only interesting, but even in a sense inspiring; and was later distressed at the aftermath.

Carnap's summary of the issue between Quine and himself was on the following lines: "Quine", he said (I am not quoting verbatim, but giving the gist as I remember it), "and I really differ, not concerning any matter of fact, nor any question with cognitive content, but rather in our respective estimates of the most fruitful course for science to follow. Quine is impressed by the continuity between scientific thought and that of daily life – between scientific language and the language of ordinary discourse – and sees no philosophical gain, no gain either in clarity or in fruitfulness, in the construction of distinct formalized languages for science. I concede the continuity, but, on the contrary, believe that very important gains in clarity and fruitfulness are to be had from the introduction of such formally constructed languages. This is a difference of opinion which, despite the fact that it does not concern (in my own terms) a matter with cognitive content, is nonetheless in principle susceptible of a kind of rational resolution. In my view, both programs – mine of formalized languages, Quine's of a more free-flowing and casual use of language – ought to be pursued; and I think that if Quine and I could live, say, for two hundred years, it would be possible at the end of that time for us to agree on which of the two programs had proved more successful".

This view of the matter might certainly be expected to be congenial to Quine, with the 'shift toward pragmatism' he signalized as one of the principal consequences of the critique in 'Two Dogmas' (Quine 1963, p. 20); and, indeed, as I recall, Quine happily assented to Carnap's diagnosis. I say that I found the discussion inspiring. The other dominating figure in the Chicago Philosophy Department at the time, besides Carnap – and in the politics of the department the single dominating figure – was Richard McKeon, who maintained that differences of philosophical principle are invariably irremediable – to be understood, in terms of a classification of the possible coherent philosophical stances, but never to be resolved: philosophers with divergent principles were doomed to talk at cross-purposes. And here, in McKeon's very presence, were two eminent philosophers who had come to an agreement not only about the character of their own disagreement, but about the conditions under which it could in fact be resolved. As to my later distress, I take it that hardly requires comment. I have never understood why Quine continued to argue his case against Carnap with no suggestion that the issue concerned the fruitfulness of a program, and not the tenability – or intelligibility – of a doctrine.

A proper assessment of Carnap's philosophy demands, first, that one understand the general character of his program; only then do there arise questions of evaluation, both of particular proposals within the program, and of its more general outlines. A first point about the general program seems in need of special emphasis – for the confusions (as I take them to be) inherent in both Quine's demand for the 'experimental meaning' of Carnap's views on logical truth, and the corresponding demand for 'verifiability' of the verifiability principle, turn on a failure to appreciate this point. In his later writings (beginning, I think, with the work on the theory of logical probability) Carnap used the term 'explication' for the activity of philosophical clarification. An explication is a proposed exact characterization of a concept. If the proposal is adopted, the concept so characterized is an explicatum – that is, an 'explicated' notion. There is a perhaps somewhat delicate issue whether the explication is to be regarded as a clarification of a notion already present before the explication has been achieved. In typical cases, something like this is so; and Carnap calls the preexisting, not fully clarified notion the *explicandum*: 'that which requires explication'. That this state of affairs involves the well-known 'paradox of analysis' is clear – indeed, this is precisely the paradigm situation of that alleged paradox; I need hardly elaborate. That on the other hand there have *actually occurred* what should quite reasonably pass for successful explications in this full sense seems to me uncontroversial; for instance, although Quine is unhappy with Carnap's account of logical truth, he is famously happy with first-order predicate logic – and would presumably agree that the exact construction of this system clarified the preexisting, insufficiently clear and precise, notion of logical inference.

Now, Carnap's distinction between 'external' and 'internal' questions, which was introduced in his paper on ontology and is deprecated by Quine, has – if one accepts it (which means: if one agrees to use it) – an obvious application to the process of explication in general. The explicatum, as an exactly characterized concept, belongs to some formalized discourse – some 'framework'. The explicandum – if such there is – belongs ipso facto to a mode of discourse outside that framework. Therefore *any* question about the relation of the explicatum to the explicandum is an 'external question'; this holds, in particular, of the question whether an explication is adequate – that is, whether the explicatum does in some appropriate sense fully represent, within the framework, the function performed (let us say) 'presystematically' by the explicandum.

In saying that the 'linguistic doctrine of logical truth' or the 'verifiability theory of meaning' is, if adopted, analytic, Carnap would be making a statement about the standing of a certain proposition within a formalized system – or, rather, in a sketch of a *family* of *projected* formal systems: devised, namely, to serve as a 'framework' for *Carnapian linguistic theory itself* (whether of 'constructed' languages or of 'natural' languages or both). That the truth of the corresponding propositions within those systems is *trivial* – imposes no restriction upon the world – is something consequent upon the characterization of the framework itself. Whether one should adopt linguistic, or theoretical, frameworks that are characterized in such a way as to embody those propositions as analytic is a question (as Carnap sees it) of quite a different type, and by no means trivial. Indeed, just such a non-trivial question is at issue in the controversy with Quine.

When I said, earlier, that the issue whether an explication is to be regarded as 'of' some presystematic explicandum is a little delicate, I had in mind not only the lack of a clear and uniform criterion for assessing external questions in general and questions of the adequacy of explications in particular, but also another point – not, I think, made by Carnap, but which I should like to propose in (or should I say 'to'?) his spirit. The question of the nature of 'presystematic notions' is obviously very complex – and somewhat vague. I don't know to what science one should say it belongs: psychology? sociology? – Quine, I dare say, would assign it to naturalistic epistemology. But I should expect Quine to agree that the notion of a presystematic notion is by its very nature vague – and perhaps usefully vague. It would be easy to cite cases in which a notion of this type – or at any rate, a word in general use – can be said to have been explicated by more than one precise explicatum. The other possibility is that a newly proposed exact concept does not correspond very well to any presystematic notion at all. It seems not to be a violation of linguistic propriety – at least, of Latin propriety – to call such a concept an 'explicatum': meaning, again, an explicated – that is, clarified (in other words, simply a clear) – concept. If one asks what such an explicatum is the explication of, more than one reply is possible. One can say that the exact characterization proposed is just the explication of the very concept in question (as a definition defines the concept whose definition it is); or that it explicates a presystematic idea, not previously in general use, but vaguely entertained by the inquirer when groping for clarity. I hope it is clear that all this is peripheral: what counts in the end – still in Carnap's view of things – is the clarity and the utility of the proposal; whether part of that utility has to do with an earlier, vaguer, general usage is distinctly a secondary matter.

This has an immediate bearing upon one aspect of the issue raised by Quine concerning the analytic. Quine was of course deeply concerned with empirical semantics – with the theory of 'natural languages'; and a rather large part of his challenge to Carnap was based upon the contention that there is no presystematic notion of analyticity in natural languages. Carnap appreciated the importance of the study of natural languages, and hoped that the semantical theories he was developing would be of use in empirical linguistics. In his reply to Quine, in the Schilpp volume, he therefore "accepted [Quine's] challenge to show that an empirical criterion for intension concepts with respect to natural languages can be given" (Carnap et al. 1963, p. 919). Observe that two questions may be distinguished here: whether such concepts are embedded in ordinary usage – so that, for instance, there might be an 'ordinary' notion of 'truth based on meanings' to serve as the explicandum for an explication; and whether such concepts can usefully be introduced as part of the technical apparatus of the theorist – as 'terms of art'. But the answers to these questions, however interesting they may be for empirical linguistics, have no bearing whatever upon either the difference in general between Quine's approach and Carnap's to philosophical issues, or the issue in particular of the viability, or utility, of the distinction between the analytic and the synthetic in formalized languages. I emphasize this because I believe that Carnap, in generously accepting the challenge posed by Quine on the empirical side, has failed to make sufficiently clear the difference between the two sorts of issues, and the important fact that the standing of his notion of the analytic as it relates to his program for scientific/philosophical explication is a matter entirely independent of the question about natural languages.

This is not to say that Carnap was 'right' about the analytic/synthetic distinction. I have so far only been trying to clarify what the crucial question is – or, rather, what it is not; for we have not yet seen what it is.

Once again, it is necessary to make some distinctions. I believe it is clear that Carnap's intense concern with questions about the analytic and synthetic derives from Kant, from Frege, and from (early) Wittgenstein. In all three of these philosophers, the primary interest of the notions is epistemological; the analytic, in particular, is that species of truth a priori knowledge of which is unproblematic, because the truth itself is trivial. In the case of Frege, to be sure, the matter is a bit ticklish, for Frege offers no criterion of analytic truth save that it is purely logical in nature – and offers no criterion of what does or does not genuinely belong to logic. One can see Wittgenstein's theory that analytic truths are tautologies as an attempt to repair this defect in Frege. But of course Wittgenstein's theory will not do, and by the time of his later writings Carnap had long abandoned it – although I do not think he ever abandoned his belief that the classification of a sentence as analytic in some sense 'explains' how we know it. My own opinion is that such purported explanations have served little purpose – that we really do not have a satisfactorily analyzed epistemological 'basis' for any department of knowledge, mathematics and logic included. We have learned – principally from Gödel, also from Skolem and others – that the notion of 'logical triviality' is highly non-trivial. The primitive view – surely that of Kant – was that whatever is trivial is obvious. We know that this is wrong; and I would put it that the nature of mathematical knowledge appears more deeply mysterious today than it ever did in earlier centuries – that one of the *advances* we have made in philosophy has been to come to an understanding of just how deeply puzzling the epistemology of mathematics really is. So on this point, I agree with Quine. (Although Quine speaks of "the obviousness [or potential obviousness] of elementary logic", he tells us that this characterization carries no explanatory value.)

But that is not the end of the matter. I said we must make some distinctions: the hope of solving an epistemological puzzle with the help of the concept of the analytic has failed; but there remain, I think, certain issues for which that notion, or at least something related to it, might still serve an important purpose.

Carnap's late view distinguished between logical truth and analytic truth – the latter being a wider concept (thus, all logical truths were, for him, 'based on meaning'; but not all truths based on meaning were truths of logic). It might be supposed that this extension was motivated by considerations of synonymy in natural languages, and by Quine's skeptical attack upon synonymy; Carnap's paper 'Meaning Postulates', which formally introduced the broader concept, appeared in 1952 and opened with a reference to Quine's example of 'bachelor' and 'unmarried' (Carnap 1952; see Carnap 1956, p. 222). However, Carnap makes clear in that very place that his central concern is another one: "Our explication", he says, "will refer to semantical language systems, not to natural languages"; and adds: "It seems to me that the problems of explicating concepts of this kind for natural languages are of an entirely different nature" (Carnap 1956, pp. 222–23). As to the real concern of the paper, that is also made plain in its introductory section; Carnap says:

> It is the purpose of this paper to describe a way of explicating the concept of analyticity, i.e., truth based upon meaning, in the framework of a semantical system, by using what we shall call *meaning postulates* . . . . It will be shown in this paper how the definitions of some concepts fundamental for deductive and inductive logic can be reformulated in terms of postulates.

In 'Carnap and Logical Truth', Quine dismisses this move rather brusquely; he writes:

> Carnap's present position is that one has specified a language quite rigorously only when he has fixed, by dint of so-called meaning postulates, what sentences are to count as analytic. The proponent is supposed to distinguish between those of his declarations which count as meaning postulates, and thus engender analyticity, and those which do not. This he does, presumably, by attaching the label 'meaning postulate'.
>
> But the sense of this label is far less clear to me than four causes of its seeming to be clear. Which of these causes has worked on Carnap, if any, I cannot say; but I have no doubt that all four have worked on his readers. (Carnap et al. 1963, pp. 404–05)

Earlier, however, Quine had shown more understanding of at least one aspect of Carnap's program. In 'Two Dogmas', after a preliminary indication of the difficulty connected with synonyms, he remarked: "I do not mean to suggest that Carnap is under any illusions on this point. His simplified model language with its state-descriptions is aimed primarily not at the general problem of analyticity but at another purpose, the clarification of probability and induction" (Quine 1963, p. 24).

That purpose remained a very important part of Carnap's concern, and of his wish to deal more adequately with what Quine calls 'the general problem of analyticity' – but here to be understood as the general problem of analyticity for what Carnap calls 'semantical language systems', which he contrasts with 'natural languages' (where, *also*, the general problem of analyticity seems to him of interest, but where it presents problems 'of an entirely different kind').

Let me try to explain what, as I understand it, the philosophical role is of the notion of a 'linguistic framework', and, relative to such a framework, of the notion of the analytic, in Carnap's later thought. By the time of his later writings, Carnap had clearly abandoned the earlier hopes for the construction of a single and permanent language that should be adequate for all of science. Even in his early work – as early as *Der logische Aufbau der Welt* – he had encouraged the exploration of alternative languages with alternative 'bases', as he called them – physical or phenomenal bases, and, if the latter, ones like his own in that book, taking as fundamental the total field of immediate experience, or ones founded rather upon sense-qualities. By the period of the centrality of 'logical syntax', this openness to alternatives had taken the 'official' form of Carnap's 'principle of tolerance', which concerned both the 'basis' and the mathematical form – the logico-mathematical strength – of the language. But at this time, the hope and indeed confidence were still there that some among the language forms to be considered would prove adequate in the strong sense I have mentioned: for all of science, and permanently. To be sure, Gödel's results had by then demonstrated that no one language could be expected to serve for all conceivable *mathematical* purposes; but Carnap's view pretty clearly was that whatever developments were required in the direction of increasing mathematical strength could just be grafted on to a stable stock or core of empirical language. But in the later writings, this hope has been essentially abandoned – or at least, very fundamentally modified. We shall presently consider a most striking piece of evidence of this change.

What I am suggesting, then, is that alternative possible 'frameworks' are alternative in a very serious sense. What sense? I would put it this way: that a linguistic or theoretical framework envisages a distinct set of possibilities for the world; that alternative frameworks are, in effect, *constitutive* of alternative notions of possibility. This set of possibilities appears in Carnap's earlier semantical works under the guise of linguistic entities, the 'state-descriptions' – sentences of maximal strength (short of logical contradiction). Under the influence of Kemeny, and with special concern for his developing theory of logical probability, Carnap replaced this formal notion by what we may reasonably call the 'ontological' notion of possible states. In terms of this notion, the semantical content of a sentence can be characterized as just the set of possible states with which it is compatible ('in which it would be true'); its logical strength, rather, by the complementary set: the set of states it excludes. Since inductive logic, as Carnap was attempting to develop it, was to be founded upon the definition of *measures of content* of sentences (or of propositions), these notions clearly were, as he says, fundamental for inductive (as well as deductive) logic.

Now, such an ontological notion of 'the possible' as I have just been discussing is of course extremely uncongenial to Quine. But it is worth pointing out that, if one agrees (under Carnap's principle) to 'tolerate' the notion so far at least as to explore its uses, there results a rather pleasant application to a point that seems, if just briefly and slightly, to have bothered Quine. In his reply, in the Quine Schilpp volume, to the article of Putnam, Quine toys with a challenge posed to him at the end of that article. Putnam referred to Quine's dismissal of the general notion of meaning, and asked how Quine can consistently hold that there is no 'fact of the matter' as to whether (for example) meanings exist (Putnam 1986, p. 425). Quine brushes this off a little breezily:

One bit that I am going to have to understand for purposes of Putnam's example [Quine says] is the bit that he renders as "Meanings exist." As already remarked, the existence of meanings poses no problem beyond synonymy; they can be taken as equivalence classes. Since I have despaired of making general sense of synonymy, perhaps Putnam is right in supposing that I make no fact of this matter and I am right in not doing so. (Quine et al. 1986, p. 430)

Quine then adds the following remark:

Dreben once put me a related but more challenging question: is there no fact of mathematical matters? For me, unlike Carnap, mathematics is integral to our system of the world. Its empirical support is real but remote, mediated by the empirically supported natural science that the mathematics serves to implement. On this score I ought to grant mathematics a fact of the matter. But how, asks Dreben, does this involve the distribution of microphysical states? What would there being a largest prime number have to do with the distribution of microphysical states?

Carnap would have said that we have here a contrary-to-fact conditional with an L-false antecedent, 'There is a largest prime number', from which anything and everything follows vacuously as consequent. That avenue is closed to me, but I can still protest that there is no coping with intensional conditionals with wildly implausible antecedents. My suggested standard for facts of the matter is directed rather at concrete situations, and pales progressively as we move upward and outward. Evidently then the upshot is that the factual and the mathematical stand apart, for me as for Carnap; but for me, unlike Carnap, the separation is a matter not of principle but of degree.

I confess this makes me slightly dizzy. Mathematics is serious, and presumably in large part true, for Quine (who does after all take his stand upon the deliverances of science). Yet mathematics occupies a region too far from the concrete for there to be a fact of the matter. Or is it that there is only a 'pale' fact? How pale? Is it too pale to be discerned at all? It is evidently too pale to be described. Saying that this is 'a matter not of principle but of degree' hardly does justice to the issue.

In any case, Quine's representation of what Carnap would have said in answer to Dreben is stated in just such formally linguistic terms as the semantics I have referred to nicely avoids. The correct late-Carnapian answer to Dreben's question, as I understand Carnap, would be that the class of microphysical states of the world admitted by the proposition that there is a largest prime number is the empty set: the 'fact of the matter' is that that proposition is ruled out by any possible facts whatever.

· The notion of a framework, then, with its *envisaged* possibilities, does at least afford us a convenient way of *formulating* statements about the 'ontological', or truth-related, bearings of sentences of a theory – including purely mathematical ones. In his paper on ontology, Carnap emphasized the relativity of this notion to the theoretical framework.

I think it has not been generally understood that, in Carnap's scheme of things, and using the terms I have quoted earlier from Gibson, semantics is fundamentally concerned with 'ontology', and not with 'methodology' or 'epistemology'. This should have been clear from the start, in view of Carnap's tripartite classification of linguistic theory: into syntax – concerned with linguistic entities alone; semantics – concerned with linguistic entities and their relations to what they refer to; and pragmatics – concerned with all the aspects of a language together, including in particular its conditions and modes of use. It should have been apparent that, under this classification, methodology and epistemology belong to pragmatics. But the point was obscured – and seems at first not to have been appreciated by Carnap himself – for two reasons. On the one hand, the liberalization that freed Carnap's philosophy from its former restriction to syntax had been made possible by Tarski's definition of truth, which showed how very general semantical notions could be characterized in a systematic way for formalized languages. There was no corresponding central concept that seemed to serve as an exact systematic foundation for pragmatics; and Carnap thought of the latter as concerned with something like *idiosyncrasies* of use in ordinary languages. On the other hand, Carnap thought – and this he seems to have continued to hold to the end of his career – that the empirical interpretation of a theory could always be achieved by specifying the *semantics* of the *empirical* part of its language. In effect, the role of pragmatics in this fundamental problem of the analysis of 'empirical content' would be restricted to the single function of distinguishing, within the language, its 'empirical part'. (I take it that the pragmatic character of this distinction is clear – assuming, of course, that the distinction is tenable at all: what part of the language is 'empirical', or what part of its vocabulary refers to the 'observable', is obviously a matter that depends upon something about the *users* of the language.)

But now, with the explicit introduction of the concept of a framework, and the implication that one of the continuing tasks of philosophy will be the examination and evaluation of alternative frameworks, it has to be clear that this activity belongs to pragmatics. Here is a statement on the subject by Carnap, taken from his reply to Charles Morris's article in the Schilpp volume:

In particular, many problems concerning conceptual frameworks seem to me to belong to the most important problems of philosophy. I am thinking here both of theoretical investigations and of practical deliberations and decisions with respect to an acceptance or a change of frameworks, especially of the most general frameworks containing categorial concepts which are fundamental for the representation of all knowledge. (Carnap et al. 1963, p. 862)

I would sum this up by saying that what in Quine appears as the distinction between concern with ontology – in the sense in which that means whatever relates to 'fact of the matter' – and concern with epistemology, is represented in Carnap as the distinction between the semantics of a framework and the pragmatics of frameworks generally (where by this last expression I mean both pragmatic questions about a single framework, and questions that involve the comparative assessment of alternative frames).

I have given one example of the application of Carnap's semantics to a Quinean question – the explication of the notion of 'fact of the matter' itself. I noted that the concept of 'possible state' involved in that application – and at the center of Carnap's semantics in general – is one Quine would find objectionable. But I must now say that I do not see how he himself can do without it. Quine – to repeat – takes his stand on science, and more particularly on physics. But the concept of the space of all states of a physical system is a central one in much of classical physics, and in all of quantum physics; and 'all states' means, of course, all possible states. If Quine rejects this notion because of the occurrence of a modality in its description, I do not see how he will accommodate physics. If he accepts this as an innocent use of the term 'possible', that innocent use is all that Carnap needs. The difference between them seems here to be that Carnap is willing to consider alternative frameworks, for alternative theories, with their alternative ontologies and their alternative conceptions of the range of the possible; whereas Quine, asserting ontological relativity, nevertheless holds that we can only conduct our discourse within one or another of competing theories – that is, we cannot find a 'framework' for semantical and pragmatic discussion itself that could serve as a kind of neutral ground. As he puts it in his reply to Gibson: "[W]hichever system we are working in is the one for us to count at the time as true, there being no wider frame of reference" (Quine et al. 1986, p. 157). That remark was made about a situation in which "we have somehow managed to persuade ourselves" that two competing systems – our own, and an 'alien jargon' – are empirically equivalent. Quine says, "[o]ur own system is true by our lights, and the other does not even make sense in our terms". He seems to imply not *only* that there is "no wider frame of reference", but that there is no possible analogue of relativistic invariance.

But the question of empirical content, and of judgments of 'empirical equivalence' of competing systems, raises the issue of Quine's second 'dogma of empiricism': the principle of verification. It must be remembered that Quine does not in fact reject the equation of meaning with (something like) verification – indeed, he says, for example, in 'Epistemology Naturalized', that "epistemology remains centered as always on evidence, and meaning remains centered as always on verification; and evidence is verification" (Quine 1969, p. 89). Rather, in rejecting the verification principle, or (as he calls it) 'reductionism', what he is objecting to is the notion that there is such a thing as the empirical meaning of a sentence. He is affirming his holism in the theory of meaning, and in epistemology – his view that "our statements about the external world face the tribunal of sense experience not individually but only as a corporate body" (Quine 1963, p. 41).

Now, I have remarked that Carnap seems never to have abandoned the view that the empirical content of a language – for him, in the case of scientific languages, always one that has been formally constructed

– can be based upon a part of that language specially distinguished as its 'observational' part; and that the empirical content of the rest of the language can then be analyzed in terms of logical relations (eventually, both deductive and inductive logical relations) to that 'observation sublanguage'. I think, to put it baldly, that this will not work; or to put it more accurately, that it does not work (for conceivably – although I doubt it – some day it will). But in my view the trouble is not the famous 'theory-ladenness' of observation terms. Whatever theory is required for ordinary life is generally quite under control by ordinary people, and what Carnap calls the 'thing language' serves very well for what we ordinarily call observation-reports; that is, there is a kind of 'minimal-theory-ladenness' that occasions no difficulties. I think the real problem is that we have no language at all in which there are well-defined logical relations between a theoretical part that incorporates fundamental physics and any observational part at all – no framework for physics that includes observational terms, whether theory-laden or not. The point can be made by contrasting the character of a typical treatise on some branch of theoretical physics, with that of a work on experimental physics. Theoretical physics can be made to look very much like mathematics; experimental physics cannot. One can argue mathematically in theoretical physics – one can deduce consequences from assumptions; but I cannot think of any case in which one can honestly *deduce* what might honestly be called an observation. What can be done, rather, is to represent (as I have put it elsewhere) 'schematically', within the mathematical structure of a theoretically characterized situation, the position of a 'schematic observer', and infer something about the observations such an observer would have. For example: in ordinary classical astronomical theory, one will represent, say, the planets – including the earth – perhaps as particles, perhaps as extended bodies; putting the schematic observer at a certain latitude and longitude of the earth, and calculating the angles between the lines from that position to the several astronomical bodies under consideration, one will infer that the observer will see those bodies along lines making the corresponding angles with one another. But that is not by any means a deduction of an observation. We have left out, for example, the light by which the observer must see; we have left out the earth's atmosphere, through which the light is refracted. Of course, serious observational astronomy must take account of atmospheric refraction, so perhaps we should put the light – that is, the electromagnetic field – and the atmosphere into our systematic representation. Are we to do so as well with the observer's telescope? the observer's eye? the observer's brain? – There are two problems here. One is sheer complexity. That one might take to be possible 'in principle' to overcome. The second is that we simply do not know enough to put in everything that would be required for an honest deduction of a genuine observation. Well, perhaps we shall know enough some day – that is why I said that Carnap's program may some day be realizable. But even if that day comes, I doubt that the program will be realizable in practice – that it 'will work' – because the complexity that might be overcome 'in principle' would still be intolerable in practice.

Now, Carnap's scheme for philosophical analysis is admirably suited to just this situation. It is exactly the theories with a highly mathematical structure – the typical theories of physics – that lend themselves, ipso facto, to construction as Carnapian 'frameworks'. The question of the empirical application of such a framework becomes appropriately a question of its pragmatics. I do not know how, systematically, a general theory of such empirical application might be made; but at least I think the problem, in the neo-Carnapian form I have just outlined, finds a suitable locus and an intelligible formulation as a problem. And I think it reasonably clear that to just the extent that we know in practice how to talk about the empirical application of specific physical theories, we can formulate what we know how to say in terms of the pragmatics of a Carnapian framework.

It does not follow from this that 'meaning postulates' must play a role in such systems. It may very well be – I am inclined to think it is – that the possibilities to be contemplated in a framework for theoretical physics as we know it today or as it is likely to develop have to be restricted by the general principles of the theory itself – principles that one would be loth to call 'analytic'. This is a serious modification of Carnap's view. It locates fundamental theory change in change of framework, and therefore outside the scope of the sort of inductive logic Carnap was trying to construct – which itself would, of course, be internal to a framework. That, it seems to me, entails a development of Carnap's views in a direction that I should characterize as 'dialectical'; for it entails a certain blurring of the distinction, dear to Carnap, between the purely cognitive, or theoretical, and the practical. Let me remind you of the passage I have quoted from Carnap's reply to Morris, in which he says that problems concerning conceptual frameworks are among the most important problems of philosophy, and adds: "I am thinking here both of theoretical investigations and of practical deliberations and decisions with respect to an acceptance or a change of frameworks, especially of the most general frameworks containing categorial concepts which are fundamental for the representation of all knowledge". If we allow these 'categorial concepts' to include categorial concepts both of fundamental mathematics and of fundamental physics – and this, in my opinion, we must do, if we are to make good use of Carnap's notions – then the full force of what I have just called the move to a kind of 'dialectic' of science appears.

There is what seems to me a very odd contrast between Carnap and Quine. Quine, with his epistemological holism, speaks of the 'web of belief', rejecting 'reductionism' with its rigid tribunal sitting in judgment on sentences. He rejects the continuing 'legislative' force of the construction even of a formalized language, insisting that as soon as the language has been created its usages must be allowed to evolve naturally (a kind of Webster's Third International doctrine of scientific language). His pragmatism, and his epistemological naturalism, seem to contrast with the rigidity of Carnap's reliance upon formal constructions and fixed rules. One can describe the contrast in terms favorable to Quine: his refreshingly relaxed manner contrasted with Carnap's more ponderous and rigid ways. An alternative description might be that where Quine in principle leaves all open to the flow of experience – and of that part of experience, in particular, that constitutes the evolution of science – Carnap, having eventually come to recognize that science develops in ways that entail revisions even of 'categorial concepts', wishes to make at least local stands in the midst of this Heraclitean flux, and endorses constructions designed to achieve the maximum possible clarity both in what we say (and our understanding of what we say), and in the basis for the decisions we make. Which is the more acceptable way of describing the contrast of course depends upon one's comparative assessment of the two approaches. But what strikes me as odd is this: it is the rigid Carnap who encouraged Quine to explore his own approach to these basic questions, and – although of course believing his own way to be the better – left it ultimately to future experience to decide. It is Quine the holist who, while denying that the issue was one in which there is a fact of the matter, continued to maintain – and to convince many philosophers – that he was *right* and Carnap *wrong*.

Moreover, if we grant that there is no fact of the matter, it is Carnap who has the clearer way of expressing the nature of the issue itself and what is at stake: it is an external question, concerning the choice of a framework, and ought to be decided by considerations – influenced by all our theoretical knowledge and the clearest understanding we can obtain of our practical/theoretical aims – of the *usefulness* of the alternatives. It is also Carnap who tells us that there is no need to make an exclusive choice – that, subject to practical constraints, it is at least to some extent possible for the same investigator to explore, and even to use, alternative frameworks.

I want in closing to call attention to a further, not unrelated, contrast; and to a feature of Carnap's late views that I have found to be little known – although it appears in a place that ought to have attracted some attention. I refer to Carnap's comments on the paper of Feigl in the Schilpp volume (Feigl, in Carnap et al. 1963, pp. 227–68; Carnap's comments, ibid., pp. 882–86). Feigl has written on physicalism – a doctrine certainly dear to Carnap's heart; and Carnap endorses his friend Feigl's views "in all major points" – but "with some qualifications". These qualifications are extremely interesting; some of them, at least, may be regarded by many as amazing. Carnap first formulates, in several alternative ways, two theses of physicalism (or rather, he formulates the first thesis in four different ways; the second in only one). What the first thesis comes to is that a language form in which all statements are intersubjectively confirmable is sufficient for expressing everything that is 'meaningful for me' – that is, for 'the knowing subject'. This is clearly a formulation of what had long been a basic – perhaps *the* basic – tenet of Carnap's empiricism. The second thesis holds that all laws of nature, including those that apply to organisms, human beings, and human societies, are logical consequences of "the physical laws, i.e., of those laws which are needed for the expression of inorganic processes".

Now, none of this is surprising from Carnap the 'reductionist'. But here is what he goes on to say. In the first place – or, rather, in the second place; but I shall cite it first – he says: "It is true that these two theses of physicalism go far beyond the present possibility of reducing extra-physical concepts and laws to physical ones. These theses do not represent firmly established knowledge but sweeping extrapolating hypotheses". Note that this statement is made not only of the thesis of reducibility of laws to those of physics, but even of the first thesis –

that all that is meaningful for the knowing subject can be expressed in a form of language in which all statements are intersubjectively confirmable. Clearly Carnap, the empiricist and (perhaps) reductionist, was in the end not a dogmatist – however much he may have been one in the flaming 1930s.

As to the second, and much stronger, thesis – that of reducibility to physical laws – Carnap says more:

> This thesis does not refer to the laws known to us at present, but to those laws which hold in nature and which our knowledge can only more and more approximate. The thesis may therefore be understood as the hypothesis that in the future it will become possible to an ever greater extent to derive known extra-physical laws from known physical laws.

When we reflect that Carnap has just glossed 'physical law' as "those laws which are needed for the explanation of inorganic processes", the thesis becomes, in effect, that whatever proves necessary for the scientific understanding of organisms and human beings will already be necessary for the scientific understanding of more elementary natural processes. This is a rather more subtle view than classical 'materialism' or 'mechanism', or their contemporary analogues. It suggests the notion of a *continuity* in nature, in which whatever functions or processes occur at 'higher' levels have their roots in the fundamental, or elementary, levels. And it is exactly such a view that Carnap goes on to present, as in his view probable, in a discussion of the doctrine of 'emergentism' with regard to mental processes: he is skeptical of emergentism not because he is convinced that mental processes are fully explicable in terms of current physics, but because he thinks it unlikely that there is a sharp boundary line in the hierarchy of natural beings.

I am sensible of having done scant justice to many of the things I have talked about, and having omitted some things I had hoped to talk about. I said at the outset that I am far from thinking Carnap altogether right; I do not know whether I have said enough to dissuade anyone from the view that he was, after all, entirely wrong. But I hope at least that I have been able to persuade some of you that there is more in his philosophy than most current representations of it imply.

REFERENCES

Carnap, Rudolf: 1950, 'Empiricism, Semantics. and Ontology', *Revue Internationale de Philosophie* **4**, 20–40, reprinted as 'Supplement A' in Carnap 1956, pp. 205–21.

Carnap, Rudolf: 1952, 'Meaning Postulates', *Philosophical Studies* **3**, 65–73, reprinted as 'Supplement B' in Carnap 1956, pp. 222–29, from which citations in the text are made.

Carnap, Rudolf: 1956, *Meaning and Necessity*, 2nd ed., University of Chicago Press, Chicago.

Carnap, Rudolf, et al.: 1963, *The Philosophy of Rudolf Carnap*, edited by Paul Arthur Schilpp, Open Court, La Salle, Illinois.

Gibson, Roger F., Jr.: 1986, 'Translation, Physics, and Facts of the Matter', in Quine et al. 1986, pp. 139–54.

Putnam, Hilary: 1986, 'Meaning Holism', in Quine et al. 1986, pp. 405–26.

Quine, Willard Van Orman: 1951a, 'On Carnap's Views on Ontology', *Philosophical Studies* **2**, 65–72.

Quine, Willard Van Orman: 1951b, 'Two Dogmas of Empiricism', *Philosophical Review* **60**, 20–43, reprinted in Quine 1963, pp. 20–46, from which citations in the text are made.

Quine, Willard Van Orman: 1960, 'Carnap and Logical Truth', *Synthese* **12**, 350–74, reprinted in Carnap et al. 1963, pp. 385–406, from which citations in the text are made.

Quine, Willard Van Orman: 1963, *From a Logical Point of View*, 2nd ed., reprinted Harper & Row, New York.

Quine, Willard Van Orman: 1969, 'Epistemology Naturalized', in his *Ontological Relativity and Other Essays*, Columbia University Press, New York. pp. 69–90.

Quine, Willard Van Orman, et al.: 1986, *The Philosophy of W. V. Quine*, edited by Lewis Edwin Hahn and Paul Arthur Schilpp, Open Court, La Salle, Illinois.

Department of Philosophy
University of Chicago
1050 East 59th Street
Chicago, IL 60637
U.S.A.

Acknowledgments

Sauer, Werner. "On the Kantian Background of Neopositivism." *Topoi* 8 (1989): 111–19. Reprinted with the permission of Kluwer Academic Publishers. Copyright 1989 Kluwer Academic Publishers. Printed in the Netherlands.

Ryckman, Thomas A. "Designation and Convention: A Chapter of Early Logical Empiricism." In Arthur Fine, Micky Forbes, and Linda Wessels, eds., *PSA 1990, Vol.2* (East Lansing: Philosophy of Science Association, 1991): 149–57. Reprinted with the permission of the Philosophy of Science Association.

Lewis, Joia. "Hidden Agendas: Knowledge and Verification." In Arthur Fine, Micky Forbes, and Linda Wessels, eds., *PSA 1990, Vol. 2* (East Lansing: Philosophy of Science Association, 1991): 159–68. Reprinted with the permission of the Philosophy of Science Association.

Haller, Rudolf. "The First Vienna Circle." In T. Uebel, ed., *Rediscovering the Forgotten Vienna Circle* (Dordrecht: Kluwer, 1991): 95–108. Reprinted with the permission of Kluwer Academic Publishers.

Jeffrey, Richard C. "After Carnap." *Erkenntnis* 35 (1991): 255–62. Reprinted with the permission of Kluwer Academic Publishers.

Wartofsky, Marx W. "Positivism and Politics: The Vienna Circle as a Social Movement." In Rudolf Haller, ed., *Schlick und Neurath— Ein Symposion* (Grazer Philosophische Studien 16/17) (Amsterdam: Rodopi, 1982): 79–101. Reprinted with the permission of Editions Rodopi BV.

Galison, Peter. "Aufbau/Bauhaus: Logical Positivism and Architectural Modernism." *Critical Inquiry* 16 (1990): 709–52. Reprinted with the permission of the University of Chicago Press, publisher.

Stadler, Friedrich. "Otto Neurath—Moritz Schlick: On the Philosophical and Political Antagonisms in the Vienna Circle." In T. Uebel, ed., *Rediscovering the Forgotten Vienna Circle* (Dordrecht:

Kluwer, 1991): 153–75. Reprinted with the permission of Kluwer Academic Publishers.

Reisch, George A. "Planning Science: Otto Neurath and the *International Encyclopedia of Unified Science.*" *British Journal for the History of Science* 27 (1994): 153–74. Reprinted with the permission of the *British Journal for the History of Science.*

Coffa, Alberto. "Carnap, Tarski, and the Search for Truth." *Nous* 21 (1987): 547–72. Reprinted with the permission of the Blackwell Publishers.

Friedman, Michael. "The Re-evaluation of Logical Positivism." *Journal of Philosophy* 88 (1991): 505–19. Reprinted with the permission of the Journal of Philosophy, Inc., Columbia University, and the author.

Richardson, Alan W. "Logical Idealism and Carnap's Construction of the World." *Synthese* 93 (1992): 59–92. Reprinted with the permission of Kluwer Academic Publishers.

Putnam, Hilary. "Reichenbach's Metaphysical Picture." *Erkenntnis* 35 (1991): 61–75. Reprinted with the permission of Kluwer Academic Publishers.

Shimony, Abner. "On Carnap: Reflections of a Metaphysical Student." *Synthese* 93 (1992): 261–74. Reprinted with the permission of Kluwer Academic Publishers.

Jeffrey, Richard C. "Carnap's Inductive Logic." In Jaakko Hintikka, ed., *Rudolf Carnap, Logical Empiricist* (Dordrecht: Reidel, 1975): 325–32. Reprinted with the permission of D. Reidel Publishing Company.

Oberdan, Thomas. "Positivism and the Pragmatic Theory of Observation." In Arthur Fine, Micky Forbes, and Linda Wessels, eds., *PSA 1990 Vol.1* (East Lansing: Philosophy of Science Association, 1990): 25–37. Reprinted with the permission of the Philosophy of Science Association.

Uebel, Thomas E. "Neurath's Program for Naturalistic Epistemology." *Studies in History and Philosophy of Science* 22 (1991): 623–46. Reprinted with the permission of Pergamon Press Ltd.

Salmon, Wesley C. "Hans Reichenbach's Vindication of Induction." *Erkenntnis* 35 (1991): 99–122. Reprinted with the permission of Kluwer Academic Publishers.

Skyrms, Brian. "Carnapian Inductive Logic for Markov Chains." *Erkenntnis* 35 (1991): 439–60. Reprinted with the permission of Kluwer Academic Publishers.

Stein, Howard. "Was Carnap Entirely Wrong, After All?" *Synthese* 93 (1992): 275–95. Reprinted with the permission of Kluwer Academic Publishers.